T0323975

PLATINUM-NICKEL-CHROMIUM DEPOSITS

PLATINUM-NICKEL-CHROMIUM DEPOSITS

GEOLOGY, EXPLORATION AND RESERVE BASE

S.K. HALDAR

Emeritus Scientist, Presidency University
Formerly Hindustan Zinc Limited,
Hindustan Copper Limited,
IMX Resources Limited, Perth

ELSEVIER

AMSTERDAM • BOSTON • HEIDELBERG • LONDON
NEW YORK • OXFORD • PARIS • SAN DIEGO
SAN FRANCISCO • SINGAPORE • SYDNEY • TOKYO

Elsevier
Radarweg 29, PO Box 211, 1000 AE Amsterdam, Netherlands
The Boulevard, Langford Lane, Kidlington, Oxford OX5 1GB, United Kingdom
50 Hampshire Street, 5th Floor, Cambridge, MA 02139, United States

Notices
Knowledge and best practice in this field are constantly changing. As new research and experience broaden our understanding, changes in research methods, professional practices, or medical treatment may become necessary.

Practitioners and researchers must always rely on their own experience and knowledge in evaluating and using any information, methods, compounds, or experiments described herein. In using such information or methods they should be mindful of their own safety and the safety of others, including parties for whom they have a professional responsibility.

To the fullest extent of the law, neither the Publisher nor the authors, contributors, or editors, assume any liability for any injury and/or damage to persons or property as a matter of products liability, negligence or otherwise, or from any use or operation of any methods, products, instructions, or ideas contained in the material herein.

Library of Congress Cataloging-in-Publication Data
A catalog record for this book is available from the Library of Congress

British Library Cataloguing-in-Publication Data
A catalogue record for this book is available from the British Library

ISBN: 978-0-12-802041-8

For information on all Elsevier publications
visit our website at https://www.elsevier.com/

  Working together
to grow libraries in
developing countries

www.elsevier.com • www.bookaid.org

Publisher: Candice Janco
Acquisition Editor: Amy Shapiro
Editorial Project Manager: Tasha Frank
Production Project Manager: Paul Prasad Chandramohan
Designer: Victoria Pearson Esser

Typeset by TNQ Books and Journals

To the Lotus Feet of Holy Mother Sri Sarada Devi.

Contents

Author Biography

S.K. Haldar (Swapan Kumar Haldar) has been a practicing veteran in the field of Mineral Exploration and Metal Mining for the past four and one-half decades. He received his B.Sc. (Hons) and M.Sc. degrees from Calcutta University and Ph.D. from the Indian Institute of Technology, Kharagpur. The major part of his career since 1966 has been focused on base and noble metals exploration/mining with short stopovers at Standard Oil (ESSO) Petroleum, Hindustan Copper Limited, and finally, Hindustan Zinc Limited, where he has undertaken various technical roles and managerial responsibilities. Since 2003, he has been Emeritus Scientist with the Department of Applied Geology, Presidency University, Kolkata, and has taught mineral exploration to postgraduate students of the department and often at University of Calcutta and Indian School of Mines, Dhanbad. He has been a consultant with international exploration entities, namely, Goldstream Mining NL/IMX Resources Ltd, Australia, and Binani Industries Limited (BIL) Infratech Ltd, India. His profession has often required visits to mines and exploration camps of Australia–Tasmania, Canada, the United States, Germany, Portugal, France, Italy, The Netherlands, Switzerland, Saudi Arabia, Egypt, Bangladesh, Nepal, Jordan, and Israel. He is a life fellow of The Mining Geological and Metallurgical Institutes (MGMI) of India and the Indian Geological Congress. Dr. Haldar is recipient of "Dr. J. Coggin Brown Memorial (Gold Medal) for Geological Sciences" by MGMI. He has authored 40 technical papers and three books:

1. Exploration Modeling of Base Metal Deposits, 2007, Elsevier.
2. Mineral Exploration—Principles and Applications, 2013, Elsevier.
3. Introduction to Mineralogy and Petrology, 2014, Elsevier.

Dr. Haldar has a unique professional blend of mineral exploration, evaluation, and mineral economics with an essence of classroom teaching of postgraduate students of three celebrated universities over the past decade.

Preface

When I was a postgraduate student at Calcutta University in the mid-1960s, I did my thesis on the Sukinda chromite deposits of Orissa, India. Mining for chromite had just begun in the country. The chromite belt was located in a valley between the Mahagiri and the Daitari hill ranges. I was put up in a small hut inside the boundary of the mining camp. The area was a dense forest, populated by wild animals and a few tribal villages. I used to do geological studies in the forest from early morning to late afternoon. No one was allowed to step outside the mine's fenced area once twilight set in. No one could open the hut door at night. Elephants regularly visited the villages for food and would destroy green banana plants. We witnessed bear families fading away into the forest. Once, I experienced the smell of a tiger and saw leftover food in a small cave. Lateritic nickel was discovered in the area by the Geological Survey of India while I was there. My three months of fieldwork in the deep forest amid nature decisively influenced me forever. Back at the University, my thesis was highly acclaimed. Thereafter, I joined metal-mining corporate companies (copper–zinc–lead). Nevertheless, I could never forget that short stay in a remote mining camp with inadequate facilities and very few modern comforts. I promised myself that I would return to the platinum–nickel–chromite industry at an opportune phase and share my knowledge with students and fellow professionals.

This first experience made a permanent and passionate impact on me, to love nature, to learn the processes of the Universe, and to understand how the mystic mother Earth hosts minerals and metals from the core to the crust for the benefit of birds, animals, and human beings. When I first visited Sukinda chromite belt in 1964, it was a sleeping hamlet, gifted by nature's love, with dense forests, mineral wealth, and peace-loving tribal people. There were three small mines separated by kilometers and supported by fewer than 100 employees. The economic interest in the area was only for small-scale production of chromite ore for the indigenous market. M/s Ferro Alloys Corporation Limited (FACOR Ltd.), a few kilometers away in the Nausahi chromium belt, produced a few tons.

I recently revisited the area, nearly after 50 years, to bring my knowledge up-to-date about the present chromium–nickel–platinum resources in the world. The present status of the project confirmed my belief that lateritic nickel could not be economically extracted due to metallurgical complexities, and was happy to learn that platinum–palladium do exist at Sukinda–Nausahi Belt. However, what I saw saddened me. There are more than 25 surface and underground working mines separated by common boundaries. The forests have totally disappeared. The majestic Mahagiri range had been reduced greatly by bulldozing. Series of newly born hills appear parallel to Mahagiri range due to mine-waste dumping. Unpaved roads serve the movement of more than 1000 trucks daily for transporting ore to distant ferrochrome factories and seaports. The sky is gloomy throughout the day due to the mine and road dust. Social evils, crime, alcoholism, and other such abuses have increased. This is the other side of the coin. We have to balance between good and evil through proper self-protective education, counseling, and training. Making stringent mine-safety rules is not the solution—it has to be implemented in the right spirit. Otherwise, what will we leave for future generations?

Five decades have passed between the two visits. The demand and supply for chromite has increased manyfold to cater to the needs of domestic and international markets. The lateritic nickel resources are still under scientific and technological research and development without any commercial breakthroughs. The presence of platinum–palladium has been reported. This was how the present project was conceived. I was interested in comparing with world-class deposits like Bushveld chromite–platinum group metals (PGMs), South Africa, Sudbury nickel–copper, Canada, and a few new discoveries in Australia.

The project was initially supported by the Government of India's Department of Science and Technology (DST) under its Utilization of Scientific Expertise of Retired Scientists Scheme HR/UR/29/2007. The book has been partially funded by the financial support of DST, New Delhi.

The book is primarily intended for academic students of Applied Geology, faculty members, and researchers, geoscientists of exploration companies, and investors. It is divided into 11 chapters. Chapter 1 is a general introduction to precious metals including the PGMs–Nickel–Chromium. Chapter 2 deals with the host affinity of these metals. Chapters 3–8 describe the occurrences/deposits/mines of six continents: Africa, North America, South America, Asia, Oceania (including Australia), and Europe. The remaining Chapters 9–11 give an overview of genetic aspects, exploration approach, and status of in situ reserves and resource base including characteristics of global deposits.

I visited all possible platinum group of elements (PGEs)–Ni–Cr-bearing areas in India for reconnaissance studies with Mr. Finn Barrett, Exploration Geologist, IMX Resources Ltd., Perth. It is my privilege and pleasure to acknowledge my academic association with Finn over a decade that has inspired me to complete this mission. I am deeply indebted to Finn for helping me with information and innumerable publications throughout the writing of this book.

Dr. Tom Evans, Executive Manager Exploration at Lonmin Plc, United Kingdom, Avinash Sarin from Washington, D.C., and Soumi Haldar from Los Angeles enriched the pages with wonderful mineral images. I am delighted to work with Marisa LaFleur and Frank Tasha, Elsevier, and thankful for their very positive attitude in this journey.

My heartfelt thanks to Mr. R. Venkatesan, Paul Prasad Chandramohan and team of Elsevier for unique support to shape the book awesome.

I am thankful to Presidency University, faculty, and students for providing an academically enriching environment. I extend my sincere thanks to all the mining camps: FACOR India Ltd., Orissa Mining Corporation Ltd., Tata Steel, and Department of Mines and Geology, and the states of Tamil Nadu and Karnataka for extending support during my field visits to Sukinda–Nausahi, Sittampundi, and Byrapur. All figures are drafted by my dear students: Subham Sarkar, Bidisha Dey, and Promita Bhar.

My humble gratitude to all the past and present monks of Ramakrishna Math, Barisha, Kolkata for their blessings in this journey.

I am short of words in expressing my emotions for my wife, Swapna, for her continuous inspiration, particularly at times of mental distress, and for bringing me back to a working mode. My family team—Srishti and Srishta (grandchildren), Soumi (daughter), and Suratwant (son-in-law) are always my source of motivation and happiness.

> I shall be telling this with a sigh
> Somewhere ages and ages hence:
> Two roads diverged in a wood, and I–
> I took the one less traveled by,
> And that has made all the difference.
> *Robert Frost, 1920*

Presidency University
Kolkta, April 19, 2016   **S.K. Haldar**

List of Acronyms

GENERAL

BIC	Bushveld Igneous Complex
Com.	Complex
DMG	Department of Mines and Geology
DST	Department of Science and Technology
FACOR	Ferro Alloys Corporation Ltd
GS	Geological Survey
g/t	grams/tonne
IMFA	Indian Metals & Ferro Alloys Limited
IBM	Indian Bureau of Mines
Int.	Intrusive
MECL	Mineral Exploration Corporation Limited
ML	Mining Lease
Mt	Million tonnes
MYB	Mineral Year Book
PGE	Platinum Group of Elements
PGM	Platinum Group of Metals
PL	Prospecting License
ppb	parts per billion
ppm	parts per million
OIG	Iron Ore Group
OMC	Orissa Mining Corporation Limited
ROM	Run of Mine
RP	Reconnaissance Permit
t	tonnes
tpa/d	tonnes per annum/day
SIC	Sudbury/Stillwater Igneous Complex
SUC	Sukinda Ultramafic Complex
USGS/USBM	United States Geological Survey/United States Bureau of Mines

METAL AND MINERALS

Ag	Silver
Au	Gold
Co	Cobalt
Cu	Copper
Chr	Chromite
Cr	Chromium
Ir	Iridium
Ni	Nickel
Os	Osmium
PGM/E	Platinum Group of Metals/Minerals/Elements
Pd	Palladium
Pt	Platinum
Rh	Rhodium
Ru	Ruthenium

Introduction

1.1 PREAMBLE

This book will address three primary metals and their associated metals. The primary metals are Platinum Group of Elements or Metals (PGE or PGM), nickel, and chromium. The associated metals include copper, gold, silver, and cobalt. The platinum group consists of six metals and occurs naturally in close association with one another as well as with nickel, chromium, copper, and cobalt. The group of metals represents a unique geological process having a common source of metal supply and a genetic model. The PGM is among the least abundant group of elements of the Earth's element family. The working deposits in South Africa and Russia are by far the largest for the PGM. The nickel deposits are significant in Canada, United States, Australia, Africa, Brazil, New Caledonia, Philippines, Indonesia, and the Caribbean. The largest chromite deposits are in South Africa, Zimbabwe, Kazakhstan, Turkey, Finland, and India.

The metals platinum, nickel, and chromium will be discussed in the same order throughout the book. The individual metal will often be substituted by its common ore-forming minerals for simple understanding in the geological domain.

1.2 THE METALS

Platinum and palladium are comparatively more abundant in nature and commercially the most significant precious metals of the platinum group. The group comprises six rare metals including ruthenium (Ru), rhodium (Rh), palladium (Pd), osmium (Os), iridium (Ir), and platinum (Pt). These elements stand for Groups 8 to 10 with position in the fifth and sixth rows of the periodic table directly under iron (Fe), cobalt (Co), and nickel (Ni). The position of the platinum group and associated metals (iron, cobalt, nickel, copper, gold, silver, and chromium) is given in the Periodic Table (Fig. 1.1).

H																	He
Li	Be											B	C	N	O	F	Ne
Na	Mg											Al	Si	P	S	Cl	Ar
K	Ca	Sc	Ti	V	**Cr**	Mn	**Fe**	**Co**	**Ni**	**Cu**	Zn	Ga	Ge	As	Se	Br	Kr
Rb	Sr	Y	Zr	Nb	Mo	Tc	**Ru**	**Rh**	**Pd**	**Ag**	Cd	In	Sn	Sb	Te	I	Xe
Cs	Ba	*	Hf	Ta	W	Re	**Os**	**Ir**	**Pt**	**Au**	Hg	Tl	Pb	Bi	Po	At	Rn
Fr	Ra	**	Rf	Db	Sg	Bh	Hs	Mt	Ds	Rg	Cn	Uut	Fl	Uup	Lv	Uus	Uuo
	*	La	Ce	Pr	Nd	Pm	Sm	Eu	Gd	Tb	Dy	Ho	Er	Tm	Yb	Lu	
	**	Ac	Th	Pa	U	Np	Pu	Am	Cm	Bl	Cf	Es	Fm	Md	No	Lr	

FIGURE 1.1   Position of platinum group of elements, nickel, chromium, copper, gold, silver, and cobalt in the periodic table.

All six members of the platinum group share affinity to Fe, Co, and Ni with consequential formation of metallic bonds over ionic bonds. This preferential tendency sets them in the "siderophile" (iron-loving) group in the geochemical classification of elements (Box 1.1) (Ertel et al., 2008). The PGE, in addition, tends to associate with copper (Cu), silver (Ag), and gold (Au) to form covalent bonds with sulfur in preference to ionized bonds with oxygen. This affinity places them in the "chalcophile" (copper-loving) group of elements (Mungall, 2005a).

The PGMs are naturally occurring rare metals, sparsely distributed, and relatively costlier than gold and silver. The elements have similar physical properties and tend to occur together in an identical metallogenic environment.

The metals gold, iridium, osmium, palladium, platinum, rhodium, ruthenium, and silver are most commonly considered the group of noble

---

BOX 1.1 GEOCHEMICAL CLASSIFICATION OF ELEMENTS

Geochemical classification is the subdivision or grouping of elements according to their joint concentration in preferred host phases in a particular natural system:

Lithophile (rock-loving): The lithophile elements are the fundamental source of rock-forming minerals of the Earth's crust. The lithophile elements include aluminum, astatine, boron, barium, beryllium, bromine, calcium, chlorine, chromium, cesium, fluorine, iodine, hafnium, potassium, lithium, magnesium, sodium, niobium, oxygen, phosphorus, rubidium, scandium, silicon, strontium, tantalum, technetium, thorium, titanium, uranium, vanadium, yttrium, zirconium, tungsten, and the lanthanides. The lithophile elements occur naturally as stable ions in the forms of silicates, oxides, halides, phosphates, sulfates, and carbonates.

Siderophile (iron-loving): The siderophile elements include cobalt, gold, iridium, iron, manganese, molybdenum, nickel, osmium, palladium, platinum, rhenium, rhodium, and ruthenium. The elements are high-density ferromagnetic and paramagnetic transition metals that readily dissolve in iron as a solid solution or in the molten state. The elements have no affinity with oxygen. They exhibit strong chemical bonds with arsenic [e.g., Sperrylite ($PtAs_2$), Chloanthite ($NiAs_2$), Cobaltite (CoAsS), and Loellingite ($FeAs_2$)], sulfur {e.g., [Pentlandite ($FeNi)_9S_8$], and Molybdenite (MoS_2)}, carbon, phosphorus, and nitrogen. In nature, platinum occurs mainly in the elemental state. Siderophile elements often bond with chalcophile elements.

The elements are bound through a metallic bond with iron in the Earth's solid inner core. Most of them are known for their rarity in the Earth's crust and recognized as precious metals.

Chalcophile (ore/copper/bronze-loving): The chalcophile elements predominantly represent sulfide ores. The elements include antimony, arsenic, bismuth, cadmium, copper, gallium, germanium, indium, lead, mercury, polonium, selenium, silver, sulfur, tellurium, thallium, tin, and zinc. Chalcophile elements have low affinity for oxygen except for tin [e.g., Cassiterite (SnO_2)]. The natural form of the occurrences of chalcophile elements is as sulfides, selenides, and tellurides. The elements silver, copper, arsenic, sulfur, and bismuth often occur in the elemental state.

Atmophile (gas/volatile-loving): Atmophile elements occur as liquid or gaseous state and mostly remain on or above the surface or in atmosphere. The elements include all the inert gases from helium to radon, nitrogen, and hydrogen.

metals. These noble metals are largely resistant to oxidation and corrosion in moist air, unlike most base metals such as iron, nickel, lead, zinc, and copper. They are considered precious due to their rarity in the Earth's crust. The most common primary metals that are generally associated with platinum group are copper, nickel,

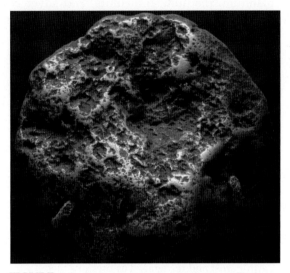

FIGURE 1.2   Native metals occur in their metallic forms, either as pure or as an alloy in nature—singly and/or in alloy with lead, copper, gold, platinum, nickel, and cobalt. Native platinum is one of the rarest and most precious metals in the Earth's crust, and was discovered in the 18th century. The platinum–palladium metals immensely contribute in pollution control of automobiles and marks the "Day of Love in Couple's journey in life." *Avinash Sarin.*

and/or chromium. The value-added secondary metals are copper, gold, silver, and cobalt.

1.2.1 Platinum Group of Metals

Platinum with symbol (Pt) and an atomic number of 78 in the Periodic Table is considered "rich man's gold." The metal was discovered by Antonio de Ulloa, a Spanish explorer, in 1735. The name "platinum" is derived from the Spanish word "platina," which literally means tiny form of "plata" or "little silver." The new metal (Fig. 1.2) was found as large placer deposits during the 16th century Spanish conquest of South America. It was called "platina del Pinto" after the Rio Pinto of Columbia. Platinum is silver-white to steel-gray or dark-gray in color and is lustrous. It is the most ductile among the pure metals. It is a malleable transition metal (Box 1.2) with a melting point at 1768°C. Pure platinum is a little harder (4–4.5) on the Mohs scale of hardness than pure iron (4). Platinum is one of the rarest elements in the Earth's crust (0.005 g/t), precious with high specific gravity between 14 and 22. It is generally nonreactive to oxidation, wear, and tarnish. This property makes it well suitable for fine jewelry. A famous native platinum nugget was recovered from the perfect circular, 8-km diameter Kondyor Massif intrusion in Russia. Platinum is often found in mining of secondary alluvial deposits in Colombia and the Ural Mountains, Russia.

BOX 1.2 TRANSITION METALS

An explicit set of metallic elements within Groups 3 to 12 in the Periodic Table contains an incomplete inner electron shell and serves as transition links between the most and the least electropositive series of elements. These metals are characterized by multiple valence, colored compounds, and an ability to form stable complex ions. The electrons they use to combine with other elements are present in more than one shell and often exhibit several common oxidation states. Iron, cobalt, and nickel are the most significant members in the transition-metal family and are the only common elements known to produce a magnetic field. The other metallic elements include platinum group, chromium, copper, gold, silver, lead, zinc, tungsten, manganese, mercury, etc.

Platiniridium is a rare naturally occurring native platinum/iridium alloy (Ir,Pt).

Palladium, the second significant member in the PGMs, has a chemical symbol of (Pd) and an atomic number of 46 in the Periodic Table. Palladium was discovered by William Hyde Wollaston in 1803 and named after the asteroid Pallas (goddess of wisdom) which was discovered at about the same time. It is a soft, malleable, ductile, silvery-white metal that resists oxidation and corrosion. Palladium has the lowest melting point at 1,555°C and lowest density (11.3–11.8 g/cm^3) among the group. Its metal dust is flammable. Palladium is usually associated with platinum, nickel, copper, and mercury ores. Palladium is commercially produced from nickel–copper deposits of South Africa and the Province of Ontario, Canada. The processing of large volumes of nickel–copper ore technically makes this extraction profitable despite its low concentration in these ores.

Iridium is a rarely occurring natural element with chemical symbol of (Ir) and atomic number of 77 in the Periodic Table. Iridium was discovered by the British chemist Smithson Tennani in 1803 and named from the Greek word "iris" meaning "Goddess of rainbow" for the brilliant color of many of its compounds. Iridium is an extremely hard and brittle metal, silvery-white in color and considered a transition element of the platinum family. It is the second-densest (22.56 g/cm^3) element after osmium (22.59 g/cm^3). It is the most corrosion-resistant metal, even at high temperatures. It has a melting point at 2410°C. Iridium is one of the least abundant elements in the Earth's crust, having an average mass fraction of 0.001 g/t in crustal rock. Metallic iridium is closely associated with platinum and other PGMs in alluvial deposits. Iridium is found in nature as free element or as natural alloys of iridium–osmium that include "osmiridium" and "iridosmine." These alloys are mixtures of iridium and osmium. The rare metals are recovered commercially as a byproduct from nickel mining and processing. Significant primary sources of iridium are in the Bushveld igneous complex in South Africa, the large copper–nickel deposits near Noril'sk in Russia, and the Sudbury Basin in Canada. Iron meteorites (Box 1.3), derived from small celestial bodies, are composed mostly of iron, nickel, gallium, germanium, and iridium (Fig. 1.3). The impact of falling on the Earth's crust causes partial melting and concentration of valuable metals.

Osmium is the least-abundant transition metal in the Earth's crust with chemical symbol (Os) and

BOX 1.3 METEORITE

Meteorites, small to extremely large size, are natural objects originating in outer space that fall onto the Earth creating a great surface impact. Most meteorites are derived from small celestial bodies as well as produced by the impacts of asteroids from the solar system. Meteorites are composed of silicate minerals and/or metallic iron–nickel. The structure of the igneous complex at the Sudbury Mining District, Canada, was formed as the result of a meteorite (1850 Ma age) impact that produced a 150–280 km multiring crater, containing 2–5 km thick sheet of andesitic melt. The immiscible sulfide liquid differentiated into Ni–PGE-dominated contact deposits by crystallization. There are over 100 deposits/mines having a total resource, including past production, of 1648 million tonnes (or 1817 million tons) at ~1% Ni, 1% Cu, and 1 g/t Pd + Pt.

Similarly, the Willamette meteorite in Oregon State is the largest in North America and 6th largest iron–nickel meteorite in the World, and contains 7.62% Ni, 18.6 g/t Ga, 37.3 g/t Ge, and 4.7 g/t Ir.

FIGURE 1.3 The iron–nickel rich meteorite contains +91% iron, 7.62% nickel, 18.6 g/t gallium, 37.3 g/t germanium, and 4.7 g/t Iridium. *Avinash Sarin.*

atomic number 76 in the Periodic Table. Osmium was discovered by the British chemist Smithson Tennani in 1803 along with Iridium during the same experiment. Osmium was named from the Greek word "osme" meaning "smell." Osmium is a hard, brittle and bluish-white metal in the platinum group. Its metallic luster remains unaffected even at high temperature. The metal has low compressibility. The metal is usually found as an un-combined trace element or as natural alloys, predominantly in platinum ores as osmium–iridium alloy. Osmium is the densest naturally occurring stable element, with a density of 22.59 g/cm³, as derived from X-ray diffraction data.

Ruthenium is a rare transition metal of the platinum group with chemical symbol (Ru) and atomic number 44 in the Periodic Table. Prof. Jędrzej Śniadecki, a Polish chemist, isolated the new element 44 from platinum ore in 1807 and named it "vestium." Later, Jöns Berzelius and Gottfried Osann confirmed the discovery of ruthenium in 1827 from the leftover residues after dissolving crude platinum in aqua regia. The metal is tin-white in color, with metallic luster, hardness of 6.5 on the Mohs scale with an average density of 12.2 g/cm³. The melting point

is high at 2334°C. Ruthenium usually occurs as a minor component of platinum ores.

Rhodium is one of the rarest and most valuable precious metals. It has the chemical symbol (Rh) and atomic number 45 in the Periodic Table. The element was discovered by William Hyde Wollaston, an English chemist, in 1803 soon after his discovery of palladium and named after the Greek word "rose." Rhodium is silver-white in color, hard, and a chemically inert transition metal and member of the platinum group. Rhodium usually occurs naturally as free-metal form or alloyed with metals of PGEs, and rarely as minerals such as Bowiete [rhodium–iridium–platinum mineral, $((Rh,Ir,Pt)_2S_3)$] and Rhodplumsite [a rare rhodium–lead sulfide mineral, $(Rh_3Pb_2S_2)$]. Rhodium is a noble metal, resistant to corrosion, found in platinum- or nickel ores together with the other members of the platinum group metals.

1.2.2 Nickel

The metal nickel reflects the technological advances of the 20th and 21st centuries, emerging as critically significant for stainless steel, a variety of specialty metal alloys, as well as currency, chemicals, pigment, and batteries. Nickel is a valuable transition metal with chemical symbol (Ni) and atomic number 28 in the Periodic Table. There is no firm date of discovery of the metal, but the unintentional use of nickel can be traced back to 3500 BC. The formal discovery of the metal in 1751 in a mineral (Niccolite or Nickeline) is credited to Baron Axel Frederik Cronstedt, a Swedish mineralogist and chemist. Apparently, he had expected to extract copper from this mineral, but instead he obtained a white metal. He called it nickel after the mineral from which it was extracted. The color is gray-white to silvery-white with metallic luster having golden tinge. Nickel is moderately hard (4–5), ductile and melting point at 1455°C. There are two main types of nickel deposits. The primary type is magmatic sulfide deposits where the principal ore mineral is Pentlandite [(Ni, Fe)$_9$S$_8$]. The secondary type is laterite hosted with principal ore minerals as Nickeliferous limonite (Fe, Ni) O (OH) and Garnierite [hydrous nickel silicate, (Ni, Mg)$_3$Si$_2$O$_5$(OH)]. There can be other types and sources of nickel such as (1) iron–nickel meteorites (Fig. 1.3), and (2) sea-bed nickeliferous copper–manganese poly-metallic nodules (Fig. 1.4).

1.2.3 Chromium

Chromium is an important commodity in the metal sector, particularly in the steel industry. The element represents its chemical symbol of (Cr) and atomic number 24. Use of the metal has been evidenced from burial pits dating from the late third century BC. However, by 1798, Louis Nicolas Vauquelin, a French pharmacist and chemist, isolated metallic chromium by heating chromium oxide in a charcoal oven, making him the discoverer of the element. Vauquelin also detected traces of chromium in precious gemstones, such as ruby or emerald. The name of the element is derived from the Greek word "χρῶμα," "chrōma", meaning "color" because many of its compounds are intensely colored. It is a steely-gray, lustrous, hard, and brittle metal which takes a high polish, resists tarnishing, and has a high melting point at 1907°C. Chromium is usually associated with PGMs in layered igneous intrusive and magmatic copper–nickel deposits.

1.2.4 Copper

Copper is one of the most functional transition metals in human society since antiquity. Copper is a chemical element with the symbol (Cu) and atomic number 29. The name is derived from the Latin word "cuprum." The metal was known to some of the oldest civilizations on record. It has a history of use in the Middle East in 9000 BC. It is a ductile metal with very high thermal and electrical conductivity and melting point at 1085°C. Copper often occurs naturally as native form, soft and malleable with a freshly exposed surface of reddish-orange color (Fig. 1.5). It occurs in native form or as minerals of copper sulfide (chalcopyrite and chalcocite), carbonate (azurite and malachite), and oxide (cuprite). Copper mineralization forms either by a Sedimentary Exhalative (SEDEX) process or as magmatic sulfide deposits associated with various noble metals.

FIGURE 1.4 Polymetallic nodules collected from the Indian Ocean bed containing 1.5% Ni, 1.4% Cu, 0.25% Co, 30% Mn, and 6% Fe. *Haldar, S.K., 2013. Mineral Exploration – Principles and Applications. Elsevier Publication, p. 374.*

FIGURE 1.5 Native copper from polymetallic sulfide deposit at Neves-Corvo mine, Portugal. *Haldar, S.K., 2013. Mineral Exploration – Principles and Applications. Elsevier Publication, p. 374.*

1.2.5 Gold and Silver

Gold is noble and precious, and documented as one of the first metals used during prehistoric culture. The Latin name for this noble metal is "aurum" with chemical symbol (Au) and atomic number 79. It is a dense, soft, malleable, and ductile transition metal with melting point at 1064°C. The color is bright yellow with metallic luster. A property of pure gold is remaining untarnished on exposure to air or water. Gold occurs mainly in native form (Fig. 1.6).

Magmatic gold–copper and gold–PGE deposits, and combined with tellurium-bearing minerals are Sylvanite [(Ag, Au) Te$_2$], Calaverite, or gold telluride (AuTe$_2$), Krennerite (Au$_3$AgTe$_8$) (Fig. 1.7), and Petzite (Ag$_3$AuTe$_2$).

Silver is a precious metal with chemical symbol (Ag) and atomic number 47. Silver is a soft, white, lustrous transition metal used for thousands of years. It possesses the highest electrical conductivity of any element and the highest thermal conductivity of any metal. The melting point is at 961.8°C. Silver occurs in native form (Fig. 1.7) and an alloy with gold ores containing arsenic, sulfur, and antimony. The common silver-bearing minerals are Argentite (Ag$_2$S), Chlorargyrite (AgCl), and Pyrargyrite (Ag$_2$SbS$_3$). The primary sources of silver include ores of copper, copper–nickel, lead, and lead–zinc.

1.2.6 Cobalt

Cobalt is one of the least-abundant elements in the Earth's crust with chemical symbol (Co) and atomic number 27. Cobalt is hard, lustrous gray metal with attractive appearance and resistance to oxidation. It is a transition metal crystallizing in the hexagonal crystal system. The name was derived from the German word "Kobold" meaning "evil spirits." The metal was used by Egyptian artisans as a coloring agent during 3000 to 2000 BC. Cobalt is one of the essential alloying elements in the Earth's crust. It has many strategic and irreplaceable industrial uses and is a central component

FIGURE 1.6 Native gold from Al Amar volcanogenic gold–zinc–copper deposit in volcaniclastic host rocks, ~180 km southwest of Riyadh, Saudi Arabia. The average grades are 7 g/t Au, 14 g/t Ag, 3.7% Zn, and 0.5% Cu. *Haldar, S.K., 2013. Mineral Exploration – Principles and Applications. Elsevier Publication, p. 374.*

FIGURE 1.7 Native silver occurs in various forms and sizes. Silver mining in the United States began on a major scale with the discovery of the Comstock Lode in 1858; 36 mines from 18 states continue silver production today in the United States. *Haldar, S.K., 2013. Mineral Exploration – Principles and Applications. Elsevier Publication, p. 374.*

of Vitamin B12. It has a high melting point (1495°C) and retains its strength to high temperature. Cobalt is frequently associated with nickel, and both are characteristic components of meteoric iron. The common cobalt bearing minerals include Cobaltite (CoAsS), Skutterudite (CoAs$_3$), Safflorite (CoAs$_2$) and Glaucodot [(Co,Fe)AsS].

1.3 THE MINERALS

Many large mafic and ultramafic-layered igneous plutonic/volcanic rocks contain thin layers enriched in PGEs, noble metals, and other copper–nickel sulfides and chromium oxides. This is a common feature in the Bushveld Igneous Complex, South Africa (Belinda et al., 2007). The metals are either in native form and/or as minerals containing the respective metals.

1.3.1 Minerals of Platinum Group of Metals

The common platinum-group minerals are Sperrylite (PtAs$_2$), Braggite [(Pt, Pd, Ni)S] or [(Pt$_5$Pd$_2$Ni)S$_8$], Cooperite [(Pt,Pd,Ni)S], Laurite (RuS$_2$), Vysotskite [(Pd,Ni)S], Merenskyite, Moncheite [(Pt,Pd)(Te,Bi)$_2$], and Isoferroplatinum [(Pt,Pd)$_3$ (Fe,Cu)].

1.3.1.1 Sperrylite

Sperrylite is widespread and the most common mineral of the platinum group and a major source of platinum. The mineral was first collected by Francis Louis Sperry, an American chemist, in 1887 from Vermillion nickel mine, Sudbury District, Ontario, Canada, and was named after him. It is an opaque platinum-arsenide mineral with chemical formula (PtAs$_2$). The color is shining tin-white with a black streak and a bright metallic luster. It is hard (6–7 on the Mohs scale) with high specific gravity of 10.6. Sperrylite in purest form contains 56.56% Pt and 43.44% As. It is predominantly associated with

FIGURE 1.8 Sperrylite crystal on weathered chalcopyrite mat from Broken Hammer copper nickel PGE deposit of Wallbridge Mining Company Limited, Sudbury Camp, North Range, Ontario, Canada. The crystal is approximately 8 mm across. *Dr Tom Evans.*

chromite ore in the Bushveld Igneous Intrusive, South Africa, and contributes as the largest source of platinum in the World. Sperrylite is closely associated with nickel ore in Sudbury, Canada (Fig. 1.8), Oktyabr'skove copper–nickel deposit of the Eastern Siberia Region, Russia and many other Cu–Ni–PGE deposits.

1.3.1.2 Braggite

Braggite is a sulfide mineral of platinum, palladium, and nickel with chemical formula [(Pt, Pd, Ni)S] or [(Pt$_5$Pd$_2$Ni)S$_8$]. The mineral was first described by William Henry Bragg and his son William Lawrence Bragg in 1932 for a sample collected from the Bushveld Igneous Complex, South Africa. The color, streak, and luster are steel-gray, white, and metallic, respectively. The mineral is soft (1.5 on the Mohs scale), and dense with high specific gravity of 10. The composition of Braggite stands between platinum-rich Cooperite and palladium-rich Vysotskite. The mineral is considered a primary economic ore containing two precious metals of the platinum–palladium group. The mineral in purest form contains 62.62% Pt, 17.08% Pd, 3.14% Ni, and 17.16% S. The mineral occurs in igneous intrusives at Bushveld, South Africa, the Stillwater Complex, Montana, United States, Lac des Iles,

Russia, The Great Dyke in Zimbabwe, and as platinum–iron nuggets in Madagascar.

1.3.1.3 Cooperite

Cooperite is a sulfide mineral composed of platinum, palladium, and nickel±copper with chemical formula [(Pt, Pd, Ni)S]. The mineral was first characterized in 1928 at Bushveld Igneous Complex by Richard A. Cooper, a South African metallurgist, and named after him. The color, streak, and luster are gray, brownish-yellow, and metallic, respectively. It is moderately hard (4.5 on the Mohs scale) with high specific gravity of 9.5. The mineral in purest form contains 62.62% Pt, 17.08% Pd, 3.14% Ni, 17.18% S, ±0.5–0.8% Cu. Cooperite is mined in sizable quantity at the Bushveld Complex, South Africa.

1.3.1.4 Laurite

Laurite is an opaque black metallic ruthenium sulfide mineral with chemical formula (RuS_2). The mineral was discovered in 1866 in Borneo and named for Laurie, wife of an American chemist. The color ranges between iron-black, white-gray, and bluish. The streak and luster are dark-gray and metallic, respectively. It is hard (7.5 on the Mohs scale) with moderate specific gravity (6.4). It occurs in ultramafic magmatic cumulate deposits and as placer types derived from the former. Laurite in pure form contains 61.18% Ru, and 38.82% S.

1.3.1.5 Vysotskite

Vysotskite is a palladium–nickel sulfide mineral with chemical formula [(Pd,Ni)S], and crystallizes in the tetragonal system. The mineral was named after Nikolai Konstantinovich Vysotskii, a geologist who found the platinum ore deposit at Noril'sk, Russia. The color ranges between silver-gray and white with a green streak. The mineral is soft (1.5 on the Mohs scale of hardness) with an average moderate specific gravity of 6.69. The average contents of elements in purest form of the mineral are 63.07% Pd, 11.59% Ni, and 25.34% S. The mineral reported from Mutoshi mine, Zaire,

North San Juan, Stillwater, Montana, United States, Mechanic Pluton, Canada, Skaergaard Intrusion, East Greenland, Talnakh Cu-Ni deposit, Noril'sk, Putoran plateau, Taimyr Peninsula, and Taymvrskiy, Eastern Siberia, Russia.

1.3.1.6 Merenskyite

Merenskyite is a rare telluride–bismuthinide-bearing platinum–palladium mineral with chemical formula [(Pd,Pt)(Te,Bi)$_2$]. The mineral was first described in 1966 for its occurrence at Merensky Reef in the Western Bushveld Igneous Complex, and named after Hans Merensky, a German–South African geologist. It is an opaque white to light gray metallic mineral. The hardness is medium at 3.5 on the Mohs scale with a moderately high specific gravity of 8.55. The average content of elements at purest form is 24.76% Pd, 5.04% Pt, 59.39% Te, and 10.81% Bi. The mineral usually occurs as inclusion in Chalcopyrite. It is a source of platinum, palladium, tellurium, and bismuth.

1.3.1.7 Michenerite

Michenerite is a rare telluride–bismuthinide bearing platinum–palladium mineral with chemical formula [(Pd,Pt)BiTe]. The mineral was named after its discoverer Charles Edward Michener, a Canadian exploration geologist. The color is silver-white with black streak and metallic luster. The hardness is low at 2.5 on the Mohs scale, and the average specific gravity is high at 9.5. The mineral in the purest form contains 17.16% Pd, 10.48% Pt, 44.93% Bi and 27.43% Te. The mineral occurs at Frood mine, McKim Township and Capre deposit (Fig. 1.9), Sudbury Camp.

1.3.1.8 Moncheite

Moncheite is a rare telluride–bismuthinide bearing platinum–palladium mineral with chemical formula [(Pt,Pd)(Te,Bi)$_2$]. It is named after the locality at Monche Tundra in Russia. The mineral is steel-gray to gray color. It has low hardness of 2–3 and moderately high specific gravity at 10. The

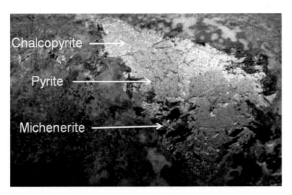

FIGURE 1.9 Michenerite [(Pd,Pt)Bi Te] occurs as intergrowth and inclusion in Pyrite and Chalcopyrite from the Capre Deposit, Sudbury Camp, Ontario, Canada. *Dr Tom Evans.*

average content of elements in its purest form is 31.21% Pt, 5.68% Pd, 40.83% Te and 22.29% Bi. It is source of platinum, palladium, tellurium, and bismuth from Monchegorsk Cu–Ni Deposit in the Kola Peninsula, Northern Region, Russia.

1.3.1.9 Isoferroplatinum

Isoferroplatinum is a gray to dark-gray metallic mineral with chemical formula [(Pt,Pd)$_3$ (Fe,Cu)]. The average content of elements in purest form is 76.13% Pt, 13.84% Pd, 2.70% Cu, and 7.27% Fe. The mineral usually occurs as rims on grains in an alluvial placer deposit, Tulameen River, British Columbia, Canada.

1.3.1.10 Skaergaardite

Skaergaardite is an intermetallic platinum mineral with the general chemical formula (PdCu). The mineral is a new mineral discovery and named after its discovery location: the Skaergaard intrusion, Kangerdlugssuaq area, East Greenland (Rudashevsky et al., 2004). The mineral name was approved by the International Mineralogical association in 2003. The color is steel gray with a bronze tint, with metallic luster and black streak. The hardness varies between 4 and 5 on the Mohs scale, and very high specific gravity at 10.64. Skaergaardite in pure form contains 59.42% Pd, 1.13% Pt, 0.29% Te, 2.27% Au, 30.09% Cu, 3.87% Fe, and rest of Zn, Pb, and Sn.

The Skaergaard mafic–ultramafic intrusion is the only known source of Skaergaardite, the most common mineraland sufficiently abundant to be mined. It is the most common source of the metal palladium.

1.3.2 Minerals of Nickel Group

1.3.2.1 Millerite

Millerite is a nickel sulfide (NiS) mineral and occurs as common metamorphic replacement of Pentlandite within serpentinite ultramafic bodies with radiating cluster of acicular needles. Millerite was discovered in 1845 and named after the British mineralogist W.H. Miller. The color is pale-brass or bronze-yellow with greenish black streak and tarnishes to an iridescent surface property. The hardness varies between 3.0 and 3.5 on the Mohs scale with moderate specific gravity between 5.3 and 5.65. The mineral possesses metallic luster, and in purest form contains 64.7% Ni, and 35.3% S. Millerite is a preferred source of nickel having the highest metal content among the remaining nickel-bearing group. It is found in Halls Gap area in Kentucky, United States.

1.3.2.2 Niccolite or Nickeline

Niccolite or Nickeline is a nickel arsenide mineral with chemical composition of (NiAs ± Fe, Sb, Co, and/or S). The unknown mineral was reported in 1751, but formally named Nickeline in 1832 and Niccolite in 1868 after Latin "niccolum." The color is distinct copper-red with pale brownish black streak and metallic luster. The specific gravity varies between 5.0 and 5.5 in Mohs scale with moderately high specific gravity of 7.3–7.7. The mineral in purest form contains 43.9% Ni and 56.1% As. Niccolite occurs in layered mafic–ultramafic intrusions at high magmatic temperature and differential segregation. The mineral is rarely used due to presence of arsenic which is deleterious to milling and smelting. However, the mineral is usually used as blending with "clean" ore which the mill and smelter can handle with acceptable recovery. The prime locations include Sudbury basin, Canada, Widgiemooltha Dome and Kambalda areas, Western Australia.

1.3.2.3 *Pentlandite*

Pentlandite is an iron nickel sulfide with chemical formula of [(Fe,Ni)$_9$S$_8$]. The mineral was discovered by and named after the Irish scientist Joseph Barclay Pentland. The color is yellow-bronze with light bronze-brown streak and metallic luster. The hardness varies between 3.5 and 4.0 on the Mohs scale with moderate specific gravity between 4.6 and 5.0. The mineral in purest form contains 22% Ni, 42% Fe, and 36% S. Pentlandite is the most common terrestrial nickel sulfide; it typically forms during cooling of magmatic sulfide melts during the evolution of parent silicate melt. Pentlandite typically concentrates within the lower margin of a mineralized layered intrusive. The best examples include the Sudbury (Fig. 1.10) deposits, Canada, Bushveld layered igneous intrusive in South Africa, the Voisey's Bay mafic troctolite (olivine, calcic-plagioclase, and pyroxene rock) intrusive complex in Canada, and Duluth gabbro intrusive in the United States. The other examples of Pentlandite deposits are Kambalda-type komatiitic ore in Western Australia, Noril'sk Ni–Cu–PGE in *trans*-Siberian Russia, and a few others in Namibia and Brazil, The supergiant Sudbury nickel deposit in Canada was formed by a large meteorite impact crater.

1.3.2.4 *Pyrrhotite*

Pyrrhotite (Fig. 1.11) is an unusual iron sulfide with variable iron content [Fe$_{(1-x)}$ S$_{(x=0-0.2)}$]. The color ranges between bronze yellow and copper-red with black streak and metallic luster. The hardness varies between 3.5 and 4.5 on the Mohs scale and an average specific gravity of 4.6. Pyrrhotite is naturally magnetic with the increase of iron content. Pyrrhotite is a common and important constituent in mafic igneous intrusive rocks such as norites. It occurs

FIGURE 1.10 The nickel-bearing massive Pyrrhotite and patches (black irregular boundary) of Pentlandite [(Fe, Ni)$_8$S$_9$] are intergrown from the Trill "offset inclusion quartz dyke" associated with the Sudbury Complex, Canada. The other nickel mineral Millerite (NiS) occurs as microlevel intergrown type. *Dr Tom Evans.*

FIGURE 1.11 Nickel-bearing bronze-brown color massive pyrrhotite with intergrowth of pale greenish Pentlandite from the Trill "offset inclusion quartz dyke" associated with the Sudbury Complex, Canada. *Dr Tom Evans.*

FIGURE 1.12 Crystalline brownish-black chromite from Kathpal mine, Sukinda Layered Igneous Complex, Orissa, India.

as segregation in layered intrusions associated with Pentlandite, Chalcopyrite, and other sulfides. The Pentlandite–Chalcopyrite–Pyrrhotite ore around the Sudbury structure formed from sulfide melts that segregated from the melt sheet produced by the meteoritic impact.

1.3.3 Minerals of Chromium

1.3.3.1 Chromite

Chromite (Fig. 1.12) is an iron chromium oxide (FeO,Cr$_2$O$_3$ or FeCr$_2$O$_4$ ± Mg). In its purest form, chromite ore contains 68% Cr$_2$O$_3$, and Cr:Fe ratio of 1.8:1. Louis Nicolas Vauquelin was the discoverer of the element "chromium," and named the mineral "ferrous-chromate-alumine" in 1798. Later, Wilhelm Haidinger named it "chromite" in 1845 with reference to its extremely high chromium content. Chromite occurs in mafic–ultramafic intrusives and as a process of early magmatic differentiation usually in layered form. The common host component is ultramafic igneous intrusive rocks. The typical host rocks are peridotite, pyroxenite, dunite, serpentinite, chromitite cumulate, and komatiite. The characteristic mineral assemblages are Olivine, Serpentine, Magnetite with low silicon, potassium, aluminum, and high to extremely high magnesium content. The commonly associated metals are nickel and PGE. The main uses of chromium are in metallurgical and chemical applications for manufacture of hard rustless steel, chrome plating, pigments, and dye. It is also used as refractory materials because of its high heat stability.

1.3.3.2 Crocoite

Crocoite is a rare lead chromate mineral having the chemical formula of (PbCrO$_4$). The mineral in the purest form contains 64.11% Pb and 16.09% Cr. It has a brilliant vivid yellow color for its best use in paints. The mineral was first reported at Berezovsk gold mines in the Urals region in 1766, and named "Crocoise" in 1832 from the Greek "Κρόκος" ("krokos" meaning "saffron"). The mineral is found in gold-bearing quartz veins traversing granite or gneisses in the presence of ultramafic rocks serving as source of chromium. The type areas include Russia, South Africa, Brazil, North America, Germany, Australia, and the Red Lead, West Comet, Platt, and a few other mines at Dundas, Tasmania.

1.3.4 Minerals of Copper

Copper-bearing minerals occur in nature by various chemical forms (carbonate, oxide, and sulfide), metal content (% Cu), and name for identification. The most important, common, and abundant copper mineral is Chalcopyrite.

1.3.4.1 Azurite

Azurite is a soft deep-blue (azure-blue) color copper carbonate hydroxide mineral [Cu$_3$(CO$_3$)$_2$(OH)$_2$]formed by weathering of copper orebodies (Fig. 1.13). The hardness varies between 3.5 and 4.0 on the Mohs scale with average specific gravity of 3.8. The mineral in the purest form contains 55.31% Cu. Azurite is seldom used as copper ore, but its presence is an excellent surface indicator and guide for exploration targeting in the vicinity and at depth. The

FIGURE 1.13 Azurite is a copper carbonate hydroxide mineral with distinct azure-blue color. It contains 55.31% copper in the purest form and its presence is an excellent exploration guide for discovery of new mineral deposits. *Soumi Haldar.*

FIGURE 1.14 Malachite is a copper carbonate hydroxide mineral with 57.48% copper in the purest form, and its presence is an excellent exploration guide for discovery of new mineral deposits. *Haldar, S.K., Josip Tišljar., 2014. Introduction to Mineralogy and Petrology, Elsevier Publication, p. 356.*

deep azure-blue color makes it suitable for use as pigment, decorative stone, and jewelry.

1.3.4.2 Malachite

Malachite is the another copper carbonate hydroxide mineral with chemical formula [$Cu_2CO_3(OH)_2$] formed by the weathering of copper orebodies in the vicinity. The color is bright green (Fig. 1.14) with light green streak. The hardness and specific gravity varies between 3.5–4.0, and 3.6–4.0, respectively. The mineral in the purest form contains 57.48% Cu. Malachite occurs worldwide including Congo, Gabon, Zambia, Namibia, Mexico, Australia, and with the largest deposit/mine in the Urals region, Russia. Malachite has been suitable for mineral pigment in green paints since antiquity, decorative vase, ornamental stone, and gemstone. Presence of malachite is an excellent surface indicator for copper exploration.

1.3.4.3 Bornite

Bornite or peacock ore is a copper-red to bronze-brown color sulfide mineral with formula (Cu_5FeS_4), and usually forms in the zone of supergene enrichment raising the metal content to 63.3% Cu in its purest form. The hardness and specific gravity varies between 3.0–3.25, and 4.9–5.3, respectively. It is a weathering and leaching product of supergene enrichment of copper

ore below the water table and above the fresh ore. Bornite occurs globally with the usual copper ore deposits.

1.3.4.4 Chalcocite

Chalcocite is a black lead-gray color copper sulfide (Cu_2S) mineral with shiny black streak. The harness and specific gravity vary between 2.5–3.0 on the Mohs scale and 5.5–5.8, respectively. Chalcocite is a characteristic mineral formed by weathering and leaching of copper sulfide minerals and concentration in the zone of supergene secondary enrichment below the zone of oxidation. The total process raises the metal content to the highest level of 79.8% Cu in Chalcocite, next to Cuprite (88.8% Cu), and the native form of 100% Cu.

1.3.4.5 Chalcopyrite

Chalcopyrite is the most important and major source of copper metal. It is widely spread copper–iron sulfide mineral with chemical formula ($CuFeS_2$). The color is typically brass to golden-yellow (Fig. 1.15), often tarnished, with greenish-black streak and submetallic luster. The hardness varies between 3.5 and 4.0 on the Mohs scale. The specific gravity ranges between 4.1 and 4.3. The mineral in the purest form contains 34.5% Cu, 30.5% Fe, and 35.0% S. Chalcopyrite also

FIGURE 1.15 Massive high-grade Chalcopyrite (brassy-gold) and Pyrrhotite (bronze-brown) ore in Chlorite–Amphibole ± Garnet schist host rock at Kolihan Section, Rajasthan, India.

contains gold, nickel, and cobalt in solid solution and may be closely associated with PGMs formed by mafic/ultramafic igneous intrusive and in greenstone belts. Chalcopyrite is the primary source of copper metal associated with many high value commodities. The other type of Chalcopyrite deposits is formed by a SEDEX process in association with lead and zinc ± gold, silver, and/or nickel.

1.3.4.6 Covellite

Covellite or Covelline is a rare copper sulfide mineral with chemical formula (CuS). The mineral was named after the Italian mineralogist N. Covelli in 1832. The color ranges between indigo-blue, brass-yellow, and deep red with lead-gray streak. The mineral is soft (hardness 1.2–2.0 on the Mohs scale), and naturally flexible in thin lamina. The specific gravity varies between 4.6 and 4.8. Covellite is a characteristic mineral formed by weathering and leaching of copper sulfide minerals, and concentration in the zone of secondary supergene enrichment below the zone of oxidation. The total process raises the metal content to the level of 66.4% Cu in Covellite.

1.3.4.7 Cuprite

Cuprite is a copper oxide mineral (Cu_2O) with cochineal-red, crimson-red, and black color and shining brown, red streak. Cuprite was discovered in 1845, and the name derived from the Latin word "cuprum" meaning copper.

The distinctive red color crystal resembles "ruby" and known as "ruby copper." The hardness varies between 3.5 and 4.0 on the Mohs scale. The specific gravity ranges between 5.85 and 6.15. Cuprite occurs as secondary mineral in the oxidized zone of copper sulfide deposits and usually associated with native copper, Azurite, Malachite, and a variety of iron oxide minerals. The mineral in its pure form contains 88.8% Cu.

1.3.5 Minerals of Gold and Silver

Gold- and silver-bearing minerals are rare in nature. The deposits occur as hydrothermal auriferous quartz veins in igneous and metamorphic rocks, volcanic exhalative sulfide deposits, associated with igneous intrusive complexes along with copper–nickel–PGMs and in consolidated to unconsolidated placer deposits. The metal contents are extremely low measured in ppm or g/t. The minerals include:

1.3.5.1 Sylvanite

Sylvanite or gold–silver telluride is the common mineral of gold–silver–tellurium with a chemical formula of $[(Ag,Au)Te_2]$. Sylvanite was first reported in Transylvania, and named after the location. The mineral mostly occurs in low-temperature hydrothermal veins. The color ranges from steely gray to almost white with silver-gray streak and metallic luster. The mineral is soft (hardness 1.5–2.0 on the Mohs scale), and moderately high specific gravity of 8.2. The mineral in purest form contains 34.36% Au, 6.28% Ag, and 59.36% Te. The mineral is reported/mined in East Kalgoorlie District, Australia, Kirkland Lake Gold District, Ontario, Rouyn District, Quebec, Canada, and Cripple Creek gold deposit, United States.

1.3.5.2 Calaverite

Calaverite or gold telluride is an uncommon telluride of gold mineral with chemical formula $(AuTe_2)$. It was discovered in Calaveras County, California, in 1861 and named for the county.

The color ranges between brass-yellow and silvery-white with green to yellow-green streak and metallic luster. The harness is between 2.5 and 3.0 on the Mohs scale and specific gravity between 9.1 and 9.3. It usually occurs as low-temperature vein-type deposit. Calaverite contains 43.56% Au and 56.44% Te. The mineral occurs in association with Sylvanite.

1.3.5.3 Krennerite

Krennerite is a gold-telluride mineral with chemical formula ($AuTe_2$) (Fig. 1.16). The mineral was discovered in 1878 at Săcărâmb gold mine, Romania, described by the Hungarian mineralogist Joseph Krenner, and named after him. The color varies from silver-white to brass-yellow with greenish-gray streak and metallic luster. It has specific gravity of 8.53, and hardness of 2.5 on the Mohs scale. Krennerite contains 43.56% Ag, and 56.44% Te. It can partially substitute gold by a relatively small amount of silver in the structure, in which case the formula changes to $[(Au_{0.8},Ag_{0.2})Te_2]$. It occurs in high-temperature hydrothermal environments.

1.3.5.4 Petzite

Petzite is a silver-gold telluride mineral (Ag_3AuTe_2) that usually occurs in vein-type deposits by high-temperature hydrothermal activity. The mineral is often associated with mercury and copper. W. Petz, a chemist, analyzed the mineral for the first time in 1845 collected from the type locality at Săcărâmb gold mine, Transylvania, Romania. The color changes between steel-gray and iron-black with grayish-black streak and metallic luster. The hardness changes between 2.5 and 3.0 on the Mohs scale and specific gravity between 8.7 and 9.14.

1.3.5.5 Argentite

Argentite is the primary source of silver, and occurs as a cubic silver sulfide mineral (Ag_2S). The mineral has been known since 1529 by local names as "glaserz," "silver-glance," and

FIGURE 1.16 Krennerite is a gold telluride ($AuTe_2$) mineral containing a relatively small amount of silver in the structure. The mineral is often found at Cripple Creek, Colorado, United States. *Soumi Haldar.*

"vitreous-silver." The name "argentite" (from the Latin "argentums" meaning silver) was used only in 1845. The color is lead-gray with shining streak, and metallic luster. Argentite is a soft mineral with hardness ranging between 2.0 and 2.5 on the Mohs scale, and specific gravity between 7.2 and 7.4. Argentite contains 87.0% Ag and 13.0% S. A related Cu-rich mineral was also reported in 1858 from Leonora Mine at Jalpa, Zacatecas, Mexico, and named "jalpaite" containing 71.71% Ag, 14.08% Cu, and 14.21% S.

1.3.5.6 Chlorargyrite

Chlorargyrite is a silver-chloride (AgCl) mineral and occurs naturally as a secondary mineral formed during the oxidation process of silver deposits. The mineral was first reported in 1877 from Broken Hill District, New South Wales, Australia, and named from the Greek, "chloros" for "pale green" and Latin for silver, "argentums." The color changes to purple or brown on exposure to light. The streak is white with adamantine luster. It is extremely soft with hardness of 1 to 2 on the Mohs scale. The specific gravity is moderate at 5.5. The mineral in purest form contains 75.26% Ag and 24.74% Cl.

1.3.5.7 Pyrargyrite

Pyrargyrite is a complex sulfosalt mineral consisting of silver sulfantimonide (Ag_2SbS_3). The name is derived from the Greek, "pyr" and "argyros," "fire-silver" with reference to color and rich silver content. The color is usually dark-red to red-black with purplish-red streak and metallic luster. The hardness on the Mohs scale is 2.5, and specific gravity is 5.8. The mineral occurs as metalliferous veins of Calcite, Aragonite, Argentiferous-galena, arsenic, and native silver. The mineral in purest form contains 59.75% Ag, 22.48% Sb, and 17.76% S. This silver ore is mined from the Comstock Lode, Nevada, United States, and many other silver mines in Mexico.

1.3.6 Minerals of Cobalt

1.3.6.1 Cobaltite

Cobaltite is a strategically significant mineral (Fig. 1.17) composed of cobalt, arsenic, and sulfur (CoAsS) crystallizing in the orthorhombic (or pseudocubic) system resembling pyrite. It contains up to 10% iron and a valuable amount of nickel. The name is derived from the German "Kobold" meaning "underground

FIGURE 1.17 Cobaltite was discovered around 1790 in disseminated sulfide from skarn-type copper–cobalt ore field (Stor Mine), Tunaberg Cu–Co Ore Field, Sweden. The mining was abundant after producing 61.97 tonnes (68 tons) of cobalt until 1905. *Soumi Haldar.*

spirit" or "goblin" in allusion to the refusal of cobaltiferous ore to smelt appropriately. The color changes from reddish silver-white, violet steel-gray to black with grayish-black streak and metallic luster. The hardness is moderately high of 5.5 on the Mohs scale, and density of 6.3. Cobaltite contains 35.5% Co, 45.2% As, and 19.3% S ± Fe and Ni. The mineral is mostly found in the Cobalt District, Ontario, Canada; Sweden; Norway; Germany; Cornwall, England; Australia; Democratic Republic of the Congo; and Morocco.

1.3.6.2 Skutterudite

Skutterudite (Fig. 1.18) is a cobalt arsenide mineral with variable substitute amounts of nickel and iron (Fig. 1.19). The general chemical formula is [(Co, Fe, Ni)As$_{2-3}$]. Skutterudite was first reported from Skuterud/Skutterud mines, Modum Cobalt mines, Buskerud, Norway, and named for the locality. The color is between steel-gray and tin-white with black streak and metallic luster. The mineral is moderately hard at 5.5 to 6.0 on the Mohs scale, and specific

FIGURE 1.18 Cluster of bright and shining crystalline form of Skutterudite with Calcite, a cobalt arsenide mineral with variable proportion of cobalt, nickel, and iron, occurs as numerous hydrothermal Quartz–Amethyst veins in Precambrian serpentine rocks extending over 35 km along fault zone near Ouarzazate Province Village, Amerzqane, Bou Azzer, Morocco. *Avinash Sarin.*

FIGURE 1.19　Well-developed nickel–Skutterudite containing 16.39% Ni, 5.49% Co, and 78.12% As, Schoenberg, Germany. *Avinash Sarin.*

gravity between 6.5 and 6.9. The general range of metal contents is 12.0–18.5% Co, 7.0–7.7% Ni, 76.1% As, 1.4–8.8% Fe, and ±120 g/t Au.

1.4 PHYSICAL AND CHEMICAL PROPERTIES

The physical properties of minerals include name, crystal system, color as it appears to the naked eye, streak by rubbing on streak plate, luster, hardness on the Mohs scale, and average specific gravity. The chemical properties comprise chemical formula and the proportion of elements in the purest form of the minerals. The information base extends with short description of mode of occurrences, and major uses of the metals and minerals. This has briefly been tabulated in (Table 1.1) for ready reference. The sequence of presentation has been intended to start with PGE, nickel, chromium, and associated metals followed by their minerals.

1.5 GLOBAL DISTRIBUTION

The major platinum group of deposits occur in close association with nickel–chromium–copper–gold–cobalt ores as sulfides ($(Pt,Pd)S$), tellurides ($Pt,BiTe$), antimonides ($PdSb$), and arsenides ($PtAs_2$). PGMs seldom occur as free elements in native form as nuggets. The common occurrences are usually in alloyed form mixed with other transition metals including palladium, ruthenium, rhodium, iridium, osmium, gold, nickel, copper, and cobalt. The wide-ranging platinum minerals include Sperrylite (platinum arsenide, $PtAs_2$) and Cooperite (platinum sulfide, PtS). The largest known resources of PGMs in the form of main/byproduct are from the mines at Bushveld chromite, in South Africa; Great Dyke complex in Zimbabwe; Noril'sk copper-nickel and Urals alluvial deposits, Russia; Stillwater, East Boulder mine, and Duluth in the United States, Lac des Iles palladium–nickel–gold–copper mine, nickel–platinum camp in Sudbury Basin and Raglan in Canada; Skaergaard gold–PGM deposit in Greenland; Munni–Munni, Panton, and Weld Range in Australia; Colombia. South Africa is the top producer of platinum with ~80% share followed by Russia at ~11%.

Palladium is found as the free metal associated with platinum and other PGMs in Australia, Brazil, Russia, Ethiopia, and North and South America, as well as with nickel and copper deposits (from which it is recovered commercially) in Canada and South Africa.

Iridium and osmium occur as the free elements and naturally alloyed with platinum and other PGMs. The most common alloys are osmiridium (Os,Ir), and iridiosmium, (Ir,Os), both containing mixtures of iridium and osmium. This rare alloy is extremely corrosion resistant, and mined in the remote rocky country around the Pieman River course on the western coast of Tasmania, Australia.

TABLE 1.1 List of Common Platinum Group of Elements (PGEs), Nickel, Chromium, and Associated Precious Metals and Their Minerals, Diagnostic Features, and Uses

Name and Formula	Crystal System	Color/Streak	Luster	Hardness*/Sp. Gr (g/cm³)	Cleavage/ Fracture	% Content of Valuable Component	Origin or Occurrence	Major Uses
MINERALS OF PLATINUM GROUP OF METALS								
Native platinum (Pt) (Fig. 1.2)	Isometric, nuggets and irregular lumps, mostly magnetic and occasionally with polarity	Whitish, steel to dark gray	Metallic	4.0–4.5 14–22 (18)	None, Hackly– Jagged, torn surfaces	100 Pt Pt alloy: 0.5–2 Pd, 1–3 Ir, 1–3 Rh, tr Os, 0.5–2 Cu, 8–18 Fe	Magmatic sulfide deposits usually in layered igneous intrusive alloyed/ in close association PGM, Cr, Ni, Cu, Co.	Catalytic converter in vehicle emission control, electrical contacts, electronics and electrodes, laboratory equipment, dentistry, medicine, jewelry, currency, trading and investment.
Native Palladium (Pd)	Granular	White, steel-gray	Metallic	4.5–5.0 11.3–11.8	None/Elastic fragments	100 Pd Pd alloy: 79.24 Pd, 18.16 Pt, 2.60 Fe		Substitute for silver in dentistry and jewelry. Pure metal is used as delicate mainsprings in analog wristwatches, in surgical instruments, and as a catalyst.
Iridium (Ir)	Isometric	White/White	Metallic	6.0–7.0 22.59	None, brittle	100 Ir Ir alloy: 52.58 Ir, 31.22 Os, 5.53 Ru, 10.67 Pt	All	Pt–Ir alloys are used in manufacturing industries for production of machine parts, containers, fountain-pen nib tip, electric contacts exposed to high temperatures and chemicals. The other uses include jewelry, electrodes of spark plugs.

Continued

TABLE 1.1 List of Common Platinum Group of Elements (PGEs), Nickel, Chromium, and Associated Precious Metals and Their Minerals, Diagnostic Features, and Uses—cont'd

Name and Formula	Crystal System	Color/Streak	Luster	Hardness*/Sp. Gr (g/cm³)	Cleavage/Fracture	% Content of Valuable Component	Origin or Occurrence	Major Uses
Osmium (Os)	Hexagonal	Steel-gray/Gray	Metallic	6.0–7.0 22.59	Perfect on (0001), brittle	100 Os Os alloy: 74.80 Os, 25.20 Ir	Trace element and natural alloy with platinum group metals	Os–Ir alloy used as electrical contacts, fountain pen tips and component of extreme durability and hardness.
Ruthenium (Ru)	Hexagonal, tabular, close packed	Tin-white/	Metallic	6.5 12.2	–	100 Os Os alloy: 44.16 Ru, 41.99 Ir, 13.85 Os	Occurs as a minor component of platinum ores.	Hard Ru–Pt–Pd alloy used for wear-resistant electrical contacts and the manufacture of thick-film resistors.
Rhodium (Rh)	Isometric	Tin-white/	Metallic	3.5 16.51	–	100 Rh Rh alloy: 61.28 Rh, 38.72 Pt	Occurs in platinum- or nickel ores together with the other members of the PGMs.	Most rhodium produced is converted to hard platinum–palladium alloy and used for wear-resistant electrical contacts and the manufacture of thick-film resistors.
Sperrylite ($PtAs_2$) (Fig. 1.8)	Isometric	Tin-white/Black	metallic	6–7 10.58	Indistinct on (001), Conchoidal	56.56 Pt, 43.44 As	Magmatic sulfide deposits	Most common and major source of platinum mineral associated with nickel ore in Sudbury, Ontario, Canada. Bushveld intrusive, South Africa, and Oktyabr'skove Cu–Ni deposit, Russia.

Mineral	Crystal system	Color/Streak	Luster	Hardness / Specific gravity	Cleavage/fracture	Composition	Occurrence	Uses
Braggite [(Pt, Pd, Ni) S] or [(Pt$_5$Pd$_2$Ni)S$_8$]	Tetragonal	Steel-gray/White	Metallic	1.5 / 10	None	64Pt, 27 Pd, 14 Ni	Magmatic segregation in layered igneous intrusion	Source for platinum, palladium, and nickel. Occurs at Bushveld, South Africa, Stillwater, Montana, USA, Lac des Iles, Russia, the great dyke in Zimbabwe and as platinum-iron nuggets from Madagascar
Cooperite [(Pt, Pd, Ni)S]	Isometric	Gray/Brownish-yellow	Metallic	4.5 / 9.5	Thin platy havits, occassionally twined.	62.62Pt, 17.08 Pd, 3.14 Ni, 17.18S, ±0.5–0.8 Cu	Bushveld igneous complex	Ore of platinum (PtS), palladium (PdS) and Nickel (NiS) ± Cu.
Laurite (RuS$_2$)	Cubic, octahedral, pyritohedral	Iron-black, white-gray. Bluish/Dark gray	Metallic	7.5 / 6.43	Parfect (111), Subconchoidal, brittle	61.18 Ru, 38.82S	Magmatic ultramafic cumulate and	Primary source of ruthenium and associated with Cooperite, Braggite, Sperrylite and other PGEs and chromite.
Vysotskite [(Pd,Ni)S]	Tetragonal	Silver gray, white/Green	Metallic	1.5 / 6.69	Irregular grains	63.07 Pd, 11.59 Ni, 25.34S	Magmatic sulfide deposits at Noril'sk region, Russia	Source of palladium and nickel
Merenskyite [(Pd,Pt)(Te,Bi)$_2$]	Trigonal	White, grayish-white	Metallic	3.5 / 8.547	Tiny grains	24.76 Pd, 5.04Pt, 59.39 Te, 10.81 Bi	Inclusion in Chalcopyrite at Rustenburg mine, Bushveld complex, South Africa	Source of telluride and bismuth with Pt and Pd.
Michenerite [(Pd,Pt)BiTe] (Fig. 1.9)	Isometric	Silver-white/Black	Metallic	2.5 / 9.5	None, granular, brittle	17.16 Pd, 10.48 Pt, 44.93 Bi, 27.43 Te	Frood mine, McKim Township and Capre deposit, Sudbury, Canada	Source of Palladium, platinum, bismuth, and telluride.

Continued

TABLE 1.1 List of Common Platinum Group of Elements (PGEs), Nickel, Chromium, and Associated Precious Metals and Their Minerals, Diagnostic Features, and Uses—cont'd

Name and Formula	Crystal System	Color/ Streak	Luster	Hardness*/Sp. Gr (g/cm³)	Cleavage/ Fracture	% Content of Valuable Component	Origin or Occurrence	Major Uses
Moncheite [(Pt,Pd)(Te,Bi)₂]	Trigonal-hexagonal	Steel-gray/ Gray	Metallic	2.0–3.0 10	Good (0001)	31.21 Pt, 5.68 Pd, 40.83 Te, 22.29 Bi	Magmatic Cu–Ni deposit at Monchegorsk, Kola peninsula, Russia.	Source of platinum, palladium, tellurium, and bismuth.
Isoferroplatinum [(Pt,Pd)₃ (Fe,Cu)]	Isometric	Gray to dark-gray/ Dark brown and tarnishes in air	Metallic	5.0 16.5	Perfect basal cleavage, malleable, deformed rather than breaking away	76.13 Pt, 13.84 Pd, 2.70 Cu, 7.27 Fe	As rims on Grains in alluvial placer deposit, Tulameen River, British Columbia, Canada.	Rich in platinum, palladium and copper metals.
Skaergaardite (PdCu)	Isometric, Hexoctahedral	Steel gray–bronze tint/ metallic	Black	4–5 10.6	–	59.42 Pd, 1.13 Pt, 0.29 Te, 2.27 Au, 30.09 Cu	Skaergaard mafic-ultramafic intrusion is the only known source.	Rich source of palladium metal.
MINERALS OF NICKEL GROUP								
Native Nickel (Ni)	Isometric	Gray-white, silvery-white/ Greyish-white	Metallic	4.0–5.0 8.90	None, Ductile	100 Ni	Magmatic sulfide deposits, laterites, meteorites and sea bed polymetallic nodules.	The primary applications of nickel include stainless steel, nickel cast iron, super-alloy in jet engines, alnico (aluminum-nickel-cobalt) alloy in horseshoe magnet, nickel plating and polishing, coinage, rechargeable batteries, electric guitar strings, special alloys and green tint in glass.

Mineral	Crystal system	Colour / Streak	Luster	Hardness / Specific gravity	Cleavage / Fracture	Composition	Occurrence	Uses
Millerite (NiS)	Rhombohedral	Pale brass or bronze yellow / greenish black	Metallic	3.0–3.5 / 5.3–5.65	Perfect, uneven, brittle	64.7 Ni, 35.3 S	Common metamorphic mineral replacing pentlandite within serpentinite ultramafic bodies with radiating cluster of acicular needles.	High grade source for Ni, used for stainless steel, super alloys, electroplating, alnico magnets, coinage, rechargeable batteries, electric guitar strings, microphone capsules, green tint in glass.
Niccolite or nickeline (NiAs)	Hexagonal	Pale copper-red / Pale brownish black	Metallic	5.0–5.5 / 7.33–7.67	Massive, reniform-columnar, Uneven, brittle	43.9 Ni, 56.1 As	Layered mafic-ultramafic intrusion at high magmatic temperature and differential segregation	Rarely used due to presence of arsenic, deleterious to smelting and milling, except blending with 'clean' ore which the mill and smelter can handle with acceptable recovery.
Pentlandite ((Fe,Ni)$_9$ S$_8$) (Fig. 1.10)	Isometric	Pale bronze yellow / Bronze brown	Metallic	3.5–4.0 / 4.6–5.0	Absent, Octahedral parting, Uneven	22 Ni, 42 Fe, 36 S	Layered mafic-ultramafic intrusion at high magmatic temperature, differential segregation	Primary source of nickel associated with PGE, tarnish resistant stainless steel, super alloys, electroplating, alnico magnets, coinage, rechargeable batteries, electric guitar strings, microphone capsules, green tint in glass.

Continued

Name and Formula	Crystal System	Color/ Streak	Luster	Hardness*/Sp. Gr (g/cm³)	Cleavage/ Fracture	% Content of Valuable Component	Origin or Occurrence	Major Uses
Pyrrhotite [Fe$_{(1-x)}$ S$_{(x=0-0.2)}$] (Fig. 1.11)	Hexagonal	Bronze-yellow to copper-red / Black	Metallic	3.5–4.5 4.58–4.64	Absent, Uneven	0.4 Fe, 39.6 S, ± Ni	Common in large mafic-ultramafic layered intrusive and sulfide deposits. The former contains Rh, Os, Ir and Ru.	No specific application other than source of sulfur for making sulfuric acid, rarely for recovering iron due to complex metallurgy and often Nickel bearing.

MINERAL OF CHROMIUM

Name and Formula	Crystal System	Color/ Streak	Luster	Hardness*/Sp. Gr (g/cm³)	Cleavage/ Fracture	% Content of Valuable Component	Origin or Occurrence	Major Uses
Chromium (Cr)	Isometric	White to steel gray	Metallic	7.5 7.2	Brittle	100 Cr		The primary applications of chromium are in metallurgy, refractory materials, dye and pigment, synthetic ruby and laser, wood preservative, tanning, and catalysts.
Chromite (FeCr$_2$O$_4$) (Fig. 1.12)	Isometric, octahedral, massive	Black, brown/ Brown	Submetallic	5.5 4.1–4.9	None, Uneven, brittle	68.0 Cr$_2$O$_3$ or 46.46 Cr, 32.0 FeO or 24.95 Fe, 28.59 O	Layered mafic – ultramafic intrusion at high magmatic temperature, differential segregation, crystallization	Primary source of chromium and applications in hard rustless steel, chrome plating, anodizing of aluminum, superalloys, refractory bricks, pigments and dyes, synthetic ruby, wood preservative, leather tanning and catalysts for hydrocarbon processing.

Mineral	Crystal system	Colour/Streak	Lustre	Hardness / Specific gravity	Cleavage/Fracture	Composition (%)	Occurrence/Association	Uses/Remarks
Crocoite (PbCrO$_4$)	Monoclinic, prismatic	Orrange, red and yellow / Yellowish orange	Adamantine	2.5–3.0 / 5.9–6.1	Distinct on (110), Conchoidal to uneven	64.11Pb, 16.09 Cr, 19.80 O	Associated with quartz veins in Russia, Tasmania, Brazil.	Low-grade source of chromium and mineral collection for brilliant magnificent colored crystal.

MINERALS OF COPPER

Mineral	Crystal system	Colour/Streak	Lustre	Hardness / Specific gravity	Cleavage/Fracture	Composition (%)	Occurrence/Association	Uses/Remarks
Native copper (Cu) (Fig. 1.5)	Cubic	Pale-rose to copper-red / Copper-red	Metallic	2.5–3.0 / 8.95	None/Hackly-jagged, highly malleable and ductile	100.00	Occurs as a natural mineral.	Source of rich copper metal with major applications in electrical wires, cables, plumbing, currency, utensils, machinery, alloy, architecture, nutritional supplements and fungicides in agriculture
Azurite [Cu$_3$(CO$_3$)$_2$(OH)$_2$] (Fig. 1.13)	Monoclinic	Azure-blue,/ Light blue	Vitreous	3.5–4.0 / 3.77–3.78	Perfect on (011), Conchoidal	55.31 Cu, 0.58H, 6.97C, 37.14 O	Azurite is often pseudomorph after Malachite and occurs together	The intense color makes it suitable for pigments, decorative/collective stone and jewelry. Presence of azurite is an excellent surface guide for exploration target searching for copper ore.
Malachite [Cu$_2$CO$_3$(OH)$_2$] (Fig. 1.14)	Monoclinic, prismatic	Bright green/Light green	Adamantine to vitreous	3.5–4.0 / 3.6–4.0	Perfect on (201), Subconchoidal to uneven	57.48 Cu, 0.91H, 5.43C, 36.18 O	Often results from weathering of copper ore	Used as mineral pigment in green paints from antiquity, decorative vase, ornamental stone and gemstone. Malachite is an excellent surface indicator for copper exploration.

Continued

TABLE 1.1 List of Common Platinum Group of Elements (PGEs), Nickel, Chromium, and Associated Precious Metals and Their Minerals, Diagnostic Features, and Uses—cont'd

Name and Formula	Crystal System	Color/Streak	Luster	Hardness*/Sp. Gr (g/cm³)	Cleavage/ Fracture	% Content of Valuable Component	Origin or Occurrence	Major Uses
Bornite (Cu₅FeS₄)	Orthorhombic	Copper-red, bronze-brown, purple/Black	Metallic	3.0–3.25 4.9–5.3	Imperfect on (111), Conchoidal	63.3 Cu, 11.1 Fe, 25.6S	In zone of secondary supergene enrichment, source of rich copper metal	Source of rich-grade copper, major applications in electrical wires, cables, plumbing, currency, utensils, machinery, alloy, architecture, nutritional supplements and fungicides in agriculture
Chalcocite, (Cu₂S)	Orthorhombic	Black lead-gray/Shiny black	Metallic	2.5–3.0 5.5–5.8	Indistinct on (110), Conchoidal	79.8 Cu, 20.2S	Zone of secondary supergene enrichment, source of rich copper metal	Source of rich copper with major applications in electrical wires, cables, plumbing, currency, utensils, machinery, alloy, architecture and nutritional supplements and fungicides in agriculture
Chalcopyrite, (CuFeS₂), (Fig. 1.15)	Tetragonal	Brass-yellow, often tarnished/Greenish black	Metallic	3.5–4.0 4.1–4.3	Indistinct on (011), Uneven, brittle	34.5 Cu, 30.5 Fe, 35.0S	Large massive, irregular veins, disseminated and porphyry deposit at granitic/dioritic intrusive and SEDEX type	Primary source of copper metal with major applications in electrical wires, cables, plumbing, currency, utensils, machinery, alloy, architecture, nutritional supplements and fungicides in agriculture

Mineral	Crystal system	Color/Streak	Luster	Hardness	Specific gravity	Cleavage/Fracture	Composition	Occurrence	Uses
Covellite or coveline (CuS)	Hexagonal	Indigo-blue, brass-yellow, deep red / Lead gray	Submetallic, resinous, dull	1.5–2.0	4.6–4.8	Perfect on (0001), Flexible in thin lamina	66.4 Cu, 33.6 S	Zone of secondary supergene enrichment, source of rich copper metal	Natural superconductor chips, electrical wires, cables, plumbing, currency, utensils, machinery, alloy, architecture, nutritional supplements and fungicides in agriculture, insecticide
Cuprite (Cu$_2$O)	Isometric-plagiohedral	Cochineal-red, crimson-red, black / Shining brown, red	Adamantine, submetallic, earthy black	3.5–4.0	5.85–6.15	Interrupted on (111), Conchoidal, uneven	88.8 Cu, 11.2 O	Zone of oxidation, secondary enrichment, source of rich copper metal	Applications in electrical wires, cables, plumbing, currency, utensils, machinery, alloy, architecture, nutritional supplements and fungicides in agriculture.

MINERALS OF GOLD AND SILVER

Mineral	Crystal system	Color/Streak	Luster	Hardness	Specific gravity	Cleavage/Fracture	Composition	Occurrence	Uses
Native gold (Au) (Fig. 1.6)	Isometric	Golden shining yellow/Gold yellow	Metallic	2.5–3.0	19.3	Dense, soft, malleable, and ductile	100 Au	Quartz veins and alluvial deposits	Bar and coinage for standard international monetary exchange, investment, jewelry, electronics, dentistry and medicine.
Sylvanite or silver gold telluride [(Ag, Au) Te$_2$]	Monoclinic	Silver-gray, silver-white / Silver-gray	Metallic	1.5–2.0	8.2	Perfect on (010), Massive, uneven	34.36 Au, 6.28 Ag, 59.36 Te	Low-temperature hydrothermal veins	Minor source of gold, silver, and tellurium.

Continued

TABLE 1.1 List of Common Platinum Group of Elements (PGEs), Nickel, Chromium, and Associated Precious Metals and Their Minerals, Diagnostic Features, and Uses—cont'd

Name and Formula	Crystal System	Color/Streak	Luster	Hardness*/Sp. Gr (g/cm³)	Cleavage/Fracture	% Content of Valuable Component	Origin or Occurrence	Major Uses
Calaverite, or **Gold telluride** ($AuTe_2$)	Monoclinic	Brass-yellow to silver-white/Green to yellow green	Metallic	2.5–3.0 / 9.1–9.3	None, bladed, platy	43.56 Au, 56.44 Te	Usually low temperature vein type deposit	Minor source of gold.
Krennerite ($AuTe_2$) (Fig. 1.16)	Orthorhombic	White to blackish yellow/Greenish gray	Metallic	2.5 / 8.53	Perfect	43.56 Au, 56.44 Te, ± Ag	Occurs in high temperature hydrothermal environment.	Minor source of gold
Petzite (Ag_3AuTe_2)	Cubic	Steel-gray to iron-black/Grayish-black	Metallic	2.5–3.0 / 8.7–9.14	None, sub-conchoidal	25.39 Au, 41.71 Ag, 32.90 Te	Vein-type gold deposit forms under hydrothermal activity	Minor source of gold and silver.
Native silver (Ag) (Fig. 1.7)	Isometric	Silver-white tarnishes dark yellow to black/gray	Metallic	2.5–3.0 / 9.6–12.0	None/Hackly, ductile and malleable.	100 Ag	In sulfide ore deposits.	Coinage, ornaments, jewelry, high-value tableware and utensils, electrical contacts and conductors, photography, dentistry and medicine, investment.
Argentite (Ag_2S)	Cubic, octahedral	Lead-gray/Shining	Metallic	2.0–2.5 / 7.2–7.4	Traces, sub-conchoidal	87.0 Ag, 13.0 S	Galena and other sulfide association.	Primary source of silver, jewelry, photo-processing, currency and investment
Chlorargyrite (AgCl)	Isometric, hex-octahedral	Purple gray, green, white, colorless/White	Adamantine	1.0–2.0 / 5.55	None, Massive to columnar	75.26 Ag, 24.74 Cl	Oxidized part of silver minerals as secondary mineral phase	Minor source of silver ore.
Pyrargyrite (Ag_3SbS_3)	Trigonal	Dark red to red-black/Dark cherry red	Adamantine	2.5 / 5.8	Occasionally distinct, conchoidal	59.75 Ag, 22.48 Sb, 17.76 S	Metalliferous veins of calcite, aragonite, lead, arsenic, and native silver.	Popular source of silver and antimony. Bright red color is attractive for mineral collection as "ruby silver".

MINERALS OF COBALT

Mineral	Crystal system	Color/Streak	Luster	Hardness*	Specific gravity	Cleavage/Fracture	Composition	Occurrence	Uses
Cobaltite (CoAsS) (Fig. 1.17)	Orthorhombic, pseudo-cubic	Reddish silver white, violet steel gray / Grayish black	Metallic	5.5	6.0–6.3	Perfect on (001), Uneven	35.5 Co, 45.2 As, 19.3 S	High-temperature hydrothermal and contact metamorphic deposit with Magnetite, Sphalerite	Industrially useful metal, high-temp superalloy, steel tools, lithium cobalt oxide battery, pigments and coloring, radio-isotope and electroplating owing to its attractive appearance, hardness and resistance to oxidation.
Skutterudite [(Co, Fe, Ni)As$_{2-3}$] (Fig. 1.18 and Fig. 1.19)	Isometric-octahedral-pyritohedral	Tin-white, lead-gray / Black	Metallic	5.5–6.0	6.5–6.9	Distinct on (100), Conchoidal, uneven	12.0–18.5 Co, 7.0–7.7 Ni, 76.09 As, 1.4–8.8 Fe, ±120g/t Au	Hydrothermal ore found in moderate- to high-temperature veins with other Ni-Co minerals	Strategically and industrially useful, high-temp super-alloy, steel tools, lithium cobalt oxide battery, pigments and coloring, radio-isotope and electroplating owing to its attractive appearance, hardness, and resistance to corrosion

*Mohs hardness scale.

Haldar, S.K., 2013. Mineral Exploration – Principles and Applications. Elsevier Publication, p. 374; Haldar, S.K., Josip Tišljar., 2014. Introduction to Mineralogy and Petrology, Elsevier Publication, p. 356.

Ruthenium and rhodium occur as the free metals and are often associated with platinum, osmium, and iridium in South Africa and in North and South America. It is also associated with copper and nickel deposits in Canada from which it is recovered commercially. There are rare minerals of ruthenium and very few minerals of rhodium.

The strength and quality of steel increases many fold by adding small quantities of nickel into the system. The important nickel-bearing minerals are Pentlandite and Pyrrhotite. The major nickel deposits are typically associated with copper-PGMs and widely distributed in Australia, Canada, Russia, and South Africa. Lateritic nickel deposits are located in New Caledonia, the Philippines, Indonesia, Western Australia, Colombia, Cuba, Venezuela, China, Tanzania, and Brazil. Major production sites include the Sudbury igneous complex, Canada; New Caledonia; and Noril'sk in Russia. Nickeliferous limonite/laterite (oxide form) has been identified in the overburden of chromite orebodies in Sukinda valley, Orissa, India. The commercial viability could not be established due to complex metallurgy.

Chromium is not found as the free metal in nature. The most important (or only commercially viable) ore is chromite ($FeCr_2O_4$). Chromite deposits are widely located in South Africa, Zimbabwe, Kazakhstan, Turkey, Finland, India, Brazil, China, United States, Albania, Iran, Madagascar, Russia, Southern Rhodesia, Transvaal, Cuba, Japan, Pakistan, and the Philippines. The largest producers of chromium ore have been South Africa (44%), followed by India (18%), Kazakhstan (16%), Zimbabwe (5%), Finland (4%), Iran (4%), and Brazil (2%). The ore is reduced with either aluminum or silicon to produce ferrochromium or charge chromium. Pure chromium is produced by removing iron from the chromite in a two-step roasting and leaching process.

The global distribution of PGMs, nickel, and chromium deposits are illustrated in Fig. 1.20 following Naldrett (2004), Zientek et al. (2005), Zientek (2012), Mungall (2005b), and Pariser

(2013). The distribution pattern indicates that this group of metals coexists and is widespread in different proportions in all six continents of Africa, America (South and North), Asia, Australia, and Europe.

1.6 DEMAND AND SUPPLY

The past and present consumption and future projection of demand for metals rise with the increase in world population. In contrast, the supply or production tends to drop with diminishing mineral resources. The gap between demand and supply primarily depends on availability of ore reserve base, nature of the orebodies, mining technology, mineral processing (beneficiation, smelting, and refining), and price of the metals. The demand-and-supply gap is minimized by adopting advanced technology in all aspects of the mineral industry and balanced by import/export. Part of the demand is substituted for by routine recycling or recovery of metals from scrap.

1.6.1 Platinum Group of Metals

There are few significant platinum–palladium mining companies in the world. These two metals have the highest economic importance, and are found in the largest quantities in the PGEs.

South Africa has been the largest producer of platinum from the Bushveld Complex since 1925, with palladium and rhodium as byproducts. The Bushveld Igneous Complex has the resources to supply the demands of the World for the next century. Platinum was discovered in the rocks of the Great Dyke, Zimbabwe, in 1918, but significant output from this extensive resource only began in the 1990s. The fresh supply of platinum in decreasing order is contributed by South Africa, Russia, North America including Canada and other countries. In addition, a bulk amount is received from recycling of the scrap which is about 25% of the fresh supply.

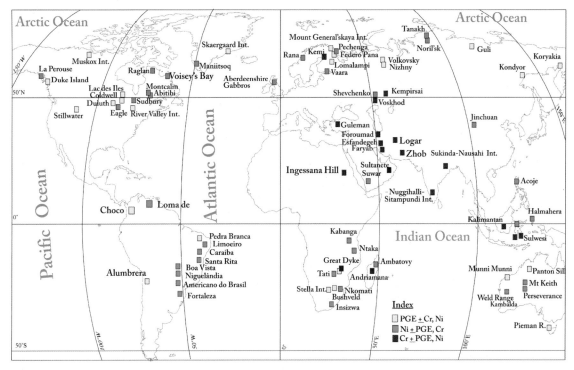

FIGURE 1.20 The global geographic distributions of major and important Platinum-group-metals, nickel, and chromium deposits exist in all six continents of Africa, America (South and North), Asia, Australia, and Europe. The location of deposits is approximate with an objective to overview the distribution pattern in the whole world.

The global supply of platinum between 1990 and 2015 is given in Fig. 11.11. The overall demand of platinum is expected to rise by 3.5% per annum due to increasing need for catalytic converters, autocatalysts, jewelry, and coins.

Russia is the largest producer of palladium. Stillwater in the United States is a palladium-rich mine which began underground mine production in 1987. The palladium demand is expected to rise by 3.5% per annum and the supply drops in the last decade. This gap in demand and supply may result increase in price of palladium over platinum.

The other four metals (rhodium, ruthenium, iridium, and osmium) are produced as co-products of platinum and palladium. The annual production of these metals depends on the production of platinum and palladium. There have been

significant improvements in mining and processing of rhodium over the years. Therefore, there has actually been a growth in rhodium supply (2.6%), whereas demand for the same has risen much higher. The current annual production and consumption of Iridium is only 3 tonnes (or 3.3 tons) per annum. The annual mine production of ruthenium is about 20 tonnes (or 22 tons). The annual production of osmium is low at less than 1 tonne (or 1.1 ton).

1.6.2 Nickel

Nickel occurs as oxides, sulfides, and silicates, with nickel ores mined, smelted, and/or refined in about 25 countries. The most important producing countries are Russia, Canada,

New Caledonia, the Philippines, Australia, Indonesia, and Cuba. There are basically four types of nickel supply—primary nickel, ferronickel, nickel in pig iron (NPI), and nickel scrap. Noril'sk nickel began producing copper and nickel in 1935 with palladium and platinum as byproducts. Canada has produced platinum metals byproducts of nickel mining since 1908.

The top five producing countries account for 64% of global mine production, whereas the three largest producers account for 40% of production. The largest is the Norilsk mine in Russia, followed by Broken Hill Proprietary (BHP) Billiton and Vale. The next seven largest producers provide 28% of the supply with 32% coming from smaller producers.

The global nickel supply is accelerating at a faster pace with growth rising at 1.6% per annum during the 1980s, 2.9% in the 1990s, and finally 3.5% since 2000 (Fig. 11.12). In 2011, nickel production was around 1.59 million tonnes (or 1.75 million tons)— which was relatively small compared with 20 million tonnes (or 22 million tons) of copper and 45 million tonnes (49.6 million tons) of aluminum. The emergence of nickel supply from pig iron is a significant development in the mineral industry. Sky-high nickel prices in 2006 and 2007 forced China to find alternative means of getting nickel units, boosting supply to around 275,000 tonnes (303,000 tons) last year from 5000 tonnes (5500 tons) in 2005.

1.6.3 Chromium

The world ore reserve base of chromite stands at 9215 million tonnes (10,160 million tons) out of which 90% was concentrated in South Africa, Zimbabwe, Kazakhstan, and Turkey. US chromium resources are mostly in the Stillwater complex in Montana. The other countries having sizeable resources are Finland, India, Brazil, China, Russia, and Albania. The global supply of chromium witnessed a steep rise from 1960. The average growth of supply in the twenty-first Century stands at 6.0% (Fig. 11.12).

TABLE 1.2 London Metal Exchange Prices of Noble, Ferrous and Nonferrous metals as of July 2015

Metal	Amount in US$
Noble metals	(US$/oz)
Platinum	1,470
Palladium	864
Rhodium	1,100
Iridium	600
Ruthenium	70
Osmium	365
Gold	1,199
Silver	
Ferrous and nonferrous metal	(US$/tonne)
Nickel	1,876
Ferrochrome	2,500
Copper	6,975
Cobalt	30,400
Zinc	2,182
Lead	2,138
Iron (steel billet)	385

1.7 METAL PRICE

The metal price of noble and associated metals as reported by London Metal Exchange is given in Table 1.2. This will depict the value of the various metals under common uses.

1.8 USES

1.8.1 Platinum Group of Metals

• The major use of Platinum is in catalytic converters (~50%), an automobile emission control device, that converts the toxic pollutants as exhaust gas to less toxic pollutants by catalyzing redox (reduction–oxidation) reaction. Catalytic converters are used in internal combustion engine fueled by either gasoline or diesel. The PGEs are highly resistant to wear and tarnishing, making them well suited internationally for fine jewelry (30%). The other uses are in chemical and petroleum refining plants, laboratory equipment, platinum resistance thermometers, currency, and investment. The chemical inertness and refractory properties of these metals find applications in electrical contacts, electronics, electrodes, and dental and medical fields.

- Palladium is used as a substitute for silver in dentistry and jewelry. The pure metal is used as the delicate mainsprings in analog wristwatches. Palladium is used as catalytic converters, which convert up to 90% of harmful gases from auto exhaust (hydrocarbons, carbon monoxide, and nitrogen dioxide) into less-harmful substances (nitrogen, carbon dioxide, and water vapor). The metal is also used in surgical instruments, electronics, hydrogen storage, coinage, photography, jewelry, and investment.
- The most important compounds of Iridium in use are the salts and acids it forms with chlorine. Iridium is an extremely hard and too brittle an element to be used in the pure state. Therefore, it is often used as an alloy. Platinum–iridium alloys are used in the manufacturing industries for the production of machine parts, containers (crucibles), fountain-pen nib heads, and electric contacts that may be exposed to high temperatures and chemicals. The other uses include jewelry and electrodes of spark plugs.
- Osmium is rarely used in its pure state. The metal is used as alloys with platinum, iridium, and other platinum group metals in electrical circuit components, fountain-pen tips, and other applications in which extreme durability and hardness are necessary.

- Most Ruthenium produced is converted to hard platinum–palladium alloy and used for wear-resistant electrical contacts and the manufacture of thick-film resistors.

1.8.2 Nickel

The current global consumption pattern of nickel indicates that 46% is used for making nickel steels, 34% in nonferrous alloys and superalloys, 14% electroplating, and 6% into other uses. Nickel is used in many specific and recognizable industrial and consumer products. The primary applications of nickel include stainless steel, nickel cast iron, super-alloy in jet engines, alnico (aluminum–nickel–cobalt) alloy in horseshoe magnets, nickel plating and polishing, coinage, rechargeable batteries, electric guitar strings, special alloys, and green tint in glass.

- Nickel is an exceptional alloy metal. The main applications are in the varieties of nickel steels and nickel cast irons. The other alloys are nickel brasses and bronzes, and alloys with copper, chromium, aluminum, lead, cobalt, silver, and gold.
- A superalloy "alnico" is composed of aluminum–nickel–cobalt–copper–titanium–iron, and used for making "horseshoe magnets." The ferromagnetic or ferroelectric alloy of the magnet can withstand an external magnetic or electrical field.
- Nickel is an excellent alloying agent for precious metals and capable of full collection of all six PGEs including gold from parent ore minerals by fire-assay technique (Noril'sk in Russia and the Sudbury Basin in Canada).
- Nickel is a highly corrosion-resistant metal and suitable to substitute for decorative silver articles and cheaper coinage metal.
- Nickel is used as a binder in the cemented tungsten-carbide or hard-metal industries and adds corrosion-resistant properties to cemented tungsten-carbide parts.

1.8.3 Chromium

The primary applications of chromium are in metallurgy, refractory materials, dye and pigment, synthetic ruby and laser, wood preservative, tanning, catalysts, and others.

- Chromium increases the strengthening effect of forming stable metal and strong corrosion-proof chromium–iron alloy material for high-speed-tool steels, stainless steel. Nickel–chromium superalloy is suitable for high-temperature jet engines and gas turbines, electroplating, chrome plating and polishing, and anodizing of aluminum.
- Chrome–yellow, a synthetic product of lead-chromate (mineral Crocoite) is one of the most used nonphotodegraded bright-yellow pigments for school buses in the United States, Postal Services in Douche Post, and may other applications.
- Synthetic red ruby (corundum) crystal by artificial doping of chromium acts as a base for laser technique on simulated electromagnetic emission of light from the chromium atom.
- The toxicity of chromium salts (chromate copper arsenate) preserves timbers from decay fungi.
- Chromium salts of nontoxic varieties (chrome alum and chromium sulfate) stabilize the leather by cross-linking the collagen fibers during tanning. Chromium-tanned leather can contain between 4 and 5% of chromium, which is tightly bound to the proteins.
- The high heat resistivity and high melting point makes chromium and chromium-oxide mixed with Magnesite the best suitable for high-temperature refractory applications in blast furnaces, cement kilns, molds for the firing of bricks, and as foundry sands for metal castings.

1.8.4 Copper

The current global consumption of pure copper metal indicates 60% is used in electrical wires and cables, 20% in roofing and plumbing, 15% in industrial machinery, and 5% in hard alloys as brass (copper–zinc) and bronze (copper–tin–arsenic). A small part of the copper supply is used in production of compounds for nutritional supplements and fungicides in agriculture.

- Copper remains the all-time preferred electrical conductor in all categories of electrical wiring, with the only exception being overhead electrical power transmission by aluminum cables. Copper wire is used in power generation, transmission, distribution, telecommunication, electronic circuitry, and numerous types of electrical equipment. Many electrical devices rely on copper wiring because of its multitude of inherent beneficial properties, such as high electrical and thermal conductivity, tensile strength, ductility, cold-flow and corrosion resistance, low thermal expansion, soldering ability, and ease of installation.
- Integrated and printed circuit boards, energy efficiency motors with greater conductivity.
- Corrosion resistant, weatherproof, and durable ancient architectural remains and roofing.
- Energy-saver kitchen and domestic utensils.
- Biostatic, antimicrobial applications, and wood preservation.
- Hard alloy of brass and bronze.
- Jewelry, decorative articles, and coinage.

1.8.5 Gold and Silver

The prime applications of gold are monetary, investment, jewelry, medicine, food and drink, electronics, and commercial.

- The yellow metal is universally used throughout the World as an efficient monetary exchange in the form of bullions (metal bars and ingots) and coins.
- Most popular as investment to get benefits within a specified date or time frame at minimum risk.

- Jewelry in various designs of brooches, bracelets, necklaces, rings, earrings, and many other forms used for personal adornment, particularly by women as a passion. The highest consumer of gold jewelry is by Greater China [1120 tonnes (1235 tons)] and India [974 tonnes (1074 tons)] in 2013. Possession of gold is social pride and social security.
- Popularly administered medicine by ancient physician and dentistry.
- Gold leaf, flake, and dust as food additives.
- Electronic contacts, medals, embroidery, paintings, and coloring agent in cranberry-glass and industrial uses.

The major applications of silver are currency, jewelry, dentistry, photography and electronics, mirrors and optical, industry and commercial, biology and medicine, investment, and clothing.

1.8.6 Cobalt

The primary applications of cobalt are alloys, batteries, catalysts, pigments and coloring, radioisotopes, and others.

- Cobalt-based superalloys are suitable for high-speed steels, gas turbine blades, aircraft engines, cutting tools, diamond tooling, and cemented carbide.
- Lithium–cobalt oxide heavy duty batteries with long life.
- Electroplating and enamel coats on porcelain products.
- Brilliant rich-blue (cobalt-blue) glass, surface coating and coloring on blue pottery and decorative articles.

1.9 SUBSTITUTES AND RECYCLING

Platinum can be partially substituted by palladium in catalytic converters in new technologies. PGMs can be substituted for other PGMs with loss in efficiency. Palladium can be substituted by base metals for conductive electrodes of multilayer ceramic capacitors. Nickel-free specialty steels are sometimes used in place of stainless steel within the power-generating and petrochemical industries. Titanium alloys or specialty plastics can be replacements for nickel metal or alloys in highly corrosive chemical environments. Austenitic grade nickel in construction applications is replaced by low-nickel, duplex, or ultrahigh-chromium stainless steels to offset high nickel prices. Chromium has no substitute in stainless steel, the leading end use, or in superalloys, the major strategic end use.

Recycling of these metals contained in old scraps will be beneficial from a long-term standpoint. Reduce, Reuse, and Recycle should be the slogan for sustainable use of these uncommon metals.

There is no suitable alternative for chromium in the manufacture of steel and chromium chemicals. Scrap metals that contain chromium can be recycled as an alternative source. However, the natural abundance of chromium in Earth's crust makes alternative sources unnecessary at this time.

Geology and Geochemistry

2.1 DEFINITION

Basic knowledge of geology, geochemistry, and a rock classification system is significant in search for Platinum Group of Elements (PGEs)-Ni-Cr mineralization. The study of geology includes mineralogy and petrology of favorable host and country rocks, stratigraphy, structure, mode of occurrence, and period of formation [age in Giga annum (Ga) or Million annum (Ma)]. Petrology deals with host and country rocks. Geochemistry gives emphasis on chemistry or chemical composition of rocks with respect to some significant increase or decrease of indicator elements in the composition of host rocks. Classification systems highlight various genetic characteristics and modes of occurrence.

2.2 GEOLOGY

2.2.1 Stratigraphy

In general, PGEs, nickel, and chromium coexist in close association with copper, cobalt, and gold in layered mafic to ultramafic plutonic/volcanic complexes. The host rocks originated from mafic–ultramafic intrusive and/or extrusive magma–lava. The primary rocks include magnesium-rich gabbro, anorthosite, norite, pyroxenite, harzburgite, peridotite, dunite, and serpentinite. The various types and periods of magmatic association include:

Major sulfide/oxide PGEs-Ni-Cr deposits related to massive magmatic activity all over the world during Archean Eon, such as the Bushveld Igneous Complex in South Africa, the Sudbury Meteorite impact structure in Canada, and the Sukinda–Nausahi complex in India. Mineralization occurs as concentrations and disseminations in layered igneous rocks in the form of layers, lenses, pockets, reefs, contact type, thin seams, and stringers within the host rocks.

The other stratigraphic association is Mesozoic ophiolite, such as New Caledonia; Indonesia; Cuba; and Manipur, Nagaland, Andaman–Nicobar and Jammu–Kashmir in India. The chromite deposits of Jammu–Kashmir occur in association with serpentinite and dunite within the Dras volcanics of Cretaceous age.

2.2.2 Host Rock

The favorable hosts for these minerals are intrusive/extrusive mafic to ultramafic igneous rocks with variable mineralogical and chemical composition, their alteration products like laterite and alluvial placers. Igneous rocks are natural products of cooling, crystallization, and solidification of extremely hot mobile molten material (magma) originating from the deepest parts of the Earth. The mode of formation can be either intrusive (plutonic) or extrusive (volcanic).

Intrusive igneous rocks are formed by cooling, crystallization, and solidification of magma within the Earth's crust surrounded by preexisting country rocks. These rocks are naturally medium to coarse grained, and designated according to shape, size, and relationship with the existing formation, as abyssal (deep seated), hypabyssal (near surface), batholithic (large felsic/intermediate massive plutons), stocks (massive plutons), laccolithic (concordant plutonic

sheets between sedimentary layers), sills (concordant tabular plutonic sheets within volcanic/sedimentary/metamorphic rocks), and dikes (plutonic sheets cut discordantly across existing rocks) (Fig. 2.26).

Extrusive igneous rocks are formed at the surface of the crust by emplacement and cooling of mantle-derived hot magma. The molten rocks, with or without suspended crystals and gas bubbles, erupts outside the crust due to lower density and spread as lava flows. Volcanic eruptions into the air or the ocean are termed "subaerial" and "submarine," respectively. These rocks cool and solidify very quickly and are, in general, naturally fine grained. The midoceanic ridges and ocean-floor flood basalts are examples of submarine volcanic activity.

The chemical composition of magma and lava is complex. The elements include oxygen, silicon, aluminum, iron, calcium, sodium, potassium, magnesium, and titanium. It also contains different amounts of water vapor mixed with easily volatile components in the form of gases and vapors, such as hydrogen sulfide (H_2S), hydrogen fluoride (HF), hydrogen chloride (HCl), carbon dioxide (CO_2), sulfur dioxide (SO_2), hydrogen, nitrogen, and sulfur. The viscosity of magma depends primarily on its chemical composition and temperature.

2.2.2.1 Crystallization of Minerals— Bowen's Reaction Series

Laboratory experiments demonstrated that certain types of minerals tend to form at specific temperatures out of molten igneous rock, occur together, and are collectively known as Bowen's Reaction Series. Two types of reaction series, continuous and discontinuous, explain the different minerals that form various rocks.

The Continuous Reaction Series is on the right side of Bowen's Reaction Series and encompasses plagioclase feldspars. Highest-temperature plagioclase contains calcium and lowest temperature it contains sodium. In between, these ions mix in a continuous series from 100%

Ca and 0% Na to 0% Ca and 100% Na at lowest temperature.

The Discontinuous Reaction Series is a group of mafic or iron–magnesium-bearing minerals: Olivine, Pyroxene, Amphibole, and Biotite. These minerals react discontinuously to form the next mineral in the series. Each higher-temperature mineral will stop forming, and the next lower-temperature mineral will begin to crystallize with decreasing temperature, provided enough silica is present in the cooling magma. The proportion of silica will increase in minerals down Bowen's Reaction Series as the magma becomes increasingly depleted in the components that went to form earlier minerals.

2.2.2.2 Cooling of Magma After Crystallization

The rock mass will be at relatively higher temperature even after crystallization of magma or lava. The rock mass and its mineral components slowly cool to ambient temperature. In the process, its volume reduces and the mass cracks (Fig. 2.1). Such an originally compact rock mass breaks over time and separates into pieces of various sizes and shapes, such as plates, three-sided, four-sided, five-sided, or six-sided prisms, cubes, spheres, or completely irregular bodies. This jointing of a cooling mass, both intrusive and extrusive, manifests remarkably distinguished "columnar structure," most commonly displayed in basalt (Fig. 2.2). Such cracking and separation must be strictly distinguished from cracks and crushing of rocks caused by tectonic movements. The style of cracks plays a decisive role in breaking and processing stone.

2.2.2.3 Classification of Igneous Rocks

Igneous rocks are classified according to their mode of formation and mineralogy (chemical composition). Mineral constituents of igneous rocks are classified according to their proportion in composition based on major, important, minor (accessory), and secondary categories.

Major minerals control the classification system. These are essential ingredients for rock and make it different from other rocks. Quartz, Potassium (K)-feldspar, and Biotite are essential minerals that compose granite. Minor minerals are not important to the rock in which they occur. Their amount is small, typically <1%. Zircon and Rutile are minor minerals in granite. Secondary minerals are formed later by weathering processes or changes in the primary minerals. The most common secondary minerals are Kaolinite (chemical weathering of Feldspar), Chlorite (weathering of Biotite, Pyroxene, and Amphibole), Sericite (weathering of feldspar), and Serpentine (hydrothermal modification of Olivine).

Igneous rocks are divided into four types based on the content of SiO_2 in chemical composition (Box 2.1):

The PGM, nickel, and chromium deposits are the main target and exclusively occur in mafic–ultramafic rocks. The following description will be focused on mafic–ultramafic rocks.

2.2.2.4 Mafic Plutonic and Volcanic Rocks

Mafic–ultramafic rocks differ in shape, form, and size depending on cooling and crystallization of magma deep in the lithosphere or on the

FIGURE 2.1 Plan view of the polygonal cracks/joints in basalt developed during crystallization and cooling of magma at Albert Hill, Mumbai, India (Haldar and Tišljar, 2014).

Earth's surface. Mafic–ultramafic magmas forming deep-seated plutonic rock types typically have their corresponding volcanic equivalents. The rocks and major mineral component are arranged in Table 2.1 in descending order of temperature with respect to Bowen's Reaction Series.

2.2.2.4.1 MAFIC INTRUSIVE IGNEOUS ROCKS

Gabbro is a mafic intrusive coarse-grained rock with allotriomorphic texture. Gabbros contain low silicon (no Quartz or Alkali feldspar)

FIGURE 2.2   The columnar structure developed in Deccan Trap flood-basalt lava flow. The unique structure formed at the Cretaceous/Tertiary boundary during crystallization and cooling of magma, and separation over time at Albert Hill, Mumbai, India (Haldar and Tišljar, 2014).

and essentially of ferromagnesian minerals and Plagioclase feldspar rich in calcium. The amount of ferromagnesian minerals equals or exceeds that of Plagioclase. Gabbros are plutonic rocks formed by cooling and crystallization of molten magma trapped under the Earth's surface and chemically equivalent to extrusive basalt. The ferromagnesian minerals are Pyroxene (Diopside or Diallage, Augite, and Hypersthene), Hornblende, and Olivine, occurring either together or singly. The Pyroxene is mostly Clinopyroxene (Diopside or Augite) with or without Orthopyroxene (Hypersthene). The feldspar is chiefly calcic Plagioclase, generally 50–60% Labradorite [(Ca,Na)(Al,Si)$_4$O$_8$], and Plagioclase composition of Bytownite to Anorthite. Gabbros are mostly dark in color, ranging between dark gray and greenish black because of a high proportion of ferromagnesian minerals (Fig. 2.3). Gabbro with Olivine is called olivine gabbro. Olivine gabbro without Pyroxene and enriched with calcium Plagioclase and Olivine, is known as **troctolite**. **Norite** is a variation of gabbro with Ca Plagioclase and predominantly orthorhombic Pyroxene (Hypersthene). Norite with Olivine is olivine norite.

Plagioclases undergo processes of "saussuritization" with action of hot solutions and extensively change into a dense, compact mixture of

---

BOX 2.1

CHEMICAL CLASSIFICATION OF IGNEOUS ROCKS

"Felsic" igneous rock refers to light color, low specific gravity, and high-silica minerals, magma, and rocks. The most common felsic minerals are Quartz, Orthoclase, and sodium-rich Plagioclase feldspar and Muscovite. The most common felsic rocks are granite and rhyolite containing ~63% SiO$_2$.

"Intermediate" igneous rocks contain SiO$_2$ between 52% and 63% with common examples being andesite and dacite.

"Mafic/basic" igneous rocks have low silica (between 45% and 52% SiO$_2$) and typically composed of minerals with high iron and magnesium content such as Pyroxene and Olivine. The most common rocks are gabbro and basalt.

"Ultramafic/ultrabasic" igneous rocks contain less than 45% SiO$_2$, >18% MgO, high FeO, low K, and generally 90%+ of mafic minerals. The most common rocks are peridotite and dunite.

TABLE 2.1   Mineral Composition of Major Mafic–Ultramafic Igneous Rocks

	Intrusive Rocks	Extrusive Rocks	Main Minerals
Mafic	Gabbro	Basalt	Ca-plagioclase (40–70%)
		Diabase	Pyroxene (Augite, Hypersthene),
		Spilite	Small quantities of Hornblende and Biotite, with or without Olivine
	Norite	Basalt	Ca-plagioclase, Pyroxene (Hypersthene) with or without Olivine
	Anorthosite	–	Ca-plagioclase (90–100%) with Pyroxene, Ilmenite, Magnetite (0–10%) ± Olivine
Ultramafic	Dunite	–	Mostly Mg-Olivine with little Pyroxene
	Peridotite	Komatiite	Olivine, one or more Pyroxene
	Lherzolite	–	Olivine, Bronzite, Diallage
	Serpentine	–	Serpentine derived from Olivine
	Pyroxenite	–	Monoclinic pyroxene ± Olivine or Amphibole (Augite, Diopside, Diallage)

FIGURE 2.3   Gabbro is dark gray to black-color plutonic rock, and chemically equivalent to volcanic basalt. It mainly contains Ca-plagioclase and ferromagnesian minerals such as Pyroxene (Augite and Hypersthene or Diallage) ± Olivine (Haldar and Tišljar, 2014).

Zoisite, Epidote, Albite, Quartz, Muscovite, and Actinolite. The new rock is known as "sosirite." In similar conditions, Pyroxenes undergo "uralitization" processes to form dense clusters of Actinolite and the modified product known as "uralite."

Gabbro texture (Fig. 2.4) is formed by simultaneously extended crystallization of bright (leucocratic) and dark ferromagnesian minerals that are compact, solid, and tough. Gabbros appear as a densely homogeneous rock often fairly of same structure and composition throughout the rock mass.

Dolerite and **Diabase** are mafic igneous rocks having the same mineralogical composition, but differing in formation. The colors are dark gray, black, and green. Dolerite is a medium-grained (Fig. 2.5) intrusive equivalent of volcanic basalt or plutonic gabbro, and usually occurs as dikes, sills, and plugs. Dolerite is heavy with specific gravity ranging between 2.9 and 3.3. Dolerite dikes are often exposed to the surface and exhibit as walls in a straight line. Diabase is a subvolcanic rock equivalent to volcanic basalt or plutonic gabbro. Diabase is subsurface volcanic rock formed by injecting gabbroic magma or lava as shallow dykes and sills under the surface of Earth. Diabase is typically fine grained having a chilled margin. The main ingredients

FIGURE 2.4   Photomicrograph of thin section showing large deformed Plagioclase phenocryst embedded in finer matrix of ferromagnesian minerals in gabbro (Haldar and Tišljar, 2014).

FIGURE 2.6   Photomicrograph of thin section showing intersertal/intergranular texture of lath-shaped Plagioclase of Labradorite (rarely Bytownite) set in a finer matrix of Clinopyroxene in dolerite dike (Haldar and Tišljar, 2014).

FIGURE 2.5   Dolerite is a medium-grained mafic intrusive igneous rock composed primarily of Plagioclase set in a finer matrix of Clinopyroxene ± Olivine, Magnetite, and Ilmenite (Haldar and Tišljar, 2014).

of dolerite/diabase are mafic lath-shaped Plagioclase of ~60% (Labradorite, rarely Bytownite) set in a finer matrix of Clinopyroxene (~20–30% Augite) ± Olivine (up to 10% in olivine diabase), Magnetite and Ilmenite (Fig. 2.6). The accessory minerals are Chlorite, Uralite, and Calcite. The rocks usually display intersertal/intergranular texture. The coarse-grained diabase with Pyroxene specifically alters to Uralite (uralite diabase) and Plagioclase from the Labradorite and Oligoclase types, known as "ophite," which are

characterized by a special structure known as "ophite or ophitic structure."

Norite is a mafic intrusive igneous rock with color ranging between light and dark gray and brown. The rock is indistinguishable from gabbro, other than the type of pyroxene under the microscope. It is composed of Ca-rich Plagioclase (Labradorite), Mg-rich Orthopyroxene/Hypersthene (Enstatite), and Olivine. The rock occurs in close association with mafic gabbro and ultramafic layered-intrusive igneous complex, e.g., Bushveld and Stillwater complexes with large platinum group of deposits, and layered igneous complexes with large deposits of chromite at Sukinda and Nausahi (Fig. 2.7), India. The common usages are ornamental facing, paving, graveyard headstone at funerary rites, and kitchen countertops.

Anorthosite is typically a coarse-grained intrusive igneous rock with color varying between white, yellowish to brown, shades of gray, bluish, and smoky pigment. It is characterized by predominance of Plagioclase feldspar (90–100%), generally Labradorite, and remaining mafic components of Pyroxene, Magnetite, Ilmenite (0–10%) ± Olivine (Fig. 2.8). The fine-grained, nearly monomineralic composition and light-color anorthosite resembles both marble and quartzite in hand specimen. If the quantity of Pyroxene increases in anorthosite, the rock grades into gabbro and vice-versa. The rock can be identified with certainty under microscope with predominance of Feldspar (Labradorite) and texture (Fig. 2.9). It can be distinguished by hardness and presence of well-developed cleavage from marble and quartzite, respectively. The principal modes of occurrence are either (1) large independent intrusive mass, or (2) layers with variable thickness, as members of banded/layered gabbro lopoliths.

2.2.2.4.2 MAFIC EXTRUSIVE IGNEOUS ROCKS

The most common extrusive mafic igneous rocks are basalt and diabase (mafic) and spilite (Plagioclase-rich in albitization of basalt).

Basalts are common aphanitic igneous extrusive rocks composed of minute grains of Plagioclase feldspar (Labradorite), Pyroxene, Olivine, Biotite, Hornblende, and less than 20% Quartz. Nepheline or Leucite may associate/proxy feldspar giving rise to varieties with special names. Ferromagnesian minerals are mainly Amphibole and rarely Biotite. Basalts are usually dark gray to black color.

Basalts are formed by rapid cooling of basaltic lava, equivalent to gabbro–norite magma, from the interior of the crust and exposed at or very close to the surface. These basalt flows are thick and extensive in which gas cavities are nearly absent. In the case of thin and irregular lava flows, gas cavities are formed at the rock surface ("vesicular"), and mostly filled by secondary minerals (Zeolites, Calcite, Quartz, or Chalcedony) forming amygdules (Fig. 2.10).

FIGURE 2.7 Field photograph of coarse-grain light color norite from footwall of open-pit Chromite mine at Boula–Nausahi layered igneous complex, Orissa, India (Haldar and Tišljar, 2014).

FIGURE 2.8 Anorthosite is typically coarse-grained dark-color rock composed primarily of Plagioclase (Labradorite) with minor amount of Pyroxene, Magnetite, and Ilmenite ± Olivine (Haldar and Tišljar, 2014).

FIGURE 2.9 Photomicrograph of thin section showing large phenocryst of deformed Plagioclase (Labradorite) in fine-grained Pyroxene and Plagioclase-rich groundmass in massif anorthosite (Haldar and Tišljar, 2014).

FIGURE 2.10 Dark gray/black massive basalt showing surface cavities filled up by secondary minerals; the rock is designated as "amygdaloidal basalt" (Haldar and Tišljar, 2014).

BOX 2.2

INSOLATION

Insolation (exposure to the sun) is the total amount of solar radiation energy received on a given surface area during a given time.

FIGURE 2.11 Spheroidal/onion-skin/concentric weathering caused by penetration of groundwater along fractures, loosening and decaying rock layer by layer in lava flow over sustained periods. Process is accelerated by repeated expansion (hot days) and contraction (cold nights) leading to weakening, e.g., Albert Hill, Mumbai, India (Haldar and Tišljar, 2014).

Outcrops of basaltic lava flows are easily susceptible to mechanical and chemical weathering by penetration of groundwater along the polygonal joints (Fig. 2.1) and fractures, loosening and decaying the rock layer by layer. The surface of weathering grows more and more rounded as the process progresses into blocks resembling spheroidal shape on a larger scale. The process is accelerated by the insolation effect (Box 2.2) by repeated expansion (hot days) and contraction (cold nights) causing stresses that lead to weakening of ties between minerals, cracking, and disintegration, forming onion-skin or concentric weathering (Fig. 2.11). Basalts have almost always aphanitic or fine-grained mineral texture resulting from rapid cooling of volcanic magma.

Tholeiitic basalts are the most common eruptive rocks produced by submarine

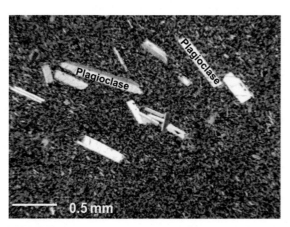

FIGURE 2.12 Photomicrograph of thin section showing Plagioclase phenocrysts in tholeiitic basalt flow (Haldar and Tišljar, 2014).

volcanism from the tholeiitic magma series, forming much of the oceanic crust and mid-oceanic ridges. Tholeiitic magma is relatively rich in silica and poor in sodium. The rock is composed of Clinopyroxene, Hypersthenes, and Plagioclase with minor iron-titanium oxide ± Olivine. Tholeiitic basalt often represents fine, glassy groundmass consisting of fine-grained Quartz. Tholeiitic basalt has fine-grained porphyritic texture, characterized by Pyroxene/Plagioclase phenocrysts (Fig. 2.12) in a fine glassy groundmass.

Alkali basalt is fine-grained dark-color volcanic rock composed of phenocrysts of Olivine, titanium-rich Augite, Plagioclase feldspar, and iron oxides. It is relatively poor in silica (unsaturated) and rich in sodium. It contains feldspathoids (Nepheline, Leucite), Alkali feldspar, and Phlogopite (magnesium mica) in the groundmass. It is typically found on updomed and rifted continental crust, and oceanic islands (the Hawaiian Islands in the North Pacific Ocean).

Flood Basalt is the product of a single or a series of giant volcanic eruptions that cover large stretches of land or ocean floor with basaltic lava. This extensive landscape resembles characteristic stair-step morphology creating plateaus and mountains, and is often called **Traps**, e.g., the Deccan Traps in western and central India covering 500,000 sq. km. The thickness can vary between 2000 m (Deccan Traps, India) and 12,000 m (Lake Superior, North America). Continental flood basalts, derived from mantle plumes rising from the deep core–mantle boundary, are major hosts for world-class large-scale Ni-Cu-PGEs ore deposits (Zhang et al., 2008). Each plume may have a complex history and varied composition. The flood basalts are favorable locations for Ni-PGE-Au signature include Deccan Traps (Fig. 2.1), Karoo, Emeishan, and Siberia. Karoo hosts magmatic ore deposits of various types, sizes, and grades.

Spilite is fine-grained sodium-rich volcanic rock formed by alteration of oceanic basalt and/or diabase in albitization processes at low temperatures in the presence of CO_2 and water rich in sodium (e.g., seawater). Albite in spilite is not caused by crystallization of lava but rather by secondary processes of Ca-plagioclase. Albite and Pyroxenes are transformed into green minerals with conversion of Ca-plagioclase into Na-plagioclase. The new minerals include Chlorites, Uralite, and Epidote. These minerals add green color to spilite. Spilite exhibits intersertal texture—typically microcrystalline. Spilite is usually found in submarine lava effusion in cushioned forms, i.e., as "pillow-lava."

2.2.2.5 Ultramafic Plutonic Rocks

Ultramafic or ultrabasic igneous rocks contain less than 45% SiO_2, >18% MgO, high FeO, and low K. The group of rocks, generally dark colored, are composed of high (+90%) magnesium- and iron-bearing mafic minerals. Modes of occurrence are usually intrusive (dunite, peridotite, and pyroxenite). It occurs as large layered-intrusive complexes hosting chromium, nickel, platinum, and palladium ± massive sulfides, gold, and cobalt. The categorization of the ultramafic group of rocks can be explained with a modal percentage diagram of Olivine, Pyroxene, and Hornblende (Fig. 2.13).

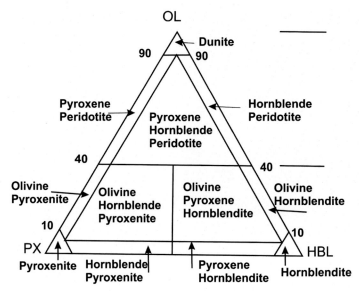

FIGURE 2.13   Modal plot of ultramafic family. *OL*, Olivine, *PX*, *Pyroxene*, and *HBL*, Hornblende (Haldar and Tišljar, 2014).

Dunite is a unique end-member peridotite family consisting almost entirely of magnesium-rich Olivine (+90%) and small share of Chromite, Pyroxene, and Pyrope. This igneous plutonic rock is coarse-grained granular or phaneritic texture and often massive or layered. Color is usually light to dark green with pearly or greasy look (Figs. 2.14 and 2.15).

Dunite is an end product of differential cooling, crystallization, and solidification of hot molten ultramafic magma processed in a huge chamber within the Earth and developed into a layered igneous complex. The composition of layered igneous complexes is often shared by a large presence of Chromite ± nickel, copper, and PGE ore deposits (Fig. 2.16). Finely ground dunite is used to sequester CO_2 and mitigate global climate change, as a source of MgO as flux in metallurgical blast furnaces, refractory and foundry applications, filtering media, and fillers.

Peridotites, the general name for ultramafic intrusive rocks, are dark green to black color, dense and massive form, with compact and coarse-grained crystal habits, and often occur as layered igneous complexes. They are primarily composed of ferromagnesian minerals (>40%), high proportion of magnesium-rich Olivine, Clinopyroxenes and Orthopyroxenes, Hornblende, and less than 45% silica (Figs. 2.17 and 2.18). Ore minerals are Chromite, Magnetite, nickel, copper, and PGEs. Rock-forming minerals constitute entirely single minerals or in combination of various proportions. The components are branded on the basis of the minerals present such as: Peridotite, Kimberlite, Lherzolite, Harzburgite, Hornblendite, Dunite, and Pyroxenite. Peridotite is the most dominant constituent of the upper part of the Earth's mantle. A special version of the peridotite is Kimberlite rock composed predominantly of Olivine, Phlogopite, Orthopyroxene, and Clinopyroxene in which a secondary valuable gem component is Diamond. Peridotites are very susceptible to changes due to the low stability of Olivine. The changing process is serpentinization of Olivine in the fibers and/or sheet clusters of Serpentine and monoclinic Pyroxene (Diallage) in Uralite in the Urals region. In this way, a new rock serpentinite is formed (Table 2.1). The layered-intrusive variety is the most suitable

FIGURE 2.14 Dunite is typically coarse-grained, light-to dark-green color, and consists almost entirely of magnesium-rich Olivine and minor share of Chromite, Pyroxene, and Pyrope (Haldar and Tišljar, 2014).

FIGURE 2.16 Alternate layers of Chromite and Olivine at Sukinda complex, that represent +90% of Chromite resources producing ~4Mt/a in India, second largest in the world after South Africa.

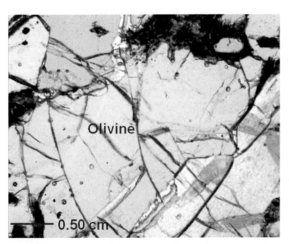

FIGURE 2.15 Photomicrograph of thin section of dunite composed entirely of Olivine (Haldar and Tišljar, 2014).

FIGURE 2.17 Peridotite is a dense coarse-grained dark green to black-color intrusive rock, often layered, composed primarily of ferromagnesian minerals (magnesium-rich Olivine and Pyroxene) and less of silica. Regular secondary minerals are Chromite, Magnetite, nickel, copper, and platinum group (Haldar and Tišljar, 2014).

host rock for chromium, nickel, copper, and platinum–palladium orebodies, and the glassy green type as gem and ornamental stones.

Lherzolite is a type of peridotite containing idiomorphic Olivine with equal share of orthorhombic Orthopyroxene-Bronzite (with irregular grains), and monoclinic Clinopyroxene-Diallage.

Harzburgite is a type of peridotite with no or very little monoclinic Pyroxene and consisting only of Olivine and orthorhombic Orthopyroxene Bronzite.

Pyroxenites are ultramafic intrusive igneous rocks composed almost entirely of dark Pyroxene group minerals rich in iron and magnesium, e.g., Augite, Diopside, Hypersthene, Diallage, Bronzite,

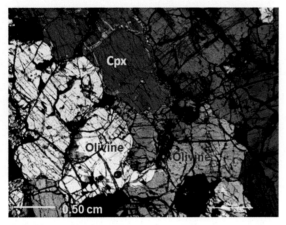

FIGURE 2.18 Photomicrograph of thin section of peridotite rock shows intercumulus texture between Olivine and Clinopyroxene (Cpx) resulting from the settling of a crystallizing magma (Haldar and Tišljar, 2014).

and Enstatite, unlike other mafic minerals Olivine and Amphibole. Absence of Feldspar and Olivine makes it different from gabbro-norite and dunite, respectively. Accessory minerals are Chromite, Magnetite, Garnet, Rutile, and Scapolite. Pyroxenite is dense and coarse grained (Figs. 2.19 and 2.20) with dark green, gray, and brown color. Pyroxenites are classified into clinopyroxenites, orthopyroxenites, and websterites that contain both the Pyroxenes. It occurs either as cumulates at the base of intrusive chamber or as thin layers within peridotites and/or xenoliths in basalt.

Pyroxenites are a source of MgO used as flux in metallurgical blast furnaces, refractory and foundry applications, filtering media, fillers, building materials, and sculptures, and often host deposits of Cr-Ni-Cu-PGEs.

Hornblendite is a rare plutonic rock composed of amphibole hornblende.

Serpentinite is composed of one or more serpentine group minerals formed by hydration and low-temperature metamorphic transformation of ultramafic rocks. Serpentinization is a good indicator in mineral exploration for platinum–nickel–chromium mineralization.

FIGURE 2.19 Pyroxenite is a ultramafic layered igneous rock consisting essentially of minerals of the Pyroxene group, such as Augite and Diopside, Hypersthene, Bronzite or Enstatite.

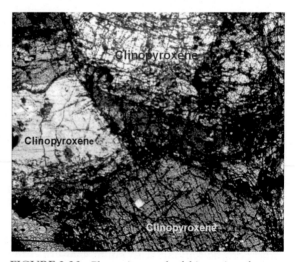

FIGURE 2.20 Photomicrograph of thin section of coarse-grained pyroxenite composed entirely by Clinopyroxene (Haldar and Tišljar, 2014).

Chromitite is an intrusive igneous cumulate (Box 2.3 and Fig. 2.21) rock composed primarily of mostly Chromite (Fig. 2.22). Chromitite is essential part of the layered intrusion containing PGMs-Ni-Cr mineralization hosted by mafic to ultramafic suits of rocks that include peridotites, pyroxenite, serpentinite, and dunite. Chromitite is the primary economic source for Chromite mining (Bushveld Igneous Complex, Great Dyke, Stillwater Igneous Complex, and Sukinda–Nausahi Igneous Complex).

Komatiite is a type of mantle-derived ultramafic volcanic rock, uncommon in occurrence and primarily of the Archean greenstone family. Komatiite was named for its occurrences in the type locality along the Komati River in South Africa. It is extremely rich in magnesium and low in silicon, potassium, and aluminum. It consists largely of forsteritic-Olivine,

---

BOX 2.3

CUMULATE ROCK

Cumulates are the typical product of accumulation of early-formed minerals by action of gravity from a fractionating cooling magma chamber. The process of accumulation is by settling or floating and typically concentrates early-formed minerals at the base of a large magma chamber. Cumulate texture is distinctive features based on condition of formation for the group of plutonic rocks. Chromitites and komatiites are typical examples of magnesium-rich mafic magma.

Orthocumulates form when the intercumulus liquid is compositionally different from earlier-formed cumulate crystals and precipitates different intercumulus minerals.

The effect of trapped liquid during crystallization and solidification of cumulus minerals in a layered complex has been studied by Barnes (1986). He concluded that in a cyclic grading from a pyroxenitic base to an anorthositic top, crystallization of a uniform proportion of trapped liquid will result in an apparent iron-enrichment trend from bottom to top of the cycle, as observed in Upper Critical Zone of Bushveld Complex.

FIGURE 2.21   Chromitite occurs as alternate layers of chromite and mafic rock formed by fractional crystallization of magnesium-rich ultramafic magma, part of Sitampundi layered igneous complex, India, a future potential host rock for PGE (Haldar and Tišljar, 2014).

FIGURE 2.22 Euhedral solid crystals of Chromite settling downward to form chromitite layers from a fractionating cooling magma, Sukinda igneous intrusive, India.

Calcic-chromium-pyroxene, Anorthite, and Chromite. The primary composition resembles that of peridotite, but, unlike the deep-seated, coarse-grained peridotite, it shows clear signs of having been erupted. It often exhibits cumulate (layered) and spinifex (large crisscrossing platy crystals of Olivine) texture. Komatiites are associated with massive nickel sulfide deposits at Kambalda, Western Australia. Komatiite-hosted nickel-copper-gold sulfide mineralization accounts for ~14% of the world's nickel production, mostly from Australia, Canada, and South Africa.

Ophiolites or ophiolite complexes are a distinct package of mafic to ultramafic plutonic (gabbro, peridotite, pyroxenite, and serpentinite) (Figs. 2.23 and 2.24), and volcanic rocks (spilite, keratophyre, pillow and sheeted basalts). An ophiolite is a section of the Earth's oceanic crust and the underlying upper mantle that has been uplifted and exposed above sea level and often emplaced onto continental crustal rocks. Ophiolites are closely associated with most of the orogenic belts in the world. The expected economic commodities associated with ophiolites include Asbestos, Talc, Magnesite, iron and titanium, manganese, copper, cumulate and podiform Chromite, and PGEs (Peterson, 1984). Ophiolite complexes can be possible host rocks for chromium and

FIGURE 2.23 Field photograph of mantle peridotite tectonite from Manipur Ophiolite, Eastern India. Rocks are phanerocrystalline, coarse grained, and melanocratic. Color varies between dark (less-altered) and pale green (more-altered), intensely serpentinized, layered, and foliated *Aparajita Banerjee*.

FIGURE 2.24 Photomicrograph of mantle peridotite showing crushed to fine-grained matrix with granular texture of Olivine (Ol), Orthopyroxene (Opx), and Clinopyroxene (Cpx) which resembles harzburgite. Serpentinization has taken place mostly around uncrushed Olivine grains. The ophiolite environment is for chromium and associated mineral deposits *Aparajita Banerjee*.

associated metals in future as reported in Northern Oman.

Laterite is a consolidated product of humid tropical weathering of mafic to ultramafic rocks, predominantly composed of Goethite, Hematite, Kaolinite, Quartz ± Bauxite and other clay minerals. Color varies between red, brown, and chocolate at top showing hollow, vesicular, and botryoidal structure (red ferruginous laterite/limonite). It changes progressively from a nodular iron oxide-rich zone at the top to structureless clay-rich zone (yellow limonite) and ultimately merges with partially altered (transition zone to nickel-bearing saprolite/ Garnierite/serpentinite) to unaltered bedrock. Laterites carry enriched grade of Fe, Al, Mn, Ni, Cu, Co, Ti, and V. Lateritic cover soil can turn into low-grade iron, aluminum, nickel–copper and gold deposits with increase of metal content. Lateritic nickel deposits are formed by extreme tropical weathering of Olivine-rich ultramafic rocks, namely, dunite, peridotite, and komatiite. The final weathered derivative, magnesium-rich serpentinite, often contains low-grade nickel. Limonite-type laterite (oxide) is rich in iron due to strong leaching of magnesium and silica. It is composed largely of Goethite containing 1–2% Ni. Silicate-type lateritic (saprolite/Garnierite) nickel ore formed below the Limonite zone. It generally contains more nickel and consists largely of magnesium-depleted Serpentine to host nickel. Lateritic nickel deposits are located near the surface, generally large size and low grade. Geochemical studies of laterite have been successfully used in exploration for Ni-Cu and gold deposits in Western Australia and an Ni deposit of Kansa at Sukinda Chromite belt, Orissa, India. The majority of lateritic nickel deposits occur in warm and tropical environments under extreme chemical and mechanical weathering, in particular, parallel to north and south of the equator. Large nickel lateritic deposits are reported from New Caledonia (Goro nickel project @ 1.48% Ni and 0.11% Co), Western Australia (Munni–Munni @ 1.07%Ni and 0.085% Co), the Philippines, Indonesia, South America, Russia (Ural Mountains), Oregon, and India (Sukinda Chromite @ 0.84% Ni).

2.2.3 Structure of the Earth and Source of Host Rocks

Earth is an oblate spheroid. It is composed of a number of different layers in spherical shells as determined by deep drilling and seismic evidence (Fig. 2.25 and Box 2.4). These layers are described as follows:

- The Earth can broadly be modeled as an outer solid-silicate crust, a highly viscous mantle, a liquid outer core that is much less viscous than the mantle, and a solid inner core.
- The core is approximately 7000 km in diameter and located at Earth's center.
- The mantle surrounds the core and has a thickness of 2900 km.
- The crust floats on top of the mantle. It is composed of basalt-rich oceanic crust and granite-rich continental crust.
- The location of magmatic Ni-Cu-PGE sulfide deposits is related to lithospheric structural design of subcontinent and its proximity to craton and paleocraton margins. Most large deposits are associated with intracontinental settings (Begg et al., 2010).

2.2.4 Morphology and Settings

The intrusive igneous rocks are formed within the crust. Large bodies of igneous rocks are created deep in the lithosphere by slow cooling of magma. The shapes are irregular and spread over several 1000 sq. km within preexisting stratigraphy (Fig. 2.26). Such massive intrusive bodies are called batholiths. There will be several offshoots from the main massive body and set with different morphology. They often form many smaller or larger enclaves, xenoliths, lopoliths, laccoliths, sills, dikes, and stocks.

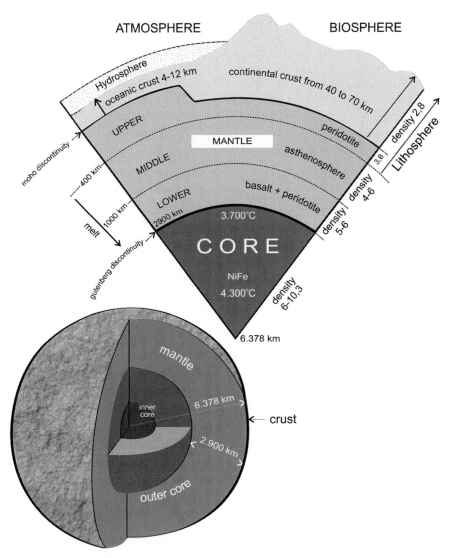

FIGURE 2.25 Schematic diagram of Earth structure representing a three-dimensional perspective (bottom) and a sectional view portraying from central core to outer surface (top) (Haldar and Tišljar, 2014).

Deep-seated mafic/ultramafic magma moves upward through fissures (volcanic neck) and erupts with high pressure as volcanoes. Extrusive mafic igneous rocks cool and solidify more quickly than intrusive igneous rocks with increasing viscosity due to loss of steam and gases. This process is particularly accelerated in sudden outbursts of lava and in explosive eruptions at volcanoes, causing a sudden solidification of lava in the form of volcanic glass. Extrusive igneous rocks are smooth, crystalline, and fine grained. Basalt is a common extrusive igneous rock and forms lava flows, lava sheets, and lava plateaus (Fig. 2.26).

Volcanic rock formed in outpourings of lava on the seabed, because of pulsating pouring

---

BOX 2.4

SOURCE OF HOST ROCKS

The **core** is rich in iron and nickel and divided into two layers. The inner core is solid with a density of $10.3\,g/cm^3$ in comparison to Earth's average density of $5.52\,g/cm^3$ and a radius of ~1220 km. The outer core is fluid with average density of ~$6\,g/cm^3$. It surrounds the inner core and has an average thickness of ~2250 km.

The **mantle** is ~2900 km thick in several different layers and occupies ~83% of the Earth's volume. The upper mantle exists from the base of the crust downward to a depth of ~400 km. This region of the Earth's interior is composed of peridotite, an ultramafic rock consisting of Olivine and Pyroxene. The middle layer, 400–1000 km below surface, is the **asthenosphere**. This layer has physical properties that are different from the upper mantle. Rocks are comparatively rigid and brittle due to cooler temperatures and lower pressures. The lower mantle extends from 1000 to 2900 km below the Earth's surface. This layer is hot and plastic. Higher pressure in this layer causes formation of basalt and peridotite.

The **lithosphere** includes the crust and the upper portion of mantle. This layer is ~400 km thick and has ability to glide over rest of the upper mantle. The deeper portions of the lithosphere are capable of plastic flow over geologic time due to increasing temperature and pressure. The lithosphere is a favorable zone of earthquakes, mountain building, volcanoes, and continental drift.

The crust represents the topmost part of the lithosphere. The materials are cool, rigid, and brittle. Two types of crust can be identified: Oceanic and Continental. Crustal rocks are less dense than those of the underlying upper mantle. Oceanic crust is thin (4–12 km) and composed of basalt with density of ~$3\,g/cm^3$. Continental crust is 40–70 km thick and composed of lighter granite with average density of ~$2.8\,g/cm^3$. These crusts compose numerous "tectonic plates" that float on top of the mantle. Convection currents within the mantle cause these plates to move slowly across the asthenosphere.

---

and mixing with seawater, contains spherical or cushion shapes known as pillow lavas.

2.2.5 Textures

Intrusive and extrusive mafic ultramafic igneous rocks depict distinct and different mineral textures. The properties are characterized by the grain size, shape, boundary relationship, and distribution pattern of certain minerals. Type of texture depends on speed and degree of crystallization of magma, lava, gases and vapors, and temperature.

Intrusive mafic–ultramafic rocks are characterized by a high degree of crystallinity due to extreme slow cooling of magma, differential crystallization, and solidification. Most minerals attain complete crystallization under a slow cooling process to form minor or major grains (Fig. 2.4), unlike amorphous masses during rapid cooling of lava. Common textures of intrusive rocks are holocrystalline (entirely crystalline), euhedral (complete bounded face), anhedral (absence of crystal face), granular, porphyritic (coarse grained), idiomorphic (entirely of euhedral crystals), allotriomorphic (entirely of anhedral crystals), and interstitial. Interstitial texture occurs when the residual mineral fills interstices between earlier crystallized grains (Fig. 2.27).

Extrusive magma will abruptly be exposed to rapid cooling after a period of slow cooling

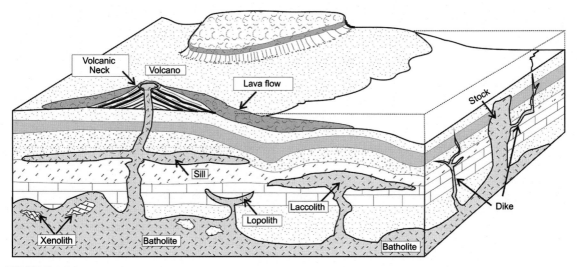

FIGURE 2.26   Conceptual diagram showing morphology and setting of igneous rocks such as batholiths, lopoliths, xenoliths, laccoliths, sills, dikes, stocks, volcanoes, volcanic necks, and lava flows (Haldar and Tišljar, 2014).

subject to high-temperature crystallization of Olivine, Pyroxene, and Ca/Na plagioclase. These minerals are large, properly developed crystals (phenocrysts) embedded in vitreous, microcrystalline core mass resulting from a sudden solidification of magma after eruption.

2.3 GEOCHEMISTRY

Geochemistry is the study of rocks with respect to major, minor, and trace elements. Some elements show close affinity to certain mineral-bearing horizons to indicate future resources. Presence of these characteristic elements acts as a pathfinder for economically viable stratigraphic horizons.

Several mafic–ultramafic rocks are the product of differential crystallization, cooling, and solidification of plutonic and volcanic magma, and often form cumulates. These rocks act as the major host in which Chromite, nickel, and precious metal-bearing sulfides (Ni-Cu-PGE-Au-Co), crystallize along with Pyroxenes, subsequent to Olivine into the interstitial spaces

of cumulates. Most komatiite-associated magmatic Ni-Cu-PGE sulfide deposits are formed from sulfide-undersaturated magmas in dynamic lava channels or magma chambers by incorporation of crustal sulfur. The process involves geochemical discrimination, crustal contamination, and depletion of chalcophile elements during cooling of magma and ore-forming processes. The discrimination is evidenced by crustal contamination, and enrichment of Thorium–Uranium–Light Rare Earth Elements (REE), depletion of Niobium–Tantalum–Titanium or sulfide segregation (Co-Ni-Cu-PGE depletion) of normal igneous fractionation or accumulation of crystals. The amount of contamination, and elemental depletion of chalcophile matter produced during the ore-forming process, depends on several factors (Lesher et al., 2001).

A large number of geochemical samples were studied from four well-known mafic–ultramafic Chromite–nickel–PGE deposits, namely, Lac des Iles, Canada; Jinchuan, China; Bushveld, South Africa; and Sudbury, Canada. Samples were analyzed for major and trace elements. The salient and characteristic compositional features

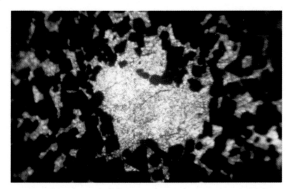

FIGURE 2.27 Interstitial texture of early-formed euhedral Chromite grains (black) showing interstices filled by late-formed and altered anhedral Olivine (light color), indicating differential crystallization and settling of the two minerals Chromite and Olivine, Sukinda intrusive complex, India.

of the best mafic–ultramafic host rocks have been observed as:

1. Abundance of MgO (up to 46%) and FeO (up to 19%),
2. Low SiO_2 (<50%), TiO_2 (<1%), Al_2O_3 (~3.2%), and CaO (~2.7%),
3. Abundance of metallic elements of economic interest, e.g., chromium (up to 8662 g/t), nickel (up to 5268 g/t), cobalt (up to 301 g/t), vanadium (up to 331 g/t), rubidium (up to 24 g/t), strontium (up to 390 g/t), Zirconium (up to 48 g/t), yttrium (up to 10 g/t), and scandium (up to 81 g/t).

Brugmann and Naldrett (1988) concluded, on the basis of geochemical studies, that the ultramafic and gabbroic intrusive complex at Lac des Iles contains PGE-bearing Ni-Cu sulfide mineralization. The chalcophile elemental patterns indicated different ore-forming processes. The mineralization of the gabbroic part is characterized by high and variable abundances of Au, Pd, and Pt, whereas the same for Ir is very low and constant. Its association with coarse gabbro and the high Pd/Ir ratios to the order of 10,000 implied a late-stage magmatic origin. Sulfide saturation is caused by mixing of gabbro–norite magma with silica and volatile-rich

components, by assimilation of older continental crust or by mixing with a volatile-rich eastern gabbro. Sulfides in the ultramafic part are characterized by much lower Pd/Ir ratios (10–30) and are typical for sulfides formed from mafic magmas.

2.4 CLASSIFICATION SYSTEM

PGEs show a natural affinity and association with chromium, nickel ± copper, gold, silver, and cobalt. The classification system for these metals can broadly be divided into three forms:

1. Genetic model
2. Morphology or mode of occurrence
3. Commercial or economic aspect

2.4.1 Genetic Model

The world-wide PGE-nickel-chromium mineralization is primarily of magmatic origin. It occurs extensively in plutonic layered magmatic rocks of mafic to ultramafic composition. Prima facie, magmatic sulfide deposits fall into two major groups based on value of their contained metals:

1. Ni ± Cu are prime valuable products, and
2. Ore enriched in PGEs elements.

The Ni-Cu group includes komatiite-(both Archaean and Paleoproterozoic), flood basalt-, ferropicrite-, and anorthosite complex-related deposits with high Mg-basalts, such as the Sudbury Ni-Cu deposit and the deposits of Ural–Alaskan-type intrusions. The Sudbury Complex is the only example related to a meteorite impact melt. PGE deposits are mostly related to large intrusions comprising both an early MgO- and SiO_2-rich magma and a later Al_2O_3-rich, tholeiitic magma. Most Ni-rich deposits occur in rocks ranging from Late Archaean to Mesozoic. PGE deposits tend to predominate in Late Archaean to Paleoproterozoic intrusions (Naldrett, 2010; Naldrett, 2011b).

The other genetic models are sedimentary, hydrothermal, and lateritic (Mungall, 2005a,b). Hydrothermal deposits are represented by a vein-like or replacement mode, accumulated by metals transported from the parent body. Sedimentary deposits are either black shale containing marine sedimentary exhalative sulfidic ore, or placer deposits derived from mafic–ultramafic magmatic sources. Lateritic Ni–Co deposits with low PGE occur in the deeply weathered profiles developed on Olivine-rich peridotites in tropical or semitropical humid climates, as seen at Goro Nickel Project on the South Pacific Island of New Caledonia, Sukinda Chromite belt, India, etc. A broad classification (Eckstrand, 2005) of PGEs and associated mineral deposits is given in Table 2.2.

2.4.2 Morphology and Modes of Occurrence

Chromite deposits are broadly classified in early 1970's as "Stratiform" and "Podiform"

according to modes of occurrence, forms, and textures. Stowe (1994) further classified Chromite deposits as "Bushveld-type" and "Ophiolitic-type" complexes based on compositions and tectonic settings. The largest deposits of Chromite (Bushveld, South Africa) and nickel (Sudbury, Canada) are identified as major sources of PGE as well. The composite PGE, nickel, and chromium industries diversified manyfold over the years. This necessitates natural microvisualization of mineralization by outlining several "forms" such as "Reef-type," "Contact-type," and others. The classification based on morphology and modes of occurrence is given in Table 2.3.

2.4.3 Commercial Classification

Chromite ore is classified based on chemical composition and physical properties for commercial uses and valuation (Table 2.4).

TABLE 2.2   Classification of PGE and Associated Deposit Types Based on Genetic Model

Genetic Model	Deposit Type	Association	Examples
Magmatic (plutonic layered mafic–ultramafic complex and volcanic flood lava)	Basal sulfide concentration	Cr-Ni-Cu-PGE	Sudbury (Canada); Bushveld (South Africa); Stillwater, Montana (USA)
	• Reef (Sulfide ± chromite/ magnetite) • Contact type • Both	PGE-Cr-Ni	• Merensky Reef (South Africa), J-M Reef (USA), Great Dyke (Zimbabwe) • Duluth, Coldwell (USA) • Stillwater (USA), Stella (South Africa) and Fedorova–Panna (Russia)
Hydrothermal	Polymetallic veins	PGE (Ni, Zn, Co, Ag, Bi)	Sukhoi Log, Russia; Nicholson Bay, (Canada)
Ophiolite	Ophiolitic	Cr	Kempirsai (Kazakhstan), Northern Oman
Laterite	Lateritic	Ni-Co	New Caledonia, the Philippines, Indonesia, Western Australia, Yubdo (Ethiopia), Sukinda (India)
Placer	Placer	PGE, Au	Deposits of Russia; Nicholson Bay, Saskatchewan (Canada)

PGE, platinum group of element.

TABLE 2.3 Classification of PGE-Nickel-Chromium Deposits Based on Morphology and Modes of Occurrence

Modes of Occurrence	Descriptions	Examples
Stratabound	Mineral deposits in stratabound formats are exclusively restricted within a single specific stratigraphic unit. Stratabound deposits will include various orientations of mineralization representing layers, rhythmic, stratiform, veinlets, stringers, disseminated, and alteration zones, strictly contained within the stratigraphic unit, but that may or may not be conformable with bedding. The shapes of plutonic–volcanic igneous host rock include batholiths, lopoliths, laccoliths, stocks, sills, dices, and lava flows. The relationship with country rock is conformable and/or unconformable.	Major primary platinum group of metals, nickel, and chromium deposits occur mostly in greenstone belt and mafic to ultramafic host rock. The significant examples are: Bushveld intrusive (South Africa), River Valley, Sudbury (Canada), Volkovsky (Russia); Stillwater PGE intrusive (USA); and Sukinda–Nausahi (India).
Stratiform (Fig. 2.28)	Hydrothermal, plutonic, and plutonic mafic to ultramafic, and "SEDEX"-type mineralization closely resembles stratification of sedimentary formation, formed by upward moving metal-bearing magma/solution through tectonic openings or porous aquifers, and deposits ore minerals in the overlying pile of sedimentary strata of shale and carbonates. Plutonic magma/volcanic lava undergoes differential crystallization to form stratification of alternate early crystallized metallic and late solidified nonmetallic minerals in numbers of rhythmic cycles.	Chromitite association: UG1 and UG2 (Bushveld), Layer-A (Stillwater) and Lode-1 to 8 (Sukinda). Sulfide association: Merensky Reef (Bushveld), J-M reef (Stillwater), Reef (Northern Finland), Main/Lower Zone (Great Dyke), Platinova (Skaergaard), Sonju Lake (Duluth). Magnetite association: Upper zone (Bushveld and Coldwell).
Layered (Fig. 2.29)	Layered features are developed during the differential crystallization, solidification, segregation, and gravitative settling of mafic to ultramafic magma in a huge chamber below the Earth's crust over a long time. These layered intrusions host magmatic ore deposits containing most of the world's economic concentrations of platinum group elements, nickel, and chromium ± copper, gold, and cobalt. These deposits are mined primarily for their platinum, palladium, rhodium, nickel, and chromium contents. Magmatic ore deposits contain crystals of metallic oxides, immiscible sulfides, or oxide liquids that formed during the cooling and crystallization of magma.	Three largest layer-type mafic–ultramafic igneous intrusive are Bushveld and Stillwater complexes and The Great Dyke (The Big Three). These type deposits host the bulk of the PGE and chrome resources in the world. These deposits are distinguished by higher proportion of magnesia-rich ultramafic to mafic rocks. The chromite bodies occur as layered and stratified units (UG-1 and UG-2 in Bushveld) within ultramafic sills. The deposits are late Archean to early Proterozoic in age and occur within stable continental shields in intracratonic rift settings.
Reef	PGE reefs are stratabound platinum-enriched lode mineralization in mafic to ultramafic layered intrusions. The term is initiated by the South African and Australian mining community for this style of PGE-enriched sulfide mineralization with distinctive texture or mineralogy (Naldrett, 2004). The Merensky Reef represents a mineralized zone within or closely associated with an unconformity surface in the ultramafic cumulate at the base of the cyclic unit. The reef-type mineralization is persistent laterally and extends along strike over km within the intrusive rock layer. The mineralized interval is thin relative to the stratigraphic thickness of layers. Primary metals of economic interest in reef type mineralization include platinum, palladium, and rhodium. The value-added metals recovered as byproduct are copper, nickel, ruthenium, iridium, osmium, and gold. Associated mineral deposits are stratiform Chromite, titanium–vanadium and magmatic sulfide-rich nickel–copper-gold.	Reef-type PGE deposits are mined in the Bushveld Complex (Merensky Reef and UG2), South Africa; the Stillwater Complex (J-M Reef), USA; and the Great Dyke (Main Sulphide Layer), Zimbabwe. The other reef-type deposits are Stella Intrusion, South Africa; Pedra Branca, Brazil; Munni–Munni, Weld Range, and Panton Sill, Australia; Muskox, Canada; and Skaergaard, Greenland.

Continued

TABLE 2.3   Classification of PGE-Nickel-Chromium Deposits Based on Morphology and Modes of Occurrence—cont'd

Modes of Occurrence	Descriptions	Examples
Contact	PGE-enriched sulfide mineralization also occurs near the contacts or margins of layered mafic to ultramafic intrusions (Zientek, 2012). Thickness of contact-type mineralization is large characterizing disseminated to massive concentrations of iron–copper–nickel–PGE-enriched sulfide zones. Mineralization occurs in the igneous intrusion and often in the surrounding country rocks. PGE-enriched contact-type deposits are mined in the Bushveld Complex. The other deposits are undeveloped and still under exploration. The primary product from contact-type deposits includes copper, nickel and platinum group elements with gold as value added byproduct.	PGE-enriched contact-type deposits are only mined in the Bushveld Complex. The other deposits are undeveloped; some are still under exploration. The exclusive contact-type deposits are Duluth, Coldwater, River Valley, and La Perouse Intrusive in North America and Canada (Zientek, 2012).
Reef and contact	Many of the platinum group deposits show the reef and contact type features. Reef-type and contact-type deposits, in particular those in the Bushveld Complex, South Africa, are the World's primary sources of platinum and rhodium.	Reef-type and contact-type deposits include Bushveld, South Africa; Stillwater Complex, North America; Federova-Pana, Russia; and Finnish Intrusive, Finland.
Shear zone	Shear is extreme rock deformation generating fractures, intense foliation, and microfolding due to compressive stress displaying wide zone of shearing in crushed rock mass with width varying between a few cm and several kilometers. Interconnected openings of shear zone serve as excellent channel-ways for mineral-bearing solutions and subsequent formation of deposits	Shear zones in orogenic belts host Chromite–Magnesite veins developed in shear zone at Sindhuvally, Karnataka, India.
Breccia zone (Fig. 2.30)	Fault breccias are tectonically deformed rocks formed by localized deformed zone of brittle deformation. Breccias are clastic rocks of large sharp-angled fragments embedded in fine-grained matrix of finer particles or mineral cement, generated by folding, faulting, magmatic intrusions, and similar forces (tectonic breccias). Tectonic breccia zones represent crush, rubble, cracked, and shatter rock mass. Breccia/conglomerate differs in shape of larger particles due to transportation mechanism. Igneous/flow/pyroclastic breccias are rocks of angular fragments of preexisting igneous rocks of pyroclastic debris ejected by volcanic blast or pyroclastic flow.	Intrusive gabbroic magma with sharp-angled fragments of Chromite embedded in fine-grained gabbroic matrix containing platinum group of elements within preexisting layered ultramafics at Boula–Nausahi, India.
Podiform	Podiform is a "type-form" in deposits of massive Chromite that occurs as pods, nodules, lenses, and layers within ophiolitic ultramafic packages. It is associated with dunite, serpentinite, and harzburgite emplaced into Mesozoic–Tertiary geosynclines along the core of folds. The primary commodity is chromium in expected association with platinum group elements osmium, iridium, and ruthenium. The deposits are more aluminous and Cr:Fe ratio exceeds 2:1.	The example of podiform Chromite deposits are Masinloc, the Philippines,

Crystalline (Fig. 1.12)	A crystal or crystalline solid is a solid material, the constituent atoms, molecules, or ions for which are arranged in an orderly repeating pattern extending in all three spatial dimensions. During crystallization, in which there is enough space for their uninterrupted growth in all directions, the crystals can have a regular polyhedral shape. That is, for example, often the case with minerals Chromite (cubic), Quartz (6-sided prism), Garnet (rhombic dodecahedral), etc. The crystallinity can be very fine to extremely coarse grained.	Most of the Chromite deposits exhibit well-developed crystalline form due to slow cooling and differential crystallization of intrusive magma. The examples can be major deposits like Bushveld Complex, South Africa; Sukinda–Nausahi Intrusive, India.
Massive (Fig. 2.31)	Massive deposits are characterized by substantial share of ore-forming minerals with very few gangue minerals. +60% sulfides in Volcanogenic Massive Sulfide (VMS), Volcanic-Hosted Massive Sulfide (VHMS), Sedimentary Exhalative (SEDEX), and plutonic mafic to ultramafic magma types show massive base metals, Chromite, and Magnetite deposits.	Chromite deposits of Bushveld, South Africa; Sitampundi in Tamil Nadu, India, etc.
Discordant strata	The intrusive deposits often exhibit cutting relationships with country rocks.	Dunite pipes (Bushveld); Robie Zone, (Lac des Iles); and Nizhny Tagil (Urals, Russia)

PGE, platinum group of element.

TABLE 2.4 Commercial Classification of Chromite Ore Based on Physical and Chemical Specifications

Use Type	% Cr_2O_3	Cr: Fe Ratio	% Al_2O_3	% Maximum Impurities	Characteristic
Metallurgical grade-A	+64	+2.5	Low	4–9.5C, 3.0 Si, 0.06 S, 0.03 P	Lump ore, non-friable
Metallurgical grade-B	56–64	1.8–2.5	Low	4–6 C, 8–14 Si, 0.04 S, 0.03 P	Lump ore, non-friable
Metallurgical grade-C	46–55	(–) 1.8	Low	6–8 C, 6 Si, 0.04 S, 0.03 P	Lump ore, non-friable
Refractory	30–40	2.0–2.5	22–34	12 Fe, 5.5 CaO, 1.0 MgO	>57% Cr_2O_3 + Al_2O_3, hard, dense, nonfriable lump ore
Chemical	40–46	1.5–2.0	Low	5 SiO_2, low Mg	Fine-grained, friable ore

Adapted from Mosier, D.L., Singer, D.A., Moring, B.C., Gallaway, J.P., 2012. Podiform Chromite Deposits – Database and Grade and Tonnage Models. Scientific Investigations Report 2012-5157 USGeological Survey, p. 54.

FIGURE 2.28 Stratification of Chromite and Olivine/Magnesite is a characteristic geomorphic expression produced by differential crystallization of cooling plutonic ultramafic magma at Sukinda layered complex, India (field photograph taken in 1964).

FIGURE 2.29 Layered ultramafic intrusive complex at left (peridotite, gabbro, and norite) showing tectonic breccia zone (center and right) at Boula–Nausahi open-pit mine, Orissa, India (Haldar and Tišljar, 2014).

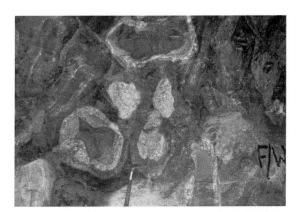

FIGURE 2.30 Irregular fragmented Chromite (black with white alteration rims) in matrix of Pt-Pd-bearing gabbro from tectonic breccia zone, Boula–Nausahi underground mine, Orissa, India (Haldar and Tišljar, 2014).

FIGURE 2.31 Massive Chromite orebody exposed to surface near Karungalpatti village at Sittampundi Complex, Namakkal District, Tamil Nadu, India (Finn Barrett, Exploration Geologist, Goldstream Mining NL, Australia, during reconnaissance field study) (Haldar, 2013).

Deposits of Africa

Platinum-Nickel-Chromium Deposits
http://dx.doi.org/10.1016/B978-0-12-802041-8.00003-1

63

3.1 BACKGROUND

Africa is the world's second-largest and second most-populous continent surrounded by Atlantic and Indian Ocean, and Mediterranean sea. It has 54 sovereign states or countries, nine territories, and two de facto independent states. Its climate ranges from tropical to subarctic. The African continent has large metallic and nonmetallic mineral resources including uranium, diamond, gold, silver, iron ore, copper, bauxite, chromite, platinum group of elements (PGEs), nickel, and cobalt.

The African continent ranks "ONE" for reserve base and annual production of PGEs, and chromite ore. The major share of PGEs, chromium, and nickel deposits are located in South Africa, Zimbabwe, and Tanzania. Other contributing countries are those of East Africa, Zambia, Botswana, Sudan, and Madagascar. The copper belt in the former Katanga Province of the Democratic Republic of the Congo and in Zambia, diamond mines in Sierra Leone, Angola, and Botswana are well known for abundant, rich products. Mining of huge and varied mineral reserves plays a vital role and is the backbone of the African economy.

3.2 SOUTH AFRICA

South Africa holds the largest share of global reserves and is the undisputed largest producer of chromite and platinum in the world. Three of the five largest chromite–PGE deposits occur near the stratigraphic middle of large layered intrusions: Merensky and UG2 Chromitite Reefs of the Bushveld Complex, South Africa, and the Main Sulfide Zone of The Great Dyke, Zimbabwe (Cawthom, 2005a). Nkomati nickel mine is the only primary nickel producer with an annual capacity of ~10%. The majority of South African nickel production is a co-product of PGE mining. The largest co-product nickel producers are Anglo Platinum and Impala Platinum Holdings, which produce ~25–30 tonnes (28–33 tons) of nickel annually from refineries based at Rustenburg and Johannesburg.

3.2.1 Bushveld Igneous Complex

The Bushveld Igneous Complex (BIC) is the largest chromite–PGE deposit of the "Big Three." The Stillwater Complex in the United States, and "The Great Dyke" in Zimbabwe are the other two in descending order. BIC is located in the northeast part of South Africa, and was discovered in 1897 by Gustaff Molengraff, a Dutch geologist, biologist, and explorer. He discovered the Bushveld Complex while mapping the Transvaal area as a working state geologist. The structure represents a huge saucer-shaped layered mafic–ultramafic intrusion of Paleoproterozoic age (2060 Ma) forming a great geological basin. BIC extends over 450 km east–west, and

250 km north–south. The surface area of potential host rocks covers over 66,000 sq. km, and has a thickness up to 9 km as preserved today.

BIC was intruded along the axis of a 500-km long north-northeast to south-southwest (NNE–SSW)-trending Thabazimbi–Murchison tectonic lineament (2500 Ma) within basement Archean granite gneisses (>2900 Ma), and Proterozoic sediments of the Transvaal Supergroup. The area is divided into four prominent limbs, lobes, or structures: Northern, Eastern, Western, and Far WesternBushveld. The former three structures have the shape of semicircular basins (Fig. 3.1), distinctly detached from each other, and separated by younger granite intrusions.

The Eastern and Western lobes show a full succession of rocks, and a type of mineralization starting from the Marginal Zone at the base to an Upper Zone at the top. This is evidenced by subsequent exploration drilling. The southern part of the Western Lobe lacks Marginal and Lower Zones. The Far Western structure is an oval-shaped body with one or two minor localities depicting preservation of only marginal rocks. The Northern lobe is separate from the other structures having a different stratigraphy.

3.2.1.1 Geology

The country rocks are presented by Archean basement composed of granite, gneisses, and

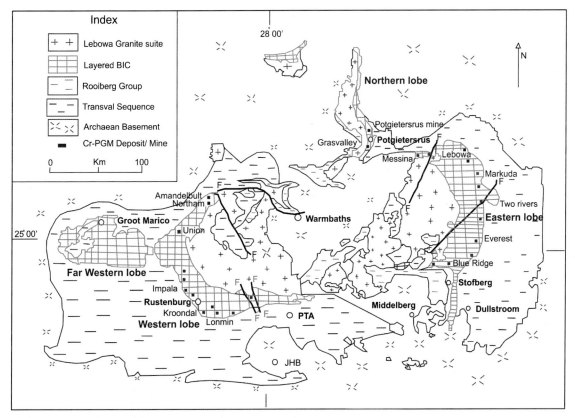

FIGURE 3.1 Schematic surface geological map of Bushveld Igneous Complex (BIC), South Africa, showing various limbs/structures and chromite–platinum (Cr-PGM) deposits/mines *Haldar, S.K., 2013. Mineral Exploration – Principles and Applications. Elsevier Publication, p. 374.*

greenstones. The basement rocks are overlain by metasediments during the Late Archean to the Early Proterozoic period. These rocks form Transvaal Supergroup (Pretoria Group) comprising argillite, shale, quartzite, banded-hematite–quartzite (BHQ), and dolomite. The total overlying sequence is covered by Early Proterozoic volcanic flows and pyroclastics (Rooiberg Suite), a precursor of the ultramafic intrusion forming the roof of the Igneous Complex. BIC comprises three main intrusive suites: Rustenburg Layered Suite, Lebowa Granite Suite, and Rashoop Granophyre Suite. The first phase of magmatism (Rustenburg) within an intracratonic setting created the largest chromite- and PGE-bearing mafic–ultramafic cumulate in the world. The rock type comprises ultramafic (peridotite, chromitite, and harzburgite) in the lower portion to mafic (gabbro, norite, and anorthosite) in the middle, and a felsic phase (granite) in the top section. Bushveld is known primarily for layered mafic–ultramafic rocks extremely enriched with chromite and PGE. The area has undergone many geological processes during Mesoproterozoic time presented by undifferentiated felsic rocks covered by a younger sedimentary sequence, a granitic intrusion, and a subvolcanic granophyre. A suite of mafic–ultramafic sills intruded into the floor rocks of Archean granites and the Transvaal Supergroup in the Bushveld Complex illustrates "half-graben" geometry and the importance of the Thabazimbi–Murchison Lineament as a structural feature. The lateral extent of different mineralized zones increases from bottom up in the N–S direction (Kinnaird, 2014). The total stratigraphic sequence is summarized in Table 3.1.

The magmatic episodes and formation of suites in the total package of Bushveld stratigraphy is described as four distinctive igneous suites. The first suite consists of early mafic sills (Cawthorn et al., 2002). The second suite is recognized as "Rooiberg Group" and is

TABLE 3.1 Generalized Geological Succession in the Bushveld Igneous Complex, South Africa

Formation	Period	Rock Type
Lebowa suite	Mesoproterozoic (2000–1400 Ma)	Younger sedimentary cover rocks, granite intrusion, and subvolcanic granophyre.
Rustenburg layered intrusive suite (BIC)	Paleoproterozoic (2500–2000 Ma)	Mafic and felsic intrusive rocks: norite–gabbro–pyroxenite–harzburgite–chromitite hosting PGE and chromium.
Rooiberg Group	Early Proterozoic (2060 Ma)	Felsite composed of volcanic flows and pyroclastics.
Transvaal Supergroup (Pretoria group)	Late Archean to Early Proterozoic (2500–2000 Ma)	Metasediments composed of argillite, shale, quartzite, banded-hematite quartzite (BHQ), and dolomite.
Basement	Archean (>2500 Ma)	Granite, gneisses, and greenstones.

composed of felsites, forming the roof of much of the layered igneous series. The felsite suite was followed by intrusion of layered mafic–ultramafic rocks of the third suite (Rustenburg Layered Series, 2060 Ma), along with accompanying dikes and sills. The third suite attained a thickness of ~10 km of mafic–ultramafic sequence through repetitive major influxes of magma into a huge subvolcanic chamber over an enormous span of time. Cooling and differential crystallization was a slow process forming subhorizontal shallow-level layers from the base of the chamber. Melting of roof rocks resulted in a series of granophyres which occur at intervals along the upper margin of the layered series. The economic aspects of the Bushveld area are confined within the mafic–ultramafic intrusive (Rustenburg Layered Intrusive Suite) which is responsible for the largest resources of precious and noble metals in the world. The fourth and final igneous suite of the Bushveld igneous event is the Lebowa Granite Suite that intruded into the center of the mafic–ultramafic rocks. Seismic and gravity data indicate that this has the form of a mushroom-like body with a central stock of granite pillowing out over the mafic rocks (Naldrett, 2004).

3.2.1.2 Mineralization

BIC contains little sulfide mineralization. Merensky Reef and Bastard pyroxenite contain minor sulfides in the immediate hanging-wall rocks indicating sulfide undersaturation (Cawthorn, 2005a,b).

The huge thickness of mafic–ultramafic host rock was achieved by intermittent replenishment, addition, and mixing of existing and new magma. The process produced repetition of layers presented by metallic minerals, concentration of mineralized bands, and host rocks. Some individual layers or groups of layers can be traced for hundreds of km.

There are more than 20 chromitite horizons in Rustenburg Layered Suite. Rustenburg Layered Suite is divided into five broad zones in ascending order (Barnes et al., 2004; McDolald and Holwell, 2011):

1. Marginal Zone composed of norite and pyroxenite sills,
2. Lower Zone comprising ultramafic suite of pyroxenite and harzburgite,
3. Critical Zone composed of the mafic–ultramafic suite (norite–pyroxenite–harzburgite and chromitite),
4. Main zone of mafic suite (anorthosite–gabbronorite and norite), and
5. Upper Zone of mafic suite (diorite–magnetite–ferrogabbroic gabbronorite, and anorthosite).

The different sills or group of sills, mineralized zones, and units display variable lithologic units, thicknesses, bulk chemical signatures, and mineral concentration arising from different

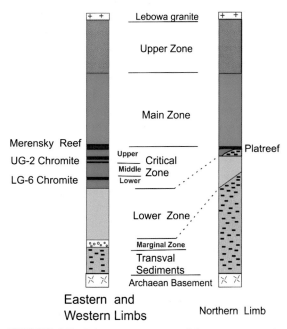

FIGURE 3.2 Schematic succession of five major mineralized horizons in eastern, western, and northern limbs at the BIC, Rustenburg Layered Suite, South Africa.

influxes of new magma and the effects of wall-rock contamination in the complex. The lower zones vary in thickness partly due to the transgressive nature of the new magma. All chromitite horizons are enriched in PGE to varying degrees. The five major mineralized horizons, from bottom to top, within the stratigraphy of the Layered Suite are conceptualized in Fig. 3.2. The mineralized zones, in general, dip toward the center of the complex.

The thickness of the bottommost Marginal zone varies between 0 and 800 m and is composed primarily of norite ± pyroxenite. The Lower zone, with thickness between 800 and 1300 m, is dominated by bronzitite, harzburgite, and dunite. The ultramafic sequences of the Lower zone occur beneath the Platreef in the Northern limb, and are separated by intervals of country rocks (Fig. 3.2) and thin sills of fine-grained norite and pyroxenite of the Marginal zone. The Lower zone extends deep beneath the

Marginal zone. The emplacement of the Lower zone occurred as multiple magma batches that differentiated, forming cyclic units with compositional reversals at their bases. The presence of olivine and orthopyroxene shows multiple reversals within the Lower zone. The style of reversals is consistent with a gradual mixing of fresh magma in the chamber (Yudovskaya et al., 2013).

The overlying Critical zone, with thickness between 1300 and 1800 m, is marked by incoming of cumulus chromitite at its base. The Critical zone is overlain by norite, gabbro, and anorthosites of the Main zone (3000–3400 m thick). Upper zone is composed of iron-bearing gabbros and diorites with thickness between 2000 and 2800 m.

The study of microstructure supported by major and trace element concentrations in whole-rock, cumulus, and intercumulus minerals from the Lower and the Lower Critical zones of the BIC provided an understanding of the formation of cumulate and to establish possible parental liquid. The primary cumulus (orthopyroxene and olivine) and intercumulus (clinopyroxene and plagioclase) minerals show relatively constant composition with respect to major and trace element throughout the lithosequence. It suggests constant composition of parental magma, and a similar crystallization path or evolution history throughout the package (Godel et al., 2010).

The Critical zone is most significant in hosting the PGE and chromium system with respect to the world's resources. The zone is divided into two broad groups: Lower and Upper with an overlapping Middle Group (Fig. 3.3). The Lower Critical zone consists primarily of bronzitites, chromitites and harzburgites that occur in unsystematic order. The Upper Critical zone is composed of anorthosites, norites ± bronzites in the lower part and chromitite, harzburgite, bronzitite, norite, and anorthosite in the upper part that made up the regular succession of rock types.

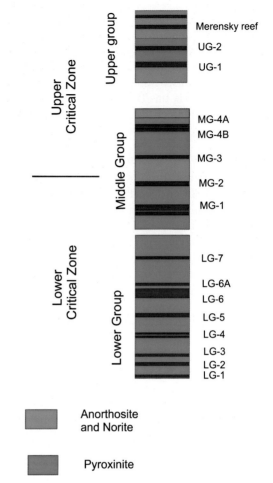

FIGURE 3.3 Schematic vertical section of the Critical Zone showing the distribution of chromitite layers (lode in red) in the BIC. The average PGEs range between 0.2 and 2.1 g/t in the Lower Critical Zone, 0.9–5.5 g/t in the Middle Group and 1.2–8.0 g/t in the Upper Critical Zone.

The chromite reefs in the Critical zone of the Eastern and Western Bushveld are named from stratigraphic bottom to top as:

1. Lower Critical Group: LG1 to LG6, LG6A, and LG7 containing 0.2–2.1 g/t Pt.
2. Middle Critical Group: (MG1 to MG3, MG4A, and MG4B containing relatively higher grade varying between 0.9 and 5.5 g/t Pt.

3. Upper Group: UG1, UG2, Merensky Reef containing grades between 1.2 and 8.0 g/t Pt.
4. The chromite content is up to 43.5% Cr_2O_3.

The PGE elements are closely associated with Fe–Ni–Cu sulfides in Merensky Reef, and Platreef. The three major, distinct, and rich orebodies in BIC are Merensky Reef, Upper Group-2 (UG2), and Platreef. Merensky and UG2 Reefs can be traced over 300 km in two separate arcs, and Platreef extends over 30 km.

3.2.1.2.1 MERENSKY REEF

Merensky Reef was an outstanding discovery of a platiniferous orebody in the Eastern Bushveld in 1924 by Dr Hans Merensky and Andries Lombaard. Merensky Reef, a stratiform mineralization of sulfide association, occurs at the top stratigraphic layer of the Upper Group (Upper Critical Zone) and fully developed both in the Eastern and Western limbs with thickness varying between 30 and 90 cm. The Reef rests on anorthosite (Fig. 3.4), and is overlain by pegmatitic feldspathoid–pyroxenite. Host rock is norite with extensive chromitite and sulfide layers or zones containing rich ore. Merensky Reef is fully developed in the Eastern and Western Bushveld. Ore zones are significantly confined to a pegmatoid layer (coarse crystals of bronzite and olivine) typically occurs at base of Merensky cyclic unit. The concentration of PGE is predominant in the vicinity of two chromitite layers positioned at the top and bottom of the pegmatoid. The highest metal values are associated with the upper chromite layer.

The concentrations of PGEs, cobalt, rhenium, gold, and silver are predominantly present in base-metal sulfide. PGEs occur (essentially as Pt–Fe alloy, Pt–Pd sulfide, and Pt–Pd bismuth telluride) at the contact between the base-metal sulfides and silicates or oxides, and included mostly within pentlandite and pyrrhotite, and less significant in chalcopyrite. The multidisciplinary analysis using three-dimensional (3-D) X-ray tomography, and in situ fluid inclusions

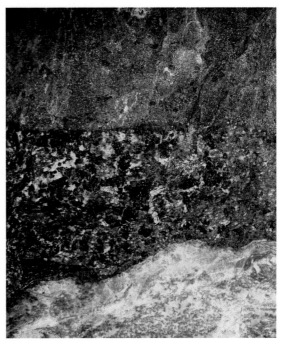

FIGURE 3.4 Stratiform chromitite mineralization (black) of the Merensky Reef rests on anorthosite (light gray) at the Karee Mine, No. 4 shaft of Lonmin, Bushveld Complex. The concentration of PGMs is predominant in the vicinity of two chromitite layers positioned at the top and bottom of pegmatoid. The highest metal values are associated with the upper chromite layer. *Dr Tom Evans.*

analysis by inductively coupled plasma mass spectrometry (ICPMS) indicate that PGEs are found either in solid solution in base-metal sulfides or as closely associated with base-metal sulfides. The PGEs were collected by a base-metal sulfide liquid initially and interacted with silicate magma rich in PGEs. The sulfide liquid then percolated downward due to compaction of the cumulate pile and redistributed at the footwall of the Reef (Godel et al., 2007, 2008).

3.2.1.2.2 UPPER GROUP-2 (UG2) REEF

UG2 chromitite is known for enrichment of PGEs, and extends for nearly the entire 400 km length of the Eastern and Western

FIGURE 3.5 Typical fine-grained massive chromitite (black) resting on anorthosite (light color) with sharp contact at the footwall of the UG2 orebody *Dr Tom Evans.*

FIGURE 3.6 The concentration of the platinum group of minerals with large sperrylite crystal is predominant in UG2 Chromitite Reef, making it the largest single PGE–Chromitite resource in the world. *Avinash Serin.*

limbs (Mondal et al., 2006). The UG2 horizon represents typically fine-grained massive and stratiforn chromite mineralization having a sharp contact with anorthosite (Fig. 3.5). It is the world's largest single PGE–Chromitite orebody and even surpassed Merensky Reef for platinum (Fig. 3.6). UG2 occurs below Merensky and above UG1 stratigraphic layer forming part of the Upper Group (Upper Critical zone). Chromitite rests at the base of the UG2 cyclic unit. The layer has sharp lower and upper contacts. The thickness of the main chromitite lode ranges between 40 and 130 cm. The platiniferous UG2 chromitite layer is underlain by anorthosite at Karee Mine. The interface between these two rock types shows evidence of irregularly distributed potholes, possibly formed due to remelting, diapirism, impact-generated dimpling, gas escape, and interference rippling (Merwe and Cawthorn, 2005).

UG2 chromitite layer is divided into 10 distinct sublayers from bottom to top based on cryptic variation of chromite mineral chemistry. Each sublayer is defined by an upward decrease of Mg combined with increase of Cr and TiO_2. These trends reflect magmatic differentiation of individual chromitite entities that unite to form a massive chromitite seam. Nickel and copper sulfides (pentlandite and chalcopyrite), interstitial to chromite grains, are rare. The associated PGE/minerals are Pt–Fe alloy, laurite, and cooperite–braggite. Platinum, osmium, iridium, and ruthenium occur as discrete metals. Palladium and rhodium are present in the crystal lattice of pentlandite that reach maxima of 2.2% Pd and 3.0% Rh. Junge et al. (2014) suggested a succession of chromitite sublayers formed due to magmatic differentiation and segregated in sequence on top of each other, and finally forming a massive chromitite layer. PGE-rich magmatic sulfides segregated simultaneously with chromite grains. These sulfides were removed mostly by reaction with chromite, thereby upgrading the PGE contents of the remaining sulfides.

3.2.1.2.3 PLATREEF

In 1925, Dr Hans discovered another promising orebody at the base of the Northern Limb, close to the Thabazimbi–Murchison Lineament, and named it "Platreef." Platreef is one of the largest and most valuable Ni–Cu–PGE

FIGURE 3.7 The cooling of plutonic magma is slow due to the extremely large volume of the subsurface magma chamber. Slow crystallization forms large sperrylite crystals in the Platreef (Mogalakwena Platinum mine, Potgietersrus, northern part of the BIC, South Africa). *Soumi Haldar.*

(Fig. 3.7) orebodies in the world. It is present in the Lower, Critical, Main, and Upper zones. The Layered suite extends over 110 km, thickness between 50 and 400 m, and dips at 30–40 degrees toward the west.

Platreef mineralization is stratabound, not stratiform, and placed in direct contact with country rock. Platreef represents a complex group of sills intruded into basement granite gneiss and sediments of the Transvaal Supergroup. It was formed because of interaction between a new gabbroic parental magma of the Main Zone, a suite of sulfur-bearing sediments, and preexisting Lower Zone cumulates arising from different inputs of magma and the effects of local wall-rock contamination. Pyroxenes from the pyroxenite facies of the Platreef at Sandloot Mine show $\delta^{18}O$ values that are up to 2.4% higher than pyroxene from the Upper and Main Zone. This difference is explained by additional contamination and assimilation of country rock (dolomite) which is in direct contact with the intrusion (Harris and Chaumba, 2001). Platreef is formed by introduction of PGE-rich sulfide droplets with intruding Platreef magma. The chromitiferous ore is rich in Fe–Ni–Cu sulfides and PGEs.

Concentration of disseminated sulfides in the upper parts of basement gneisses is caused by downward migration of residual sulfide liquid and fractionation out of the remaining platinum elements. Sulfide-rich zones with massive sulfides accumulating at the footwall are due to the permeability of host rocks. There is fundamental floor-rock control on the distribution of PGEs from Platreef into the footwall rocks. Sulfide liquids produce an intimate PGE–Base Metal Sulfide (BMS) association. Xenoliths and irregular bands of chromitite within Platreef are common features (Holwell et al., 2006).

Platreef unit comprises a diverse package of pyroxenites, peridotites, and mafic rocks with Ni–Cu–PGE mineralization. Ni–Cu sulfide mineralization is usually abundant and the overall grades of PGE are lower in Platreef than in other Bushveld deposits. It is general believed that the PGEs are closely associated with sulfide mineralization. But detailed study by laser ablation ICPMS on drill cores through Platreef suggests that a significant proportion of Pd, Os, and Ir are hosted by base-metal sulfides at the top of Platreef. In contrast, Pd [as Pd–Te–Bi–(Sb) phases], Os, and Ir (as Os–Ir alloys) occur in solid solution and as discrete inclusions within base-metal sulfide throughout the core (Hutchinson and McDonald, 2008).

3.2.1.2.4 MAIN ZONE

Main Zone gabbronorites below the Pyroxenite Marker in the Eastern Bushveld display normal upsection fractionation trends in mineral compositions and constant initial Sr-isotopic ratios of bulk plagioclase separates. There are distinct lateral variations in Mg numbers of orthopyroxenes, Anorthite (An)-contents of bulk plagioclases, and incompatible trace-element concentrations. These lateral variations primarily reflect variations in amount of trapped liquid and to a lesser extent local assimilation of country rocks (Lundgaard et al., 2006).

3.2.1.2.5 UPPER ZONE

Iridium and rhodium are enriched in the sulfide-poor Main Magnetite Layer of the Upper Zone.

3.2.1.3 Geochemistry

The parent magma responsible for chromium–platinum-hosted mafic–ultramafic intrusive at BIC has been categorized into two predominant types: "U-type" and "Tholeiitic-type." U-type magma characteristically contains high magnesium (MgO: 8–15%), high silica (SiO_2: 57–60%), rich chromium (200–1800 g/t), nickel (70–450 g/t), and low titanium (TiO_2: <0.5%). U-type (or boninitic) magma is formed by a second-stage melting process to substantiate the high content of PGEs at the Bushveld. Phenocrysts of pyroxenes and olivine are typical texture. Sr, Nd, and Pb isotopes support significant upper crustal contamination. Tholeiitic magma is relatively rich in silica and poor in sodium. The rock is composed of clinopyroxene, hypersthenes, and plagioclase with minor iron–titanium oxide, and little or no olivine. The tholeiitic magma type has undergone crustal contamination of the lower crustal type.

The geochemical compositions of two types of magma from the BIC were compared by Barnes and Maier (2002). They opined that U-type magma composition is similar to a magnesium-basaltic andesite with an estimated MgO (12.65%), SiO_2 (55.87%), TiO_2 (0.37%), Al_2O_3 (12.55%), FeO (9.15%), MnO (0.21%), CaO (7.29%), S (866 g/t), Cr (952 g/t), Ni (257 g/t), and Cu (61 g/t). Tholeiitic magma contains an estimated MgO (7.26%), SiO_2 (50.48%), TiO_2 (0.71%), Al_2O_3 (15.79%), FeO (11.61%), MnO (0.18%), CaO (10.86%), S (340 g/t), Cr (314 g/t), Ni (146 g/t), and Cu (60 g/t). Noble metals are in the range of ppb. It is evident that magnesium-basaltic andesite represent ~3 times of sulfur and 1.5–2 times of the concentration of PGEs compared to tholeiitic magma. Basaltic-andesite will crystallize olivine, orthopyroxene, and then plagioclase. Tholeiitic magma will crystallize plagioclase followed by clinopyroxene, possibly preceded by minor or no olivine.

Barnes and Maier (2002) stated that rocks of the Lower Zone and Lower Critical Zone are composed mainly of harzburgite and pyroxenite indicating the parent magma was derived from MgO-rich liquid. Rocks of the Main Zone include gabbronorite and indicate a signature similar to tholeiitic magma. Rocks of the Upper Critical Zone, particularly close to mineralized reefs, are the ultimate crystallized product of a mixture of two magmas. The proportion of tholeiitic magma in mixture increased progressively upward. The study is based on adequate Pt–Pd, and other trace elements in the Lower, Critical, and Main Zones. The analyses include trace elemental ratios of Th/Hf (0.47), $^{87}Sr/^{86}Sr$ (0.7059–0.7000), and isotopic signature εNd at (−4.92 to −6.74).

3.2.1.4 Reserve and Resources

The Bushveld Complex contains +75% of the world's platinum resources from some of the richest chromite and PGE deposits. The single lode, the UG2 chromitite unit, contributes 58% of resources (Naldrett et al., 2011a). Some layers contain 90% chromite making the rare rock type, chromitite.

Chromite reserve and resource estimated at 11,550 Mt grading between 40 and 50% Cr_2O_3, 2.87 g/t Pt, and 1.80 g/t Pd, with 5.67 g/t PGEs (Naldrett, 2004). Merensky Reef stands at 4210 Mt grading between 4.4 and 6.9 g/t PGE over mining width (Cawthorn et al., 2002). The Merensky Reef contains 3.5 and 9.5 g/t Pt in pyrrhotite, pentlandite, and pyrite (Naldrett, 2004). Merensky and UG2 reefs contain ~90% of the world's PGE resources of 9952 Mt @ 3.05 g/t Pt and 1.76 g/t Pd. Platreef has been estimated at 1597 Mt grading 1.77 g/t Pt, 2.01 g/t Pd, 0.41% Ni, and 40–50% Cr_2O_3.

Deposits have extensively been drilled and depth continuity established up to 1500 m for Platreef and 3300 m for Merensky and UG2.

3.2.2 Uitkomst Complex (Nkomati Nickel Deposit)

The Uitkomst Complex is a layered basic intrusion recognized for hosting viable magmatic base-metal sulfide deposits containing Ni–Cu–PGE–Cr mineralization in South Africa (De Waal, 2001).

3.2.2.1 Location

The Uitkomst Complex is situated close to the small resort town of Badplaas in Mpumalanga Province, 300 km east of Johannesburg, 200 km east of Pretoria, and 50 km east of the Eastern limb of the BIC. The complex is a layered mafic–ultramafic–polymetallic massive sulfide intrusion elongated parallel to the major faults. The massive sulfide deposit was discovered in 1990 and initial production established in 1997 with an annual production of 36,000 tonnes (~40,000 tons) of Ni-concentrate since 2005 (Kirste, 2009). The complex holds the only primary nickel mining along with cobalt as byproducts of copper–PGEs, and chromium in South Africa. The Nkomati nickel mine is one of the lowest-cost nickel mines in the world.

3.2.2.2 Geology

The layered mafic–ultramafic igneous complex is an intrusion with a U–Pb "sensitive high-resolution ion microprobe" (SHRIMP) age of 2044 ± 8 Ma. The deposit represents an escarpment area with a moderately flat basin and an undulation on either side. The geological setting is similar to the Bushveld Complex and portrayed in Fig. 3.8.

Basement Archean Nelshoogte plutonic granites and gneisses (3250 Ma) are well developed southeast of the complex. Rocks vary between light gray feldspar–quartz and darker biotite-hornblende gneisses showing a high grade of sericitization and silicification, and are frequently cut by chlorite and quartz veins enriched with stringers of chalcopyrite and pyrite. Overlying units are a series of sedimentary packages composed of Black Reef and Oaktree quartzite, Malmani dolomite/chert, and Timeball–Hill shale. The Uitkomst mafic–ultramafic complex intruded into the total succession of the Lower Transvaal Group along a fault zone (Fig. 3.9).

3.2.2.3 The Complex

The Uitkomst Complex is a body of tabular shape covering area of ~10 km in length oriented in the northwest–southeast direction, 800 m in width, and up to 750 m vertical thickness. The initial magma was emplaced at the base of the Malmani Subgroup and flowed as sills from southeast to northwest. Intrusion/emplacement continued upward within the regional geological setting between existing fault and fracture zones. The mafic–ultramafic complex portrays an elongated trough-shaped body that concordantly penetrated sediments of Transvaal Supergroup. The intrusion dips at a shallow angle (4–5 degrees) to the northwest. Contact with Archean basement granite is stratigraphically ~10 km below the base of the coeval Bushveld Complex (Maier et al., 2004).

The Uitkomst Layered Complex is divided into two major groups: Main and Basal Groups. These groups display large differences in vertical thickness, abundances of sulfides, and xenoliths of the country rocks. The consistent thickness of individual lithostratigraphic units of the complex along the central axis, over a distance exceeding 12 km, is remarkable. The complex is further subdivided into six/seven subunits based on lithology and mineralization (Fig. 3.9). Chromitite becomes extremely massive up to 6 m thick with typical igneous-layered structure toward the top.

3.2.2.4 Mineralization

The bulk of mineralization is concentrated in the Lower Harzburgite and Basal Gabbro units with a common form like finely disseminated, local concentrations in pegmatoidal pyroxenite, net-textured, clusters of ore enclosing olivine/orthopyroxene, and layered

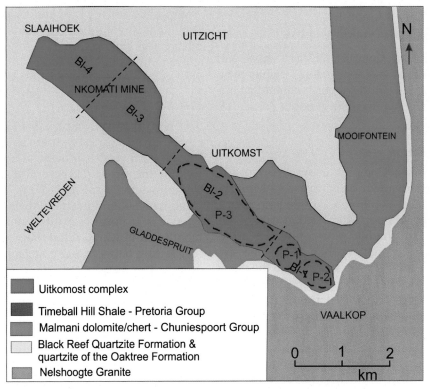

FIGURE 3.8   Simplified surface map of the Uitkomst Complex hosting low-cost nickel mining with byproduct metals: Cu, Co, Pt, Pd, and Cr. The open Pits 1, 2, and 3 are in Blocks 1 and 2 in the southeast end of the complex. *Based on/After Theart, H.F.J., De Nooy, C.D., 2001. The Platinum Group Minerals in two parts of the massive sulfide body of the Uitkomst Complex, Mpumalanga, South Africa. South African Journal of Geology 104, 287–300. Figure 1 and published with the permission of the Geological Society of South Africa.*

massive ore. The sulfides are more abundant near xenoliths of country rock and pods in the Basal Gabbro. The dominant minerals in decreasing order are pyrrhotite, pentlandite, chalcopyrite, magnetite, ilmenite, chromite, PGEs, digenite (Cu_9S_5), and pyrite. Minor minerals include violarite ($Fe^{2+} Ni_2^{3+} \underline{S}_4$), mackinawite {$(Fe, Ni)_{1+x}S$}, awaruite ($Ni_2Fe$ to Ni_3Fe), galena, sphalerite, native Cu, arsenopyrite, cobaltite, and millerite (Fig. 3.10).

3.2.2.5 *Reserve and Resources*

Nkomati is a polymetallic deposit comprising disseminated sulfide-bearing nickel, copper, cobalt, and PGEs, and a massive oxidized chromitite orebody. African Rainbow Minerals (ARM) estimated reserves and resources in two categories:

1. Measured and indicated ore reserve and resource stand at 281 Mt grading 0.34% Ni, 0.12% Cu, 200 g/t Co, and 0.86 g/t 4E (Pt + Pd + Rh + Au).
2. Massive oxidized chromitite ore of 0.23 Mt at 34% Cr_2O_3 Cut-off.

3.2.3 Stella Intrusion

The Stella Layered Intrusion was focused upon during a gold exploration program in the 1990s

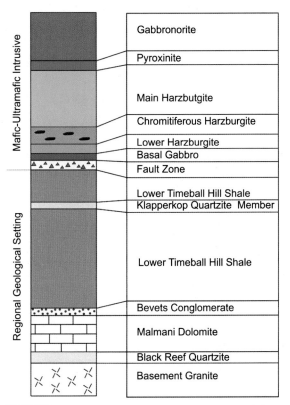

FIGURE 3.9 Conceptual stratigraphic sequence of older regional geological setting intruded by younger mafic–ultramafic Uitkomst Complex following fault/shear zones on either side. The magmatic emplacement continued upward between existing fault and fracture zones.

FIGURE 3.10 Common ore type at the Nkomati mine contains pyrrhotite, pentlandite, chalcopyrite, magnetite, ilmenite, chromite, PGEs, digenite, arsenopyrite, cobaltite, and millerite.

by Anglo American Prospecting Services in the Eastern Kraaipan Greenstone Belt, South Africa.

3.2.3.1 Location

The Stella Intrusion is located in the North West Province of South Africa near the international border with Botswana. The intrusive is 330 km west of Johannesburg, 230 km southwest of the Bushveld Complex, 45 km west of the Platgold Mine, and 25 km north of Stella Township (Fig. 3.11).

3.2.3.2 Geology

The Stella area is dominated by Kraaipan Greenstone rocks that occur within Archean cratonic granites and gneisses of the Ventersdorp Supergroup. The greenstone rocks are demarcated in two parallel belts, trending NNW–SSE with a steep dip toward the SSW. The belts are characteristically parallel to the western limb of the BIC. The rocks are poorly exposed and covered by the 3–8 m thick unconsolidated aeolian Kalahari sands. Greenstone rocks comprising deformed and metamorphosed volcano-sedimentary Banded Iron Formation (BIF), magnetite-rich quartzite, amphibolites, conglomeritic layers, and associated Archean granitoids.

The eastern belt is referred to as the Goldridge Belt hosting Kalahari gold or the Kalgold Mine. The western belt is referred to as the Stella Belt hosting the Stella Layered Intrusion (SLI), enriched with a series of platinum–gold deposits (Kalahari Platinum or Kalplats project) of Platinum Australia Limited (2014). The strike extensions of eastern and western belts are 100 and 120 km, respectively, with maximum thickness up to 5 km (Fig. 3.11).

The Stella layered complex is a sill-like magmatic package intruded within Kraaipan Greenstone rocks. The dimensions of the layered complex, estimated from magnetic bedrock signatures and borehole information, stand at 12 km long and 1.5 km wide. Rocks comprise distinct cumulate-textured gabbros, leucogabbros, and magnetite gabbros. Magnetite content

FIGURE 3.11 Location and geological map of the Kalplats project and prospecting held by PLA and its JV partners, Stella Layered Intrusive, South Africa. *Lewins, J.D., Hunns, S., Badenhorst, J., 2008. The Kalahari Platinum project. In: Third International Platinum Conference 'Platinum in Transformation', The Southern African Institute of Mining and Metallurgy, Paper 48, pp. 355–366.*

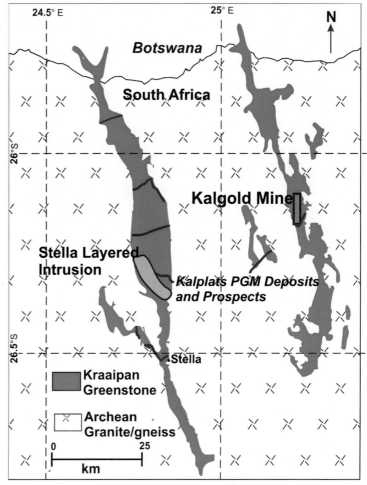

ranges from low disseminated (1–2%) to strong segregations forming magnetitite layers (50–90% magnetite). The rocks of the SLI and Kraaipan greenstone are structurally deformed resulting in overturned folding and consequent steepening of strata at 80–85 degrees to the west-south-west. The age of the SLI has been estimated at 3033 ± 0.33 Ma.

3.2.3.3 Mineralization

The platinum group of mineralization occurs in magnetite-rich gabbros about 150 m above the footwall contact with Greenstone/basement rocks. A PGE-enriched magnetite layer is 100 m thick that includes a number of laterally continuous reefs constituting the oldest mineralization of this type known on Earth (Maier et al., 2003). Footwall gabbros witnessed strong shearing and thrusting to push the mineralization contact close to the Kraaipan Greenstone and basement granites. A 20–30-m thick late-phase coarse-grained feldspar-rich gabbro (pegmatoidal) intrudes and crosscuts earlier cumulate rocks of the footwall section. The lower part of the footwall rocks are devoid of mineralization with <10 ppb Pt, Pd, Au, and Cu. Hanging-wall

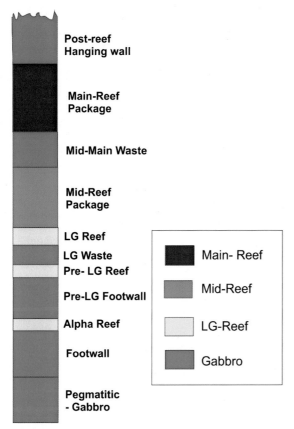

FIGURE 3.12 Generalized stratigraphy of the Stella Layered Intrusion. Mineralization is grouped into three different packages from footwall to hanging wall: Low Grade Reef (LG), Mid-Reef (MR), and Main Reef (M).

rocks are moderately homogeneous, medium-grained magnetite gabbros. Mineralized packages are exhibited at Fig. 3.12.

The mineralized horizon is grouped into three distinct packages (Fig. 3.12) from footwall to hanging wall as: Low Grade Reef (LG), Mid-Reef (MR), and Main Reef (M) (Lewins et al., 2008). The PGEs at the Stella complex are dominated by sperrylite ($PtAs_2$) and stibiopalladinite (PdSb), as well as in native form. The minerals are usually fine-grained blebs, ranging in size between 0.5 and 45 μm. The native metals are typically concentrated in cracks of Ti-magnetite, and as occlusions in

gangue minerals (amphiboles, carbonates, and chlorite).

3.2.3.4 Reserve and Resources

Surface drilling established eight deposits from north to south in an en echelon pattern as Crater, Vela, Sirius, Orion, Serpens North, Serpens South, Mira, and Crux over a 12-km strike length. Several prospects, namely, Scorpio, Tucana, Serpens East, and Pointer have been identified.

Resources of all deposits have been estimated at 75 Mt grading 1.42 g/t (Pt + Pd + Au) following South African Mineral Resource Committee code compliant (SAMREC).

3.2.4 Insizwa Nickel–Copper–PGE

3.2.4.1 Location

The Mount Ayliff Intrusive Complex (Insizwa) is located in Lesotho–KwaZulu Natal and Eastern Cape Provinces of South Africa, ~120 km west of Durban, 55 km northeast of Qumbu town, and easily accessible as N2 highway runs through the nearby Waterfall George deposit. The Mount Ayliff complex covers ~800 km² surface area and up to 1200 m of thickness. The total intrusive exposure is divided into four segments separated by volcanic and sedimentary sequence. The segments are Insizwa in the northwest, Ingeli in the northeast, Tonti in the center, and Tabankulu in the south (Fig. 3.13).

3.2.4.2 Geology

The Karoo basin can be described by two major formations:

1. Thick pile of sedimentary package at base comprising tillites (glacial Dwyka Formation), sandstones, shale, and discrete lava flows (Ecca and Beaufort Formations) accumulated between the Early Carboniferous and the end of the Early Jurassic (Lightfoot and Naldrett, 1983; Maier et al., 2013a).

FIGURE 3.13 Simplified surface map of Mount Ayliff Intrusion displaying four mafic–ultramafic segments: Insizwa, Ingeli, Tonti, and Tabankulu. *Modified after Cawthorn, R.G., Maske, S., Wet, M.D., 1988. Contrasting magma types in the Mount Ayliff Intrusion (Insizwa complex), Transkei: evidence from ilmenite compositions. Canadian Mineralogist 26, 145–160.*

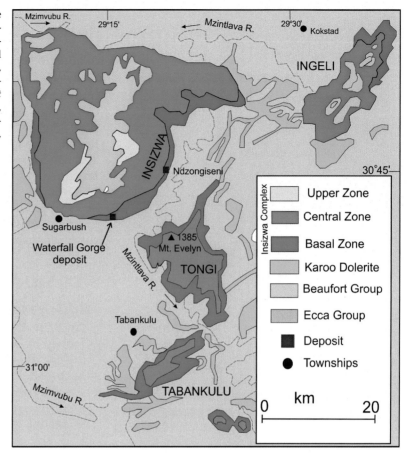

2. Plutonic and volcanic igneous events at the top that include groups of layered intrusives are considered to form part of the Karoo flood basalt province that occurred between the Middle and Early Jurassic. The deep-seated magma entered along subvertical feeder dikes forming the southern extremity of the body.

The mafic–ultramafic complex comprised from base:

1. Basal Zone: Basal gabbro showing cumulus olivine gabbro with augite, bronzite, and plagioclase, picrite-basalt (Box 3.1) showing olivine cumulate with intercumulus augite, bronzite,

and plagioclase; and troctolite showing plagioclase–olivine cumulate with intercumulus bronzite and augite.

2. Central Zone: Olivine–plagioclase cumulates with intercumulus augite–bronzite gabbro.

3. Upper/Roof Zone: Cumulus–intercumulus quartz diorites and monzonites.

3.2.4.3 Mineralization

Flood basalt and related intrusive rocks of Jurassic age are reported from Karoo province, South Africa, Mozambique, Australia, New Zealand, Antarctica, and Serra Geral Mountain Range, Southern Brazil. However, Ni–Cu–PGE mineralization has only been reported and explored from intrusive rocks of feeder conduits

---

Box 3.1

PICRITE-BASALT

Picrite-basalt or picrobasalt is a variety of high-magnesium olivine-rich basalt. It is dark with yellow-green olivine phenocrysts (20–50%) and black to dark brown pyroxene, mostly augite. It often associates with tholeiitic basalts and olivine-rich alkali basalts. Picrites and komatiites are similar chemically. Komatiite lavas are products of more magnesium-rich melts, and exhibit the spinifex texture. Komatiites are largely restricted to the Archean. In contrast, picrites are magnesium-rich as crystals of olivine accumulated in normal melts by magmatic processes.

Picrite-basalt hosts magmatic Ni–Cu ± PGE deposits at Insizwa (South Africa), Kabanga

(Tanzania), Eagle (United States), Voisey's Bay (Canada), Noril'sk (Russia), Pechenga (Russia), and Jinchuan (China).

Troctolite is a mafic intrusive rock, consists essentially of olivine, calcic plagioclase, and minor pyroxene. It is an olivine-rich anorthosite, or a pyroxene-depleted gabbro. However, unlike gabbro, no troctolite corresponds in composition to a partial melt of peridotite. Troctolite is necessarily a cumulate of crystals fractionated from melt.

Troctolite hosts some layered intrusions as Merensky Reef (South Africa), Stillwater Complex (United States), and Voisey's Bay (Canada).

---

at the Karoo flood basalt province. The mineralization occurs in disseminated form throughout the Basal and Upper zones of all four intrusive segments. The economic concentrations have been observed along the southeastern margin of the Insizwa massif at the Waterfall Gorge profile. The bulk of mineralization consists of globular, net-textured, massive, and disseminated magmatic nickel copper sulfides at the base of the lower Basal Zone. The pattern shows consistent fractionation into sheets of massive pyrrhotite–pentlandite-rich mineralization at the base and chalcopyrite-rich veins at the top hanging-wall gabbro. Disseminated mineralization at Tabankulu profiles is uneconomic.

3.2.4.4 Exploration and Resources

The Insizwa intrusive is classified under greenfield exploration with the concept of a massive Ni–Cu–PGE deposit at the base of a layered complex by BSC Resources Limited, Johannesburg. Exploration initiated with regional reconnaissance by 1500 soil and trench samples showing value of 0.20–6.0% Ni+Cu and 0.5 g/t to 7.16 g/t Pt+Pd+Au over lengths of

0.5–10 m. The reserve and resources are yet to be established.

3.3 BOTSWANA

The Republic of Botswana is a landlocked country located in southern Africa. Production of nickel, copper, gold, cobalt, and PGEs supports the Nation's economy.

3.3.1 Geology

Archean basement rocks of the Zimbabwe Craton (<3460 Ma) extends into Botswana and is composed of granitoids, schist, and gneisses that incorporate greenstone belts comprising mafic, ultramafic, and felsic volcanic rocks. Zimbabwe and Kaapvaa Cratons (3600–2500 Ma), located about 250 km in the south, are separated by the younger Limpopo formation of Archean to Proterozoic age. The Limpopo Belt, runs east–west with thrust margins on both sides and includes three major formations: Central, Southern

Marginal, and Northern Marginal zones on either sides. The Limpopo is composed of granulite facies tectonites, and intruded by a mafic–ultramafic intrusive of the Selebi–Phikwe Complex (Maier et al., 2007). The Archean–Proterozoic rocks of Botswana are covered by Phanerozoic sedimentary basins (Fig. 3.14).

The Francistown Arc Complex is represented by the Tati belt (2703±30Ma). Bagai et al. (2002) dated the Vumba greenstone belt, located along the southwestern margin of Zimbabwe craton, at 2647±4 to 2696±4Ma. The Francistown is well comparable with the Upper Bulawayan greenstones (2700Ma) in Zimbabwe. The Tati greenstone belt comprises volcanic and sedimentary rocks and is intruded by granitoids. The Lady Mary Formation, at the base of a volcano-sedimentary succession, consists of altered komatiite, komatiitic basalt, schist,

limestone, and iron formation. The overlying Penhalonga Formation consists of basaltic, andesitic, and rhyolitic volcanics, phyllite, black shales, and limestone. This is capped by the Selkirk Formation consisting of dacite and rhyolitic volcaniclastic rocks, and minor amounts of mafic volcanic rocks, quartzite, and quartz sericite schist. The Selkirk Formation hosts Phoenix, Selkirk, and Tekwane meta-gabbronoritic intrusions and the Sikukwe metaperidotite intrusion (Fig. 3.14).

The Selebi–Phikwe complex, located between the Kaapvaal and Zimbabwe cratons, forms part of the Limpopo Belt. The Phikwe Complex is located within the Limpopo Central Zone and consists of Archean gneisses. It contains the Selebi–Phikwe belt of mafic–ultramafic intrusions (Fig. 3.14) that is hosted by granite–anorthosite–gneiss.

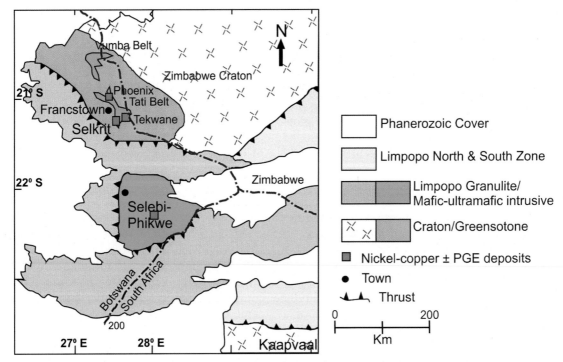

FIGURE 3.14 Simplified geologic map of eastern Botswana showing the Ni–Cu–PGE intrusive complex *Modified after Johnson, R.S., 1986, The Phoenix and Selkirk nickel–copper sulphide ore deposits, Tati Greenstone Belt, eastern Botswana. In: Anhaeusser, C.R., Maske, S., (ed) Mineral deposits of Southern Africa. Geological Society of South Africa, p. 243–248.*

3.3.2 Mineralization

The Ni–Cu–PGE deposits of eastern Botswana are clustered in two distinct groups of intrusive complexes. The first group of intrusions occurs within and on the periphery of the Tati greenstone belt. The second group of deposits is located about 200 km south of the Tati Belt and is hosted in the Selebi–Phikwe mafic–ultramafic intrusive complex within the Central Zone of Limpopo.

3.3.2.1 Tati Nickel–Copper–PGE Belt

The Phoenix, Selkirk and Tekwane deposits were discovered in 1963 based on mapping, regional geochemical soil sampling, and stream sediment geochemistry. Host rocks are gabbroic–troctolitic intrusions (2700 Ma).

3.3.2.1.1 PHOENIX NICKEL–COPPER–PGE DEPOSIT

The Phoenix deposit is located at the northern periphery of the Tati greenstone belt, 45 km east of Francistown. The host rocks are *meta*-gabbronorites intrusive into tonalitic paragneisses. Magmatic sulfide mineralization occurs in disseminated, massive, and veins in gabbronorites. Ore minerals are pyrrhotite, pentlandite, chalcopyrite, merenskyite, hollingworthite, and sperrylite.

3.3.2.1.2 SELKIRK NICKEL DEPOSIT

The Selkirk nickel deposit is located ~45 km east of Francistown and composed of layered metagabbro norites and anorthosite. The orebody is 250-m long and a 20-m thick lens of massive sulfides. The ore minerals are disseminated pyrrhotite, chalcopyrite, and lamellar pentlandite.

3.3.2.1.3 TEKWANE

The Tekwane intrusive is a 2.7-km long, 1.3-km wide troctolite, olivine-rich gabbronorite. Sulfide minerals are pyrrhotite, pentlandite, chalcopyrite, sperrylite, michenerite, native gold, melonite, hessite, and cobaltite (Maier et al., 2007).

3.3.2.2 Selebi–Phikwe Nickel–Copper– PGE Belt

The Selebi–Phikwe intrusive complex was discovered between 1963 and 1966, and hosted mafic–ultramafic intrusions within gneisses of the Central Zone of the Limpopo.

3.3.2.2.1 PHIKWE NICKEL DEPOSIT

The Phikwe forms part of the Selebi–Phikwe belt. Sulfide ores occur as disseminated and massive to semimassive forms throughout amphibolite, forming thick concentrations up to 38 m. Pyrrhotite, pentlandite, and chalcopyrite are the main sulfide minerals.

3.3.2.2.2 DIKOLOTI–LENTSWE–PHOKOJE

The Dikoloti, Lentswe, and Phokoje deposits are located a few km to the west of the Selebi–Phikwe intrusions. Sulfide minerals occur in disseminated form along basal and upper contacts of the intrusions. Pyrrhotite and pentlandite are the main sulfide minerals.

3.3.3 Reserve Base

The Reserve Base at Selkirk stands at 124 Mt grading 0.23% Ni, 0.27% Cu, and 0.57 g/t PGEs. The same at Phoenix has been estimated at 100 Mt @ 0.28% Ni and 0.21% Cu. The Phikwe deposit is estimated at 31 Mt @ 1.36% Ni and 1.12% Cu.

3.4 MOROCCO

The mineral industry is the main foreign exchange-earning sector in Morocco. There are few occurrences of chromium–nickel–PGE mineralization viz. the Beni Bousera Cr–Ni–PGE mineralization and the gold–palladium in the Bou Azzer ophiolite.

3.4.1 Beni Bousera Chromium–Nickel– Platinum Group Mineralization

The Beni–Bousera Cr–Ni mineralization formed at the intercontinental margins of Africa (Morocco) and Europe (Spain) signifies the coexistence of maucherite ($Ni_{11}As_8$) with pyrrhotite, pentlandite, and chalcopyrite in the form of globules (Piña et al., 2013). This close association of pyrrhotite, arsenide, and sulfide minerals accounts for the bulk of PGEs and chalcophile elements in ultramafic massif host rocks.

The partition coefficients of PGEs and chalcophile elements between arsenide and sulfide phases have been computed from mineral samples of the Amasined mine, Beni–Bousera massif. Piña et al. (2013) concludes that "PGE, Au, Bi, Te, and Sb are strongly compatible into arsenide, Re and Ag, and Se is moderately incompatible. The values are consistent with distribution of PGE and chalcophile elements in a number of Ni-Cu-PGE ore deposits containing arsenide-rich zones and underline the potentially important role of arsenide minerals as collectors of these elements. The implications for exploration are considerable: formation of arsenide melts can lead to formation of As-PGE-rich horizons or lenses associated with S-rich and PGE-poor mineralized zones in magmatic sulfide deposits." A number of factors such as initial PGE content of sulfide melt, timing of arsenide segregation/crystallization, and efficient concentration of As–PGE-bearing melt will influence formation of an economic PGE deposit.

3.4.2 Bou Azzer Gold– Palladium Mineralization

The Bou Azzer ophiolite complex in the central Anti-Atlas of Morocco is known for its Co–As–Ni–Cr and low-grade palladium deposits.

Geological setting of northwestern Morocco can be addressed in three major episodes: basement, ophiolite series in the middle, and volcano-sedimentary cover. Precambrian basement marks the northern border of the West African Craton. The NW segment is occupied by the Pan-African orogenic belt (750–540 Ma). Continental margin domain is located to central part of Anti-Atlas comprising ophiolite. Bou Azzer ophiolite is 4–5 km thick, and trends 120°NE–120°SW. The complex consists of serpentinized peridotites and harzburgite in the core (2 km), layered pyroxenite–gabbros (500 m) on either side, large stocks of quartz diorite, partly pillowed basaltic lavas, and poorly developed dike swarms. It is a slice of Neoproterozoic oceanic crust that obducted onto the continental margin of the West African craton.

The ophiolite complex hosts a number of small lenticular bodies of massive, podiform, and disseminated chromitite. Co–Ni arsenide mineralization is structurally controlled and concentrated within quartz–carbonate veins. Chromitite occurs as lenticular bodies of variable dimensions, <30-m long mainly in central and eastern parts of the Bou Azzer Belt. The high magnesium (11–16% MgO) and high chromium spinel reveals complex alteration patterns of successive enrichment of chromium (48–62% Cr_2O_3) and manganese in the core and a loss of aluminum toward the rim (Ghorfi et al., 2008). Rh-bearing laurite and Os–Ir–Ru-rich alloys are in great abundance as a complex intergrowth texture within chromium spinel. Platinum and palladium concentrations are low as is characteristic of ophiolite chromitite. Minerals include millerite, pentlandite, nickeline, maucherite, skutterudite, and löllingite. Cobalt arsenide occurs in quartz–carbonate lenses along the borders of serpentinite massifs and grades laterally through talc–carbonate zones into serpentines.

3.5 NAMIBIA

Mining is the biggest contributor to the nation's economy. There is no mining of chromium–nickel–platinum group metals in Namibia. Several mining companies commenced exploration for magmatic nickel sulfides in the Kunene Complex during late 1990. The Anglo American Company recognized mafic–ultramafic satellite intrusions, located at the periphery of the Kunene troctolite–anorthosite complex for nickel exploration campaign during 2001.

3.5.1 Kunene Anorthosite Complex

The Kunene Anorthosite Complex is one of the largest single-anorthosite massif suites of ~18,000 sq. km that spreads across Namibia and Angola. The anorthosite complex represents a semidomal structure to the south (Zebra Mountain lobe) and a long elongated body to the north. Several small mafic–ultramafic intrusions are scattered along the southern periphery of the main troctolite–anorthosite massif.

Regional geology comprises granite-gneiss basement (early Proterozoic to late Archean), and supracrustal rocks of the Epupa Complex (Paleo–Mesoproterozoic). The Epupa Complex is composed of ortho- and paragneisses, metasediments, and volcanic rocks. The basement rocks are intruded by the Kunene Complex of mafic rocks in two phases. The early phase is dominated by deformed white anorthosite, mainly occurring in the Angola region. The late phase is unaltered leucotroctolite. Zebra Mountain in Namibia region is an interlayered mafic sequence of leucotroctolite and white anorthosite with a total thickness of ~16 km. The age of the Kunene intrusion is debatable and 1470 ± 25 Ma (U–Pb, Maier et al., 2013b) would be a close representation with possibly extending magmatism to a time span of 250 Ma. The mafic–ultramafic intrusions exist as satellites to main Kunene Complex to the southern and western periphery of Zebra Mountain. The lithocomposition includes dunite, harzburgite, pyroxenite, gabbro, troctolite, and anorthosite.

Maier et al. (2013b) conducted whole-rock geochemistry from Zebra Mountain and satellite intrusions. The sample values from the Zebra lobe vary between 46.31 and 50.70% SiO_2, 0.29 and 8.39% MgO, 9 and 271 g/t Cr, and 4 and 235 g/t Ni. The same for the Ohamaremba Satellite Intrusion varies between 36.05 and 45.61% SiO_2, 6.71 and 24.42% MgO, 30 and 220 g/t Cr, and 247 and 1610 g/t Ni. The relative enrichment in MgO, Cr, and Ni of Kunene magmatism suggests that the primary magmas were predominantly mantle-derived picrites or basalts. The formation of the huge white anorthosite mass was due to early crystallization of ascending magma followed by downward draining of residual liquid. The interlayered dark troctolitic sills within the anorthosite plutons and the younger mafic–ultramafic satellite intrusions in the periphery of the Zebra anorthosite is the outcome of subsequent repetitive magma pulses.

Ni–Cu–PGE sulfide mineralization, forming minor deposits, occurs in some of the small satellite intrusions that include the Ohamaremba troctolite, the Oncocua pyroxenite, and the Ombuku peridotite–gabbronorite. The latter contains a massive chromitite layer. The Ohamaremba sulfides have high tenors of Ni. This is due to relatively high Ni contents of the parental tholeiitic magmas.

3.6 SUDAN

Sudan is the 17th fastest-growing country in the world and the rapid development is largely from the oil profile. The government has begun privatizing several of its mineral assets. Podiform chromite occurrences are reported in 1930s and small-scale mining has been in progress since the 1950s. Lateritic profiles with anomalous nickel values have been found associated with ophiolite mafic–ultramafic rocks at Ingessana Hills, Jebel Rahib, and Nuba Mountains.

3.6.1 Ingessana Hills Chromite–Nickel Deposits

The Ingessana hills area of Blue Nile State lie 80 km southwest of Ed Damazin, known for gold, chromium, and lateritic nickel occurrences, and promising for further mineral exploration. All basic infrastructural supports for import and exports are well established along with rich Savanna–Subtropical climate. The Ingessana massif is a large Late Proterozoic mafic–ultramafic complex representing the main and central parts of the Ingessana–Kurmuk Ophiolite Belt. It extends from the Nile to northwest of the Red Sea Hills as major thrust slices. Mafic–ultramafic units are extensively serpentinized dunite, harzburgite, and overlain by peridotite–pyroxenite.

Chromite occurs in serpentinized lower dunite–harzburgite and associated talc carbonates. Orebodies are lenticular and range in size between 10 and 50 m long, 1 and 3 m thick, and 30 and 100 m down-dip. Ore occurs in podiform type, massive, compact, often sheared, and consists of chromite spinel. Lateritic and serpentinized ultramafic capping indicates an anomalous discontinuous zone with grade up to 2.8% Ni over an extensive length of 200 km aligned to the NE–SW of the Ingessana Hills. Ni-enriched zones are small remnants of the lateritic weathering profile. The weathering profiles are significant potential geochemically explored targets for lateritic nickel deposits in the region.

The reserve base has been estimated at 1 Mt @ 58.57% Cr_2O_3, 3.04% Fe_2O_3, 11.42% Al_2O_3, and 0.22% SiO_2. Chemically, the ore is of metallurgical grade, with a high Cr/Fe ratio. Systematic chromite mining started in 1963 by a state-owned company from underground operation with an annual capacity of ~10,000 tonnes (~11,020 tons).

3.6.2 QaIa En Nahal Area

Chromite has been reported at Qala En Nahal and Umm Saqata areas, hosted by mafic–ultramafic rocks. The area is located to the northeast of the Blue Nile–Dinder–Rahad Rivers. Chromite occurs as small pods, lenses, and massive in serpentinized dunite–harzhurgite units.

Orebodies vary in length between 3 and 6 m, thickness between 1.5 and 2 m, and concentration ranging between 25 and 37.8% Cr_2O_3.

3.6.3 Jebel Rahib Area

The Jebel Rahib area is located in northern Darfur State, and dominated by rift-type metasediments and metavolcanics. The metavolcanic rocks are the southern continuation of the Jebel Rahib Late Proterozoic ophiolite complex. Three chromite zones have been identified. The major one trends NNE–SSW over 1 km and ~200 m wide. Chromite mineralization is exclusively hosted by serpentinized dunite–harzburgite taconites and occurs as pods up to 50 m in length and 2–3 m in thickness. The ore is compact, massive, and coarse @ 55% Cr_2O_3. Laterization took place in late Proterozoic time covering ~1000 sq. km with a thickness of ~25 m and contains a concentration of Ni that encourages exploration for lateritic nickel deposits.

3.6.4 Nuba Mountains

The eastern and western margins of the Nuba Mountains in South Central Sudan, over a stretch of 50 km, are predominantly covered by low-grade metavolcanic and ferruginous cherty sediments with dismembered ophiolitic bodies. The Late Proterozoic ophiolite comprises serpentinized dunite–harzburgite units that contain 22 podiform chromite lenses. The chromite lenses have an average grade of 30% Cr_2O_3, 22% Fe_2O_3, 12% Al_2O_3, and 12% SiO_2. The ore is composed of chrome spinel intimately intergrown with silicates.

The presence of anomalous copper, zinc, and nickel has been reported from discontinuous gossan outcrops. The average nickel content in the different gossans ranges between 112 g/t in the Biteria area and 5744 g/t in the Jebel Tumluk area with the highest value of 1.00% Ni.

3.6.5 Red Sea Hills Region

The Red Sea Hills Region of Eastern Sudan is an integral part of the Late Proterozoic Pan-African Arabian Nubian Shield. The region comprises plutonic equivalents and is separated by sutured ophiolites.

There are a number of chromite bodies in the Onib ophiolite occurring as small pods in the Wadi Sudi area. The most significant occurrence is reported from the Wadi Hamissana area where massive, semimassive, coarse-grained podiform chromite lenses are scattered within the serpentinized ultramafic sequences. The largest body is a lens 250 m long, 5 m average thickness, and 100 m down-dip. The average grade is 28% Cr_2O_3 with 2:5 Cr/Fe ratios.

3.7 TANZANIA

Tanzania is the fourth largest gold producer in Africa after South Africa, Ghana, and Mali. Minerals identified include gold, iron ore, nickel, copper, cobalt, silver, precious gemstones, phosphate, coal, uranium, etc. Nickel projects have been focused in the northwestern part with deposits like Kabanga, Kagera, Dutwa, Zanzui nickel oxide, Heneti Ni, and Ntaka Hill nickel sulfide. PGE mineralization is located in the Nkenja mafic–ultramafic body, Ubendian metamorphic belt.

3.7.1 Kabanga Ni Sulfide

Kabanga nickel sulfide deposit represents one of the most significant Ni sulfide discoveries of last two decades (Maier et al., 2010a). It is located 130 km southwest of Lake Victoria in the Ngra District of Kagera Region in northwestern Tanzania. The project is initiated as one of the best undeveloped greenfield significant nickel sulfide deposits. Nickel occurrence was reported in the Kapalagulu area in 1914, followed by the finding of lateritic nickel at Musongati in the

1970s. Nickel sulfide mineralization at Kabanga Main was discovered in 1976 and Kabanga North during ongoing exploration.

The basement setting is composed of Archean–Early Proterozoic granites and gneisses and covers the Tanzania craton to the east and the Congo craton to the west. The basement rocks are overlain by Mesoproterozoic rocks of the Kibaran Orogenic Belt that continues over 1500 km trending NE–SW to NNE–SSW from Katanga, Democratic Republic of the Congo, to the south, up into southwest Uganda to the north with subvertical dips on either side. The metasedimentary successions of the orogenic belt (Kagera Supergroup) comprise detrital components from nearby Archean and Paleoproterozoic basement rocks between 1780 Ma and 1370 Ma (Alonsoa et al., 2012). Maximum thickness is up to 5 km of alternating pelitic and arenaceous rocks including graphite- and sulfide-bearing pelitic rocks, thick quartzite ridges, basaltic–rhyolitic sills and lavas. The orogenic belt witnessed a long period of intracratonic intermittent depositional activity with periods of interruption of deposition, erosion, and magmatism.

The repetitive magmatic intrusions occur exclusively within the Kibaran orogenic metasedimentary formation between 1400 Ma and 1375 Ma. Kabanga nickel sulfide deposits are a part of the Kabanga–Musongati–Kapalagulu belt of nickel-bearing mafic–ultramafic intrusions. A series of sill-like large layered mafic–ultramafic intrusions extend over 500 km in the NE–SW direction from Lake Tanganyika in the south to the Uganda border in the north. The overlying Neoproterozoic basin Group comprises arenites, arkosic arenites, shales, and siltstones, and unconformably overlies the Tanzania craton (Fig. 3.15).

Sulfide minerals are hosted by harzburgite and orthopyroxenitic intrusions that crystallized from magnesian basaltic and picritic magmas (Maier and Barnes, 2010b). Mafic–ultramafic intrusions are ~4 km long and up to 1 km thick. Layered rocks are crystallized

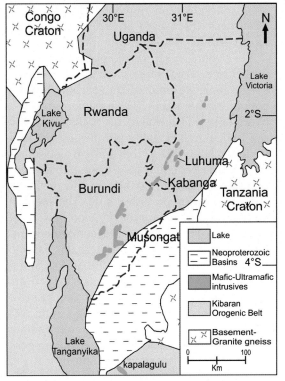

FIGURE 3.15 Regional geologic map of the northeastern Kibaran orogenic belt displaying the Kabanga–Musongati–Kapalagulu belt of nickel-bearing large layered mafic–ultramafic intrusions *Modified after Maier, W.D., Barnes, S-J., Ripley, M., 2011. The Kabanga Ni sulfide deposits, Tanzania: a review of ore-forming processes. Economic Geology 17, 217–234.*

from several pulses of compositionally distinct magma hosted by sulfide-bearing pelitic schist. The early magma pulses comprising high-magnesium siliceous basalt with ~13% MgO. It forms a network of fine-grained gabbronorite and orthopyroxenite sills with acicular texture. The magma indicates enrichment of incompatible trace elements viz. "Large-Ion Lithophile Elements (LILE)" and "Light Rare Earth Elements (LREE)" and had distinctly negative Niobium (Nb) and Tantalum (Ta). Subsequent magma pulses are less contaminated with more magnesia (14–15% MgO). The magma emplaced into earlier-formed sills creating medium-grained

harzburgites, orthopyroxenites, gabbronorite–orthopyroxenite fragments within olivine-rich matrix (Maier et al., 2011).

The contamination of parent magma formed basal olivine–orthopyroxene cumulates at the Kapalagulu intrusion. The parent magma at the Musongati intrusion is contaminated forming dunites and harzburgites at its base. The high-magnesia composition of the parental magmas and the abundance of sulfides in the host sedimentary rocks support enrichment of nickel and PGEs. Both intrusions have potential for magmatic Ni and PGE deposits. A thick weathered lateritic crust containing up to 4 g/t PGE formed at Musongati to host one of the largest Ni-laterite deposits (Maier et al., 2008).

The Kabanga intrusions contain abundant disseminated sulfides (pyrrhotite, pentlandite, minor chalcopyrite, and pyrite). Numerous thick and massive nickel sulfide layers, lenses, and veins occur in the lower portions and the immediate footwall of Kabanga Main and Kabanga North intrusions. The base of the intrusions with the host rocks forms sulfide–hornfels breccias caused by postemplacement tilting of the intrusions. This structure is suitable for solid-state mobilization of ductile sulfides into shear zones. Layers and lenses of massive sulfides extend over hundreds of meters at a thickness of >10 m. The rich mineralized horizons extend for 100 m away from the intrusions offering an important exploration target for additional nickel sulfide reserves and resources.

There are four distinct orebodies, namely, Kabanga Main, North, Block-1, and Nyanzali. Mineralization contains 1–3% Ni, 0.1–0.4% Cu, and <1 g/t platinum group of metals (PGMs) at high Ni/Cu ratios of 5–15. One unit of the Kabanga North is extremely rich with 5% Ni, 0.8% Cu, and 10 g/t PGE. The reserve base of the Kabanga nickel project stands at 37.4 Mt @ 2.63% Ni (measured + indicated). Inferred resource stands at 21 Mt @ 2.60% Ni as of December 2010 (Maier et al., 2011).

Initial mining adopted open-pit technology. The subsequent operation path developed as an underground mine due to having reached optimum economic depth. The mine practiced conventional underground mining and processing technology, and exported nickel concentrate to international markets. The anticipated mine life with current resources is ~30 years.

3.7.2 Kapalagulu Layered Intrusion

The Kapalagulu layered mafic intrusion forms part of the Kambalda mafic–ultramafic system and is located in the southern extension of the Luhuma–Kabanga–Musongat Igneous Complex (Fig. 3.16). Layered gabbroic rocks that form the upper part of the Kapalagulu Intrusion crops out on the side of a mountain range, capped by Neoproterozoic Itiaso Group quartzite, near the eastern shore of Lake Tanganyika at Kungwe Bay (Wilhelmij and Cabri, 2015). The Kapalagulu mafic–ultramafic intrusion extends over 30 km as an elongated series of ridges trending NW–SE, dipping east, and culminates at a thrusted/faulted contact with registrant quartzite and phyllite of the Itiaso Group of metasediments. The footwall of the intrusive complex rests on Ubendian basement gneiss.

The area has been explored for platinum and nickel in both lateritic and harzburgite protolith in a southeastern extension of the Kapalagulu Intrusion (Lubalisi Zone) as a Joint Venture Program between Goldstream Mining and Lonmin PLC during 2001–2009. The exploration activities include airborne and ground electromagnetic, magnetic, and radiometric geophysical surveys, grid-based soil sampling, geologic mapping, and 53,227 m of drilling. The drilling component comprised 34,041 m in 124 diamond-drill holes, 14,182 m in 729 air-core holes, 4000 m in 213 rotary air-blast holes, and 1004 m in 11 reverse-circulation drill holes.

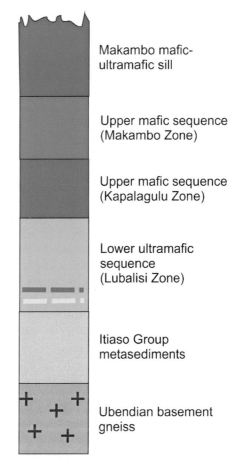

Makambo mafic-ultramafic sill

Upper mafic sequence (Makambo Zone)

Upper mafic sequence (Kapalagulu Zone)

Lower ultramafic sequence (Lubalisi Zone)

Itiaso Group metasediments

Ubendian basement gneiss

FIGURE 3.16 Generalized stratigraphic sequence of the Kapalagulu mafic–ultramafic sequence showing the PGE Reefs (yellow) and overlying nickel regolith (blue), mainly concentrated in the lower layered ultramafic harzburgite.

Mineralization occurs in harzburgite at the base of the layered gabbro complex of the Kapalagulu Intrusion, emplaced between Paleoproterozoic Ubendian basement gneiss, and overlying Neoproterozoic Itiaso Group metasediments. The high-grade mineralization of the PGEs is associated with chromitite and sulfide-bearing harzburgite within a southeastern extension of the Kapalagulu Intrusion (Lubalisi Zone) that is covered by a layer of nickel-rich laterite regolith. The poorly layered southeastern harzburgite forms part of an >1500-m thick Lower Ultramafic Sequence that continues upward into a succession of well-layered gabbroic rocks of the Upper Mafic Sequence. The layered gabbro is devoid of any platinum mineralization. The Lower Ultramafic Sequence and the laterite regolith overlie the mineralized harzburgite. The Lubalisi Zone of harzburgite is underlain by basal dunite and overlain by an interval of layered harzburgite and troctolite.

3.7.3 Kagera Ni Sulfide

The Kagera nickel sulfide project is located in the Kagera Region, Ngara District of northwest Tanzania. The primary ore minerals are pentlandite, chalcopyrite, pyrrhotite, and pyrite. The project comprises mineral rights to six properties totaling 860 sq. km area in the prospective Kabanga–Musongat mafic–ultramafic belt. The key licenses are located directly adjacent to the Kabanga Nickel Deposit.

3.7.4 Dutwa Ni Oxide

The Dutwa Laterite Nickel deposit is located in northern Tanzania, and was discovered by African Eagle Resources. The company completed a positive scoping study on the Dutwa deposit in July 2009 and is working toward a feasibility study.

African Eagle Resources estimated Joint Ore Reserves Committee (JORC)-inferred category resources for Ngasamo (west Dutwa) at 35.30 Mt grading 0.90% Ni and 0.036% Co. The same for Wamangola (Dutwa Main) is 56.80 Mt averaging 0.86% Ni and 0.030% Co. The total resources for the Dutwa deposits stand at 92.10 Mt @ 0.88% Ni and 0.031% Co.

3.7.5 Zanzui Ni Oxide

African Eagle Resources is also evaluating a second promising nickel oxide deposit at Zanzui, located 60 km from Dutwa.

3.7.6 Nkenja Platinum–Palladium–Gold

The Nkenja mafic–ultramafic body is located in southern Tanzania. The area was mapped earlier with possible indication for chromium–platinum occurrences, and explored extensively between 2005 and 2008 that revealed the existence of a significant enrichment of widespread platinum, palladium, and gold mineralization (Evans et al., 2011).

The basement geological setting is dominated by the Tanzanian craton to the north and the Bangweulu craton to the south, and composed of older Archean granite and syenite. Mineralization occurs in the central part of the Ubendian metamorphic belt. The Nkenja intrusion forms an elongate lens-shaped body covering a surface area of 4.5 km × 3 km bounded on all sides by ductile tectonic margins. These margins dip moderately toward the southwest. The Nkenja comprises several lensoidal tectonic blocks of alternating mafic (amphibolite) or ultramafic composition, separated by greenschist-facies ductile shear zones. There are three distinct ultramafic blocks: the Lower Block in the footwall, the Central Block, and the Nkenja Block in the hanging wall.

The host rocks include partly serpentinized chromitiferous dunite, wehrlite, and olivine clinopyroxenite. The rocks are tectonically intercalated with amphibolitized metagabbro. Disseminated and seam-type aluminum-rich chrome spinel occurs in dunite. The seams are thin, interrupted, and often deformed at granulite facies. PGE values are erratically distributed and are associated with trace to minor amounts of disseminated sulfides (pyrrhotite, pentlandite, heazlewoodite (Ni_3S_2), chalcopyrite, and bornite). The mineralization is primarily igneous controlled by sulfides in ultramafic rock. There has evidently been a strong metamorphic and/or hydrothermal overprint on the original magmatic concentration of PGE-bearing sulfides.

The area has been explored by Tausi Minerals, a Tanzanian subsidiary of Goldstream Mining NL (now IMX Resources Ltd.), applying surface mapping, soil and auger drill sampling (15,576 m), rotary air-blast holes (656 m), and diamond drilling (25,619 m). Exploration results indicate that the abundance and style of all PGE sulfide mineralization are consistently anomalous throughout the Nkenja ultramafic body being part of the crustal section of a dismembered Paleoproterozoic ophiolite. The ore minerals are sulfides of nickel, copper, PGE, gold, and chromium.

3.7.7 Ntaka Hill Ni–Cu

The Ntaka Hill nickel–copper deposit is located ~400 km south of Dar es Salaam in the Nachingwea district, southeast Tanzania. Oxidized copper and malachite showings have been known historically.

The country rocks comprise amphibolites, biotite gneisses, quartzite, migmatites, and mafic granulites forming a volcano-sedimentary sequence at an oceanic arc environment to the Tanzanian craton during the early to mid-Neoproterozoic magmatic events (640–596 Ma).

The Ntaka nickel–copper deposit is hosted by a mafic–ultramafic intrusive body covering 5 × 3 sq. km surface area. Host rocks include olivine-poor pyroxenite, pyroxene–olivine cumulate, peridotite, and hybrid gabbro. Pyroxenites include websterite, orthopyroxenite, and clinopyroxenite (enstatite and augite phases, respectively). The contact relationships with the host rocks are obscured by strong deformation and similarities in composition and texture between the amphibolite host rock and an amphibole-altered marginal pyroxenite. The ultramafic body has been affected by the same amphibolite-to-granulite facies metamorphism and northeasterly overturned isoclinal folding as have the enclosing meta-volcano-sedimentary rocks. It is interpreted as a high-level sill derived from high-MgO basic magma that has subsequently been folded and metamorphosed.

Three styles of mineralization are recognized, characterized by (1) abundance of disseminated, net-textured, massive sulfides, (2) tectonically remobilized sulfide veins with sharp contacts, and high-grade Ni–Cu, and (3) sulfides with graphite and biotite in tectonically modified magmatic hybrid zones at the margin of the ultramafic intrusions. Ni-rich mineralization formed from high-Mg basaltic melts indicates high temperatures of mantle melting and rapid ascent to the crust. The higher-grade deposits are hosted in smaller intrusions, and have evidently needed a multistage process to upgrade sulfide mineralization to economic levels (Evans, 2011).

The deposit was drilled by Selection Trust and Inco in the early 1950s, and BHP Ltd between 1996 and 1998. Goldstream Mining NL drilled high-grade massive sulfides in 2006. Since then, the property is being explored by Continental Nickel Ltd and IMX Resources Ltd. Infill drilling on geophysical targets at a 50-m grid interval has proceeded successfully.

The global mineral reserve and resources has been estimated by IMX Resources Ltd, as of August 2013, at 20.32 Mt grading 0.58% Ni, 0.13% Cu, and 200 g/t Co (Measured and Indicated). Inferred resources stand at 35.93 Mt grading 0.66% Ni, 0.14% Cu, and 200 g/t Co. The Ntaka Hill has targeted resource definition in support of near-surface open-pit mining opportunities.

3.8 ZAMBIA

The economy of Zambia is based primarily on agriculture, and partly on mining and energy. The Zambian economy has historically been based on a major Nkana open-pit copper mine, Kitwe. The international exploration companies are investing in a new nickel project at the Munali Ni–Cu–PGE deposit, Northwestern Province.

3.8.1 Munali Ni–Cu–Co–PGE Deposit

The Munali Ni–Cu–PGE deposit is located ~60 km south of Lusaka in southern Zambia.

It was discovered in 1969 based on a regional stream sediment geochemical survey. The deposit was under exploration between 1970 and 2006 by agencies e.g., Anglo Exploration Ltd., Apollo Mining Ltd., and Albidon Ltd.

The Munali nickel deposit represents an early phase of basic magmatism in the early Neoproterozoic Katanga rift, hosted by a composite gabbro–ultramafic body emplaced at a high level in the rift into passive margin sediments (Evans, 2012). The country rocks around the deposit are metasediments moderately deformed at low amphibolite facies. The metasediments are conglomerate, quartzite, shale, biotite–andalusite–scapolite schist, and carbonates that unconformably underlain by Mesoproterozoic basement crust (island arc granite gneiss) and metasediments of unknown age (the Nega Formation).

The Munali mafic–ultramafic intrusion consists of a relatively undifferentiated coarse gabbro core and ultramafic marginal breccia with orthocumulate textures. The intrusive unit has a surface exposure of 2.5×0.5 sq. km trending NW–SE with a steep dip between 60° and 70° to the SW. Nickel sulfide mineralization is exclusively confined within an unusual coarsely brecciated olivine–apatite–magnetite sulfide marginal zone, intruded into the main gabbro showing skarn-type mineralization. The Munali gabbro can be classified into three types: (1) Microgabbro, fine to coarse grained, with doleritic texture and altered scapolite; (2) Poikilitic gabbro, medium to coarse grained, with laminated plagioclase cumulate and altered scapolite; and (3) Olivine basalt, extremely fine-grained olivine basalt to dolerite with altered scapolite. The Munali gabbro has been dated at about 855 Ma by SHRIMP U–Pb analyses on zircons extracted from a pegmatite phase (Evans et al., 2011).

The nickel and iron–nickel sulfide mineralization occurs in two main stages: (1) a millerite–vaesite–pyrite assemblage that formed disseminations and a semimassive type in

vuggy-textured ultramafic rocks, and (2) a later millerite–bravoite–molybdenite assemblage in quartz–kyanite. The mineralization contains minor copper and trace amounts of cobalt and PGEs. A discrete zone of copper sulfides underlies a portion of the nickel sulfide zone (Capistrant et al., 2015).

The total indicated and inferred resources have been estimated at 10.30 Mt grading 1.2% Ni, 0.20% Cu, 0.07% Co, 0.6 g/t Pd, and 0.03 g/t Pt. The in situ metal inventory at the Munali stands at 123,500 tonnes (~136,000 tons) of nickel and 246,800 ounces of platinum (Albidon Ltd.). The mining right is wholly owned by Australian Company Albidon Limited and consists of two deposits: the Enterprise (Munali Phase-I) and the Voyager in the northwest extension.

3.9 ZIMBABWE

The Republic of Zimbabwe is a landlocked country located in the southern part of the African continent bordered by South Africa in the south, Botswana to the southwest, Zambia to the northeast, and Mozambique to the east. The capital and largest city is Harare (Fig. 3.17). The country's economy has been significantly based on agriculture, tourism, and mineral-based industries. The important minerals produced and exported include diamond, platinum group and base metals, chromite and ferroalloys, cobalt and nickel, and coal and coal-based methane gas.

3.9.1 The Great Dyke

The Great Dyke of Zimbabwe hosts the second largest world reserve of PGMs and chromite after the Bushveld layered complex in South Africa. The chromite mining and ferrochrome smelting industries have been active over 100 and 50 years, respectively. The chromite occurs in two geologically distinct resources: (1) small, lenticular chromite deposits associated with deformed Archean komatiitic sills, and (2) thin, laterally extensive chromitite layers within the 550-km long, late Archean Great Dyke layered intrusion (Prendergast, 2008). The platinum group and associated nickel, copper, cobalt, gold, and silver occur in the second type of layered intrusive mafic–ultramafic complex.

3.9.1.1 Location

The Great Dyke is a linear geological intrusive feature that cuts across the entirety of Zimbabwe and passes close to the west of the capital Harare (Fig. 3.17). The surface is undulating hills, and elevation rises toward the north.

3.9.1.2 Discovery History

The Great Dyke was first reported in 1867 by the explorer Karl Mauch. However, the existence of rich chromite ore deposits was not realized until around 1918. The metal platinum was discovered in a plagioclase pyroxenite ultramafic intrusion in 1925. The Grainger brothers attempted to extract platinum in 1926 in the southern Wedza Subchamber (Fig. 3.17) using primitive and inadequate techniques and could not succeed. An extensive drilling program was carried out during the late 1960 and 1970s targeting host pyroxenite occurrences. Exploration continued covering the entire length of the dike by Union Carbide, Anglo American Corporation, Cluff Minerals, Metals Mining, Broken Hill Proprietary (BHP), and others. The outcome established North and South magma chambers, centrally plunging "Y-shaped" or "boat-like" subchambers, and several mining blocks (Fig. 3.17).

3.9.1.3 Geology

The Great Dyke, in its present plane of erosion, is longitudinally subdivided into a series of narrow attached layered complexes and chambers. The northern chamber consists of the Musengezi, Darwendale, and Sebakwe subchambers. The southern chamber consists of the Selukwe and Wedza subchambers. The Darwendale and Sebakwe subchambers are known as the Hartley Complex.

The Great Dyke is a layered igneous intrusive of mafic and ultramafic rocks. It trends N15°E–S15°W and transects through the country rocks of Archean greenstone (mafic–sedimentary), granites, and gneisses forming the Zimbabwe craton. A series of satellite dikes run on either side of The Great Dyke. The dike is bounded by the Zambezi metamorphic belt in the north and the Limpopo belt to the south. The intrusive is 550 km (340 miles) in length and 4–11 km

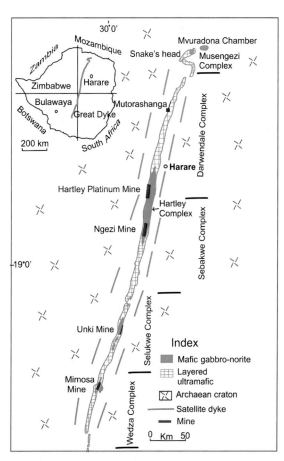

FIGURE 3.17 Geologic map of The Great Dyke of Zimbabwe showing Ni–Cu–PGE and Zn–Pb–Cu deposits/mines *Haldar, S.K., 2013. Mineral Exploration – Principles and Applications. Elsevier Publication, p. 374.*

wide. Several fractures parallel to the intrusion are identified all over the craton. The layers of ultramafic (dunite, harzburgite, and bronzite) and mafic (gabbronorite) rocks depict an essentially uniform synclinal sequence in transverse view as exposed on either side of the longitudinal axis. The plunge is to the south in the northern part and to the north in the southern part to form a series of stacked canoes.

The dike developed as a series of initially discrete magma chambers, which joined up as the chambers filled with incoming magma. The chambers coalesced below the Main Sulfide Zone (MSZ) before erosion. The dike comprises a collection of igneous rocks ranging from ultramafic to mafic. It is divided vertically into two major successions. The lower ultramafic sequence of ~3000 m thick is dominated from the base upward by cyclic repetitions of chromitite, dunite, harzburgite, olivine, bronzite, and pyroxenite–bronzitite. The upper mafic sequence of ~1200-m thickness consists mainly of plagioclase-rich rocks, such as norites, gabbronorite, and olivine gabbro. A marginal border group, consisting mainly of fine-grained gabbro, norites, and pyroxenite, exists in all subchambers at the contact of the magma and the country rock. The cyclic units of mafic rocks are 14 in number and formed as the result of periodic influxes of fresh magma of one compositional type, a high-magnesia basalt containing about 15% MgO.

The emplacement of the intrusive magma is mainly controlled by this structure. The age of emplacement has precisely been indicated as 2579 ± 7 Ma by U–Pb age determinations on zircon and rutile (SHRIMP).

3.9.1.4 Mineralization

The Great Dyke hosts a strategic economic resource with significant quantities of chromium and PGMs with associated nickel, copper, cobalt, and gold. The primary mineralized horizon is confined essentially within the layered ultramafic sequence (Fig. 3.19). There are two distinct types of mineralization, such as chromium oxide and sulfide zones containing Pt, Pd, Ni, Cu, Co, and Au.

Chromitites occur entirely in the upper pyroxenite P1 (C1–C2) and lower dunite (C3–C12) host rock as massive bands and disseminated throughout the dike. The width of the upper chromitite composite seam (C1c, C1d, and C2) is up to 4 m. The average chromitite grade varies between 35 and 51% Cr_2O_3 with Cr/Fe ratio between 2 and 3. The lower group chromitites (C5–C12) of the lower pyroxenite and dunite sequences is characterized by average thicknesses of 10–15 cm, average grade ranging between 43 and 54% Cr_2O_3, and Cr/Fe ratios of chromite between 2.7 and 3.9.

The pyroxenite of the P1 layers (bronzitite and websterite), a member of the layered ultramafic sequence, is located directly 10–50 m below the mafic (gabbronorite)—ultramafic (pyroxenite) contact. The P1 pyroxenite horizon hosts economically exploitable quantities of PGMs in the MSZ. The base metals occur as disseminated intercumulus Fe–Ni–Cu sulfides within an interval referred to as the base-metal subzone, below which is a sublayer enriched in PGMs called the PGE subzone. The base metal and PGE subzones together make up the MSZ (Fig. 3.18). The MSZ has a fine structure made up by a number of successive, geochemically distinct layers and typical vertical element distribution patterns characterized by a general upward-zoning sequence in the order Pd→Pt→base-metal sulfides (OberthÜr, 2011). Disseminated sulfides with irregular base metals and low PGE content are also present locally at the mafic–ultramafic contact. The MSZ is a lithologically continuous layer with thickness between 2 and 3 m that forms an elongated basin. The rich PGM concentrations occur at its base. The ratio between platinum and palladium varies vertically.

The sulfide content in the MSZ varies between 2.5 and 10% by volume. The sulfide minerals in decreasing order are pyrrhotite, pentlandite, chalcopyrite, and pyrite. The zone contains PGMs (Pt, Pd, Rh, Ru, Ir, and Os), Au,

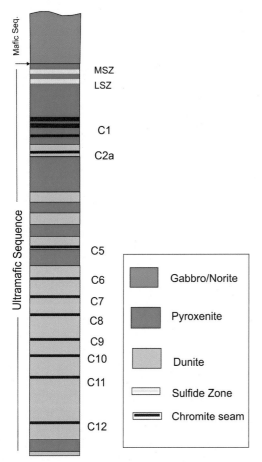

FIGURE 3.18 Schematic stratigraphy of the mafic–ultramafic sequence and distribution of chromitite and sulfide mineralization in The Great Dyke of the Zimbabwe craton. *C*, chromite seams; *LSZ*, Lower Sulfide Zones; *MSZ*, Main Sulfide Zones.

prices but remains an important potential PGE resource for the future. Platinum group minerals are generally very fine grained and occur as minute inclusions ~20 μm within sulfides. Both the Main and the Lower sulfide zones are found in all five subchambers and are essentially continuous and regularly developed throughout the preserved P1 layer. Several chromitite layers (~140 Mt) with low platinum metals below the MSZ are mined for chromium throughout the dike. Hydrothermal alteration occurred in the MSZ with the involvement of magmatic fluids and transfer of S from lower to higher stratigraphic positions in the PGE-bearing zones (Li et al., 2007).

Markwitz et al. (2010) proposed the essential ingredients for nickel sulfide mineral systems in Zimbabwe as: (1) proximity to Archean greenstone belts, craton margins, and orogenic belts; (2) large mafic–ultramafic intrusions and their feeders; (3) crustal contamination of basic–ultrabasic magma with sulfidic crustal lithology; and (4) the proximity to major fault zones that allowed fertile magma to ascend into the upper crust. The potential new nickel sulfide discoveries are from both within and outside the greenstone belts, including the Karoo flood-basalt province, the Mashonaland dolerites, and the Limpopo, Magondi, and Zambezi orogenic belts.

Chromite occurrences are traditionally classed into two principal geologic types. The large and most significant deposits are the stratiform type in layered intrusions, e.g., the Bushveld, the Great Dyke, and the Stillwater. The other type is podiform deposits in ophiolites, e.g., the Kempirsai, Kazakhstan, and the Insizwaa, South Africa. Prendergast (2008) proposed another different category based on the specific geologic features like age, magmatic association, tectono-magmatic setting, and magmatic architecture. The features include primary chromitite characteristics of large thickness, middle to upper stratigraphic location, and elongate form, as well as interpreted mode of formation, e.g., Archean komatiitic sill-hosted chromite deposits.

Ag, and Co. It occurs as stratabound zone, 1–3 m thick, located entirely within plagioclase pyroxenite. The PGE-rich MSZ will be the main target for potential PGE mining in The Great Dyke of Zimbabwe.

A second zone of sulfide-associated Pt mineralization occurs stratigraphically 10–60 m below the MSZ and is known as the Lower Sulfide Zone (LSZ). The LSZ is generally thicker (30–80 m) than the MSZ with a considerably lower metal grade. It is not economic at current

3.9.1.5 Resource and Mining

The total unmined chromite reserves and resources stand at 2574 million tonnes (2837 million tons) grading 5.42 g/t total PGEs, 0.21% Ni, 0.41% Cu, and 40% Cr_2O_3 (Naldrett, 2004). The chromite is mined throughout The Great Dyke, especially in the Darwendale, Lalapanzi, and Mutorashanga areas. The three largest chromite-mining companies are Maranatha Ferrochrome, Zimalloys, and Zimasco.

The reserve/resource potential of the PGMs is summarized in Table 3.2.

PGEs are mined at the Ngezi Mine south of Selous by Zimplats of Impala Platinum Group, at the Unki Mine near Shurugwi by Anglo American, and at the Mimosa Mine near Zvishavane by Zimasco for Impala Platinum. Oxidized, near-surface ores are a promising future source for PGMs because of their easy accessibility. They constitute a resource of 160–400 Mt, but mining has so far been hampered due to insufficient recovery rates.

The producing Ngezi and Mimosa mines are the lowest cash-cost platinum mines in the world. PGE metals with a total reserve of 333 Mt @ ~5 g/t PGE are currently mined at the Ngezi

TABLE 3.2 Summary of Reserve/Resource Potential of Platinum Group of Metals at The Great Dyke Complex

Area/Block	Reserve/ Resource in Mt	Grade g/t (4E)
Snake's Head	535	1.16
Hartley	167	4.8
Zimplats (Mhondoro project)	816	2.0
Zimplats (Seloes project)	878	2.0
Ngezi Mine	214 (reserve)	3.43
	1879 (resource)	3.63
Unki Mine	115	4.28

Oberthür, T., 2011. Platinum-Group element mineralization of the main sulfide zone, Great Dyke, Zimbabwe. Reviews in Economic Geology 17, 329–349; 4E= (Pt + Pd + Rh + Au).

mine by Zimplats, the Unki by Anglo American, and the Mimosa by Zimasco.

3.10 MADAGASCAR

The economy of Madagascar depends on natural resources and trade that includes agriculture, fishing, forestry, and minerals. The island holds one of the largest reserves of ilmenite and zircon from heavy-mineral sands, as well as important reserves of chromite, iron ore, cobalt, copper–nickel, and coal, oil, and gas. There are three potential areas of interest in Madagascar with respect to chromium, nickel–cobalt, and platinum, namely, (1) the Andriamena ultramafic complex hosting chromite, (2) the Ambatovy laterite complex hosting nickel and cobalt, and (3) the Ambodilafa nickel–copper–platinum complex. Madagascar is the 10th largest chrome producer in the world.

3.10.1 Andriamena Ultramafic Complex

Most of the chromite deposits are located in the northern part of Madagascar. The first chromite occurrence was discovered in 1948, situated close to the village of Androfia in the Tsaratanana District, Betsiboka Region. The chromite deposits are hosted exclusively by the Andriamena layered ultramafic intrusive complex.

Reserves and resources of various key deposits, namely, the Bemanevika mine (3.00 Mt), the Anengitra mine (0.16 Mt), and the Befandriana mine (0.45 Mt), assume possible PGM potentiality. The mining activity commenced only in 1968 with intermittent closure of mining activity. The majority of the production is from a number of small open-pit mines.

State-owned Societe Kraomita Malagasy (KRAOMA) is the main chromite producer with capacity of around 40,000 t/y of concentrates, 80,000 t/y of lumpy ore from the Andriamena complex, and another 20,000 t/y from the Befandriana mine. Chromite concentrates and lumpy ore (100,000 to 140,000 t/y) are the major mineral export out of Madagascar destined for China, Japan, and Sweden. Chromite concentrate grade varies between 48 and 50% Cr_2O_3 with 20–30 g/t phosphorus. The grade of lumpy chrome ore varies between 42 and 44% Cr_2O_3. The crude ore generally contains 70 g/t phosphorus.

The majority of chrome resources are located in the northern parts of the country. This calls for long-term investment in upgrading rail infrastructure critical to expanding production in Madagascar.

3.10.2 Ambatovy Nickel–Cobalt Deposits

The Ambatovy laterite nickel–cobalt complex is located 80 km east of Antananarivo, the capital of Madagascar, near the town of Moramanga. The project is well connected by main road and rail system to Antananarivo and the main seaport of Toamasina to the east coast at a distance of 208 km. The Joint Venture operates an open-pit mining operation and an ore preparation plant at the mine site. The Ambatovy nickel and cobalt mining, processing, refining, and marketing joint venture between Sherritt (owner and operator), Sumitomo, Korea Resources, and SNC-Lavalin is the largest finished-nickel and finished-cobalt operation from lateritic ore in the world.

The deposits were identified in 1960. Since discovery, the deposits have been explored by various international exploration agencies till 2004. The exploration commenced with surface mapping and pit sampling. A total of 1282 diamond-drill holes have been drilled representing 55,000 m of drilling.

The project is based on two large, weathered lateritic nickel deposits: the Ambatovy and the Analamay, located 3 km apart. Combined, the deposits constitute one of the world's largest lateritic nickel reserves and resources, covering an area of about 18 sq. km with depths ranging between 20 and 100 m (average 40 m).

Laterite nickel deposits are formed by leaching of the ultramafic portion of a large mafic–ultramafic intrusive complex. The deposits are predominantly classified as ferralite (lateritic), some of which is underlain by saprolite. A material type termed low-magnesium saprolite (LMS) lies below the ferralite and between the ferralite and saprolite, where the latter is present. Only the ferralite and LMS have been considered in the mine plan. The distribution of the saprolite is erratic; consequently, it has been classified as inferred, and is not included in the measured and indicated resource, although its exploitation may prove economic.

The combined measured and indicated mineral resource for the Ambatovy/Analamay deposits is estimated at 134 Mt grading 1.04% Ni and 0.10% Co (SNC-Lavalin, 2006). The mine life is currently projected to be 29 years.

The Ambatovy is a large-tonnage, long-life nickel–cobalt mining project in Madagascar. No blasting is required due to the soft nature of the ore and is surface mined by hydraulic excavators, and delivered to the ore preparation plant at the mine site by haul trucks. The plant separates the soft lateritic soil from rock and waste material.

The Ambatovy laterite nickel–cobalt project attained commercial production in January 2014. The operation is currently focused on optimizing project performance and bringing it up to its nameplate capacity of 60,000 t/y of refined nickel, 5600 t/y of refined cobalt, and 210,000 t/y of ammonium sulfate fertilizer for at least 29 years of mine life. This will place nickel as the top export commodity of Madagascar providing a significant stimulus to the national economy.

3.10.3 Ambodilafa Nickel–Copper– Platinum Occurrences

The Ambodilafa copper–nickel–platinum mineralization lies within the 17.5-km long and 6-km wide Vohipaha mafic–ultramafic intrusive complex located 160 km southeast of Antananarivo (Capital of Madagascar). Jubilee Platinum PLC of the United Kingdom explored at the

Ambodilafa prospect in 2009. The mineralization is associated with semi-massive, net-textured, and disseminated pyrrhotite and chalcopyrite. The host rocks comprise pyroxenite, olivine pyroxenite, peridotite, and olivine gabbro. The textures suggest an orthomagmatic origin to the mineralization due to primary magmatic sulfide segregation within a mafic–ultramafic intrusive complex. The Bureau of Geological and Mining Research (BRGM) explored the western part of the area and reported intersections of 13.9 m grading 0.45% Ni and 0.26% Cu at 243-m depth, including 8 m grading 0.57 and 0.2% Cu. There was another intersection of 3.8 m @ 0.67% Ni and 0.25% Cu at 260 m. The company has planned a drilling program at Ambodilafa.

Deposits of North America

Platinum-Nickel-Chromium Deposits
http://dx.doi.org/10.1016/B978-0-12-802041-8.00004-3

4.1 BACKGROUND

North America is a large landmass (continent) solely situated within the northern and western Hemispheres, encompasses 45 countries, plus the seemingly detached massive landmass of Greenland. The United States, Canada, and Mexico have significant and multifaceted economic systems. The continent is richly endowed with natural resources including abundant coal, iron ore, bauxite, copper, natural gas, petroleum, mercury, nickel, PGEs, gold, and silver.

4.2 UNITED STATES

The United States of America (USA), or the United States (US), is a federal republic consisting of 50 states that enjoys the largest economic power in the world due to an abundance of natural resources, a well-developed infrastructure, and high productivity. The US is a major contributor to global mining industries with iron ore, gold, zinc–lead, copper–nickel, and PGEs. The US has limited significant chromite deposit/reserves, and restricted to only a few locations, including Montana, California, and Oregon. Chromite mining is only possible by support price in a tactical emergency.

4.2.1 Stillwater Layered Igneous Complex

The Stillwater Layered Intrusive Complex is the second largest such complex in size and the richest known platinum resource of "The Big Three."

4.2.1.1 Location and Discovery History

Stillwater Complex is Neo-Archean mafic–ultramafic layered intrusion, located and exposed in the northern edge of the Beartooth Mountains, within Stillwater, Sweet Grass, and Park counties, 130 km southwest of Billings in south-central Montana. The East Boulder project is located ~20 km west of the Stillwater mine.

The adjacent areas to the east and west of the Stillwater mining counties were known to contain Cu, Ni, and Cr as early as the 1880s. Exploration and mine development history is closely tied to wartime demands for copper and chromium. The Stillwater Complex was intensively explored for chromite ore during World War I. Deposits of PGE, iron ore, and aluminum were recognized later. A.L. Howland and team members (INCO) recognized the possibility of finding reef-type PGEs mineralization in the Stillwater Complex in 1936, based on sulfide-bearing samples from the layered gabbro. A variety of exploration programs followed comprising mapping (Fig. 4.1), soil and talus geochemistry, geophysical surveys (Magnetic, Induced Polarization, and Electromagnetic), trenching, diamond drilling, and underground exploration. Johns-Manville, the exploration geologist, discovered the large platiniferous J-M Reef in 1973. The first mine on the J-M Reef, near Nye, Montana, started commercial production in 1986 and produced platinum, palladium, gold, silver, rhodium, copper, nickel, and cobalt. The East Boulder mine that accesses the J-M Reef at the western end commenced in 2002. J-M Reef ranks second with respect to reserves and resources of PGMs in the world.

4.2.1.2 Geological Setting

The Stillwater Complex is a large stratiform and layered mafic–ultramafic intrusion with many similarities to the Bushveld in South Africa, and The Great Dyke in Zimbabwe. Regional geology comprises gneisses, granites, and metasedimentary packages of Archean age (2730–2790 Ma). The layered intrusion was emplaced into what is now Montana at ~2709 Ma, and intruded by younger Archean quartz monzonite plutons, Proterozoic mafic dikes, and Tertiary felsic sills affected by several episodes of deformation (Zientek et al., 2005). The overlying unconformable cover sequence is represented

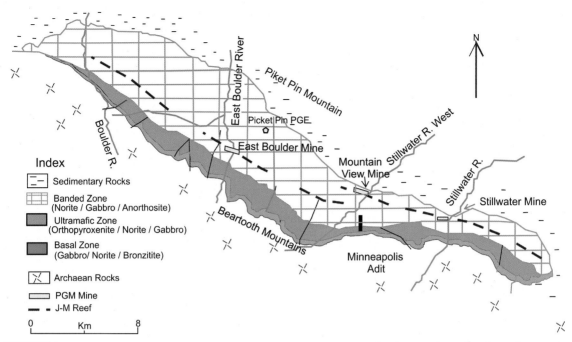

FIGURE 4.1 Simplified geologic map of the Stillwater Layered Igneous Complex showing the mafic–ultramafic intrusive sequence and the Ni–Cu–PGE mines.

by Phanerozoic sedimentary rocks and recent unconsolidated soil. The area has experienced extensive metamorphism, faulting, and folding.

The total package of layered mafic–ultramafic intrusion is exposed in a fault-bounded block along the northern edge of the Beartooth Mountains uplift. The layering can be traced for 48 km (30 mi), striking N60° W–S60° E, and dipping between 50° and 90° to the northeast. The stratigraphic thickness of the layered complex is ~6 km. The complex is extensively metamorphosed layered ultramafics (olivine > orthopyroxene) in the lower part, and a mafic (olivine–orthopyroxene-bearing gabbroic) complex in the upper part. The cumulate stratigraphy of Stillwater is divided into three distinct zones or five principal series from bottom upward: the Basal Zone, the Ultramafic Zone (Lower Peridotite and Upper Bronzitite Series), and the Banded Zone (Lower Banded, Middle Banded, and Upper Banded Series).

The Basal Zone consists of a Lower Basal chilled fine-grained gabbro, norite, and an Upper Basal feldspar pyroxenite-and-bronzitite cumulate, containing inclusions of metamorphosed Archean sedimentary rocks. The thickness varies between 14 and 400 m. The Ultramafic Zone is composed of a Lower Peridotite and an Upper Bronzitite Series with thickness varying between 2000 m in the Mountain View area and 840 m in the western part of the complex. The Lower Peridotite Series consists of a number of cyclic units of harzburgite overlain by dunite, orthopyroxenite, chromitite, and bronzite pyroxenite. The Peridotite series hosts alternating chromite seams. The Upper Bronzitite Series is massive bronzite pyroxenite.

The Banded Zone covers more than 75% of the exposed thickness of the Stillwater Complex. The Lower Banded Zone is composed of alternating layers of norite, gabbronorite, and olivine-bearing cumulates. The olivine-bearing rocks in this subzone are coarse-grained troctolite, dunite, interlayered with anorthosite, norite, and gabbronorite. The lowermost layer of Norite (I) is a distinctive marker unit along the entire length.The Lower Banded Series hosts economic platinum-bearing deposits in the J-M Reef along the entire length of the Stillwater Complex. The Middle Banded Series is composed of successive layers of anorthosite and olivine-bearing cumulates. The Upper Banded series is composed of olivine-bearing cumulates and gabbronorites with a thickness up to 4300 m. A generalized total stratigraphic column of the Stillwater Layered Intrusive Complex is given in Fig. 4.2.

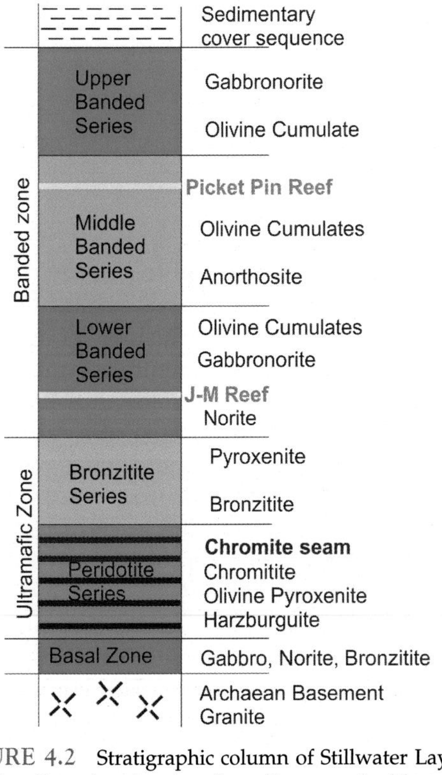

FIGURE 4.2 Stratigraphic column of Stillwater Layered Intrusive Complex showing chromite seams (red layers) in the Ultramafic Zone (Lower Peridotite Series), platinum-bearing J-M Reef (yellow band) in the Mafic Zone (Lower Banded Series), and Picket Pin Pt–Pd mineralization (yellow band) in the Middle Banded Zone.

4.2.1.3 Mineralization

The Stillwater contains magmatic mineralization variably enriched in critical commodities like chromium, nickel, and PGEs. The J-M Reef shares the major reserves and is the prime producer of PGEs in the US (Zientek and Parks, 2014). The Basal Zone contains disseminated low nickel–platinum mineralization, and occasionally changes to localized massive shape and grade.

The layers of semimassive to massive chromite orebodies are usually present within the Lower Peridotite Series of the Ultramafic Zone. Chromite-rich seams are labeled in the order of A to K from the base upward, with seam thicknesses up to 8 m, and grade between 35 and 47% Cr_2O_3. The chromite layers are enriched in IPGE (Ir, Ru and Os group). PGE-bearing minerals are laurite, pentlandite, millerite, and chalcopyrite. IPGE and Rh were originally collected by chromite, and subsequently, small quantities of base-metal sulfide liquid was added to the chromite layers from the overlying magma. IPGE and Rh in chromite diffused from the chromite into base-metal sulfides and converted some of the sulfides to laurite (Barnes et al., 2015).

The Stillwater Complex contains several stratabound and stratiform nickel–copper sulfide-bearing horizons with irregular PGEs enrichment. The most important platinum-bearing orebody, J-M Reef, is located in the Lower Banded Series of the Banded Zone. It is a continuous layer and consists of a 1 to 3 m thick pegmatitic peridotite and troctolite with disseminated sulfide minerals. Common sulfides include pyrrhotite, pentlandite, chalcopyrite, moncheite, cooperite, braggite, kotulskite, and platinum–iron alloys with an average 20–25 g/t PGEs at 2-m thickness. It has a Pd/Pt ratio of about 3.6. The J-M Reef formed from a PGE-enriched parental magma that had likely been formed at depth below the Stillwater magma chamber by interaction of the parental magma with S-rich metasedimentary rocks, followed by redissolution of sulfides in the Stillwater magma (Keays et al., 2011). The J-M Reef is similar to the Merensky Reef of the Bushveld complex.

The Picket Pin deposit represents a strat-abound disseminated-sulfide-bearing PGE zone that occurs in the upper 150 m of the Anorthosite subzone II within the Middle Banded Series. Mineralization is traceable at the same stratigraphic interval over 22 km of exposed strike length. Arsenides and antimonides are the main hosts for Pd and Pt with erratic grades reaching up to 3 g/t.

4.2.1.4 Reserve Base and Mining

The Stillwater Mining Company estimated total measured and indicated reserves at 38.1 Mt averaging 19.11 g/t Pd, 5.30 g/t Pt, 0.05% Ni, 0.02% Cu, 35–47% Cr_2O_3 (Naldrett, 2004). Reserves at Stillwater mine stand at 16.4 Mt at 21.6 g/t Pd and Pt, and 41% Cr_2O_3. Reserve base at East Boulder stands at 21.7 Mt at 18.2 g/t Pd and Pt. The resource has been estimated at 155 Mt.

Stillwater Mining Company is operating two underground mines, one at Stillwater River valley and the other at East Boulder Mine. Mine entries include adits driven from the Stillwater valley side, an internal decline, and a shaft. Three stoping methods are mechanized captive cut-and-fill, ramp-and-fill using hydraulic drill jumbos and Load-Haul-Dump vehicles (LHDs), and sublevel stoping.

4.2.2 Duluth Complex

4.2.2.1 Outline

The Midcontinental Rift system is a pronounced geological rift in the American Midwest that pulled apart the American craton during the Mesoproterozoic (1100 Ma). The rift mechanism caused a series of magmatic emplacements and volcanic episodes through the Earth's crust separating the older formations. The cooling of the magma formed mafic–ultramafic rocks in the area of the rift system.

Disseminated to massive Ni–Cu–PGE-sulfide mineralization in the Midcontinental rift is hosted by small primitive intrusions formed early in the evolution of the rift and are typically associated with transtensional structures in cratonic blocks adjacent to a main rift. The examples include the Eagle deposit in Michigan, and the Current Lake Complex in Ontario, Canada. The intrusions share the geochemical fingerprints of more primitive lowermost volcanic rocks of the Midcontinental rift system. Large-tonnage low-grade Cu–Ni–PGE deposits of the Duluth intrusion in the Great Lakes tectonic zone formed significantly later than these small primitive intrusions; their host rocks share the same lithophile trace element signature as these early intrusions (Reid et al., 2015).

The northwest and northern margins of Lake Superior are occupied by layers of the North Shore Keweenawan Volcanic Group, and the adjoining layered mafic rock formations of Duluth Beaver Bay complexes, and that of Sonju Lake. The mafic magmatic intrusive complex developed along the now southern margin of Lake Superior and formed rich nickel–copper deposits of the Eagle complex and the Bovine Intrusion.

4.2.2.2 Location and Discovery

The Duluth Complex is located at the northwest and northern margins of Lake Superior, and covers a large part of the State of Minnesota. The layered intrusive complex starts next to the city of Duluth. The complex crops out as a large arcuate body and extends over 250 km northeast to the Canadian border (Fig. 4.3). The Cu-sulfide mineralization was discovered by F.W. Childers in 1948.

4.2.2.3 Geological Settings

The Duluth Complex includes granitic basement, greenstone, volcanics, and sediments of the Archean (2700 Ma), overlying the Paleoproterozoic Animikie Group of the Biwabik Iron Formation, and the Proterozoic Virginia sedimentary Formation.

Magmatic intrusions took place ~1100 Ma along the western and northwestern margin of Lake Superior within regional Archean country

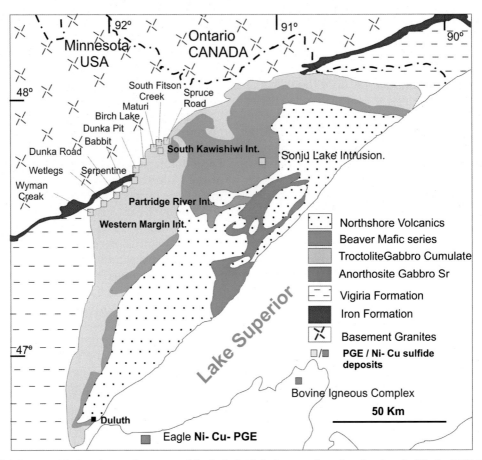

FIGURE 4.3　Simplified surface geologic map of the Duluth Mafic Igneous Complex showing the Ni–Cu–PGE deposits of Duluth, Eagle, and Bovine *Modified after Naldrett, A.J., 2004. Magmatic Sulfide Deposits – Geology, Geochemistry and Exploration. Springer Publication, p. 728.*

rocks. The intrusive sequence is composed of a felsic, early anorthositic and gabbroic series at the base, younger troctolite and gabbroic cumulates in the middle and the Beaver Series at the top. Repetitive layered cycles are present in the lower part with pegmatoidal plagioclase in sharp contact with the underlying troctolite, and formed the bottoms of the sequences. The sequences grade upward into medium-grained plagioclase cumulates and plagioclase–olivine cumulates (Naldrett, 2004). Total thickness is 1000–1200 m (Fig. 4.4). The cover sequence is irregularly bedded calcareous sandstones of the Bayfield Group.

4.2.2.4 Mineralization

The Ni–Cu–PGE sulfide mineralization in Duluth Complex is classified into three types (Ripley, 2014):

1. Disseminated Ni–Cu–PGE sulfide in gabbroic–troctolitic sheet-like intrusions.
2. Ni–Co-enrichment disseminated through massive sulfide associated with smaller, early rift conduit-related, mafic–ultramafic intrusions.
3. Low-sulfide PGE enrichment in well-differentiated tholeiite layered intrusions.

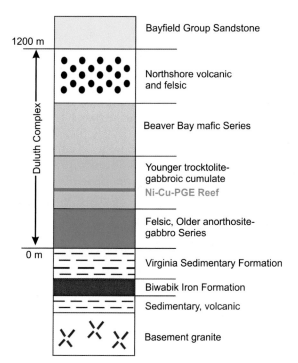

FIGURE 4.4 Complete stratigraphic sequence of the Duluth Complex showing Archean Basement Rocks, the Ni–Cu–PGE-bearing Intrusive Complex, and the younger Bayfield Sandstone.

In situ partial melt pockets and veins indicated that partial melting occurred in the granitic footwall up to 125 m from footwall-intrusion contact at temperature ~850 to 920°C based on mineral assemblage and two-pyroxene thermometry. Partial melting of the footwall granite caused sulfide liquid to sink and infiltrate into partially molten rock. In the proximal 10 m of the footwall, the sulfide assemblage is pyrrhotite and pentlandite. The distal part between 10 and 100 m from the contact contains the main minerals of chalcopyrite, bornite, millerite, PGEs, and native gold (Benkó et al., 2015).

Magmatic Ni–Cu–Pt sulfide deposit requires an addition of S from the country rocks and/or crustal contamination into the source magma. Queffurus et al. (2014), analyzed the Selenium content, S, and $\delta^{34}S$ of the magmatic sulfides of the Duluth Complex and the host-rock sediments of the Virginia Formation. It was observed that sedimentary host rocks are characterized by low S/Se ratios (~3000), close to mantle values. Therefore, sedimentary rocks are not suitable as a source of sulfide mineralization. An S-rich bedded pyrrhotite layer, located at the contact with the intrusion, has high S/Se ratios (~20,000) that are also higher than magmatic sulfides and could thus be the source of additional sulfur for formation of Ni–Cu–Pt deposits at Duluth.

The Duluth intrusion has been explored since 1951 when the first hole was drilled at the Spruce Road deposit and continues till now. The International Nickel Company (INCO) Ltd sank the first shaft at the Maturi deposit in 1967.

The mineral resources of the Duluth Complex have been estimated at + 3200 Mt grading ~0.46% Cu, ~0.13% Ni, and 0.40–1.10 g/t PGEs.

4.2.3 Eagle Complex

4.2.3.1 Location and Discovery

The Eagle mafic–ultramafic intrusive complex is located on Yellow Dog Plains in the State of Michigan's Upper Peninsula. The nearest community town is Big Bay at a road distance of 24 km. The city of Marquette ~53 km by road, provides a regional airport, rail, and shipping facilities, workshop, and commercial services. The Ni–Cu deposit was discovered in 2002 during a routine exploration program.

4.2.3.2 Geological Setting

Host country rocks are manifested by three geological eras: Archean, Paleoproterozoic, and Mesoproterozoic, separated by marked unconformities related to major regional tectonic events. Archean basement rocks are composed of gneisses, granites, metasediments, and metavolcanics. Paleoproterozoic rocks include metasediments and metavolcanics of the Marquette Range Supergroup. Mesoproterozoic rocks of the Baraga Group include quartzite, carbonates, iron formations, volcanics, and turbidities. The assemblages are capped by carbonaceous slate with 1–10% pyrite and/or pyrrhotite up to 300 m thickness.

The olivine-rich small mafic–ultramafic intrusive complex occurs as dike swarm of high-magnesium picritic magmatism in a dynamic magma conduit, and emplaced during the early stages of the Midcontinental Rift System (~1100 Ma). Eagle intrusions outcrop as two distinct peridotite bodies: Yellow Dog Peridotites/ Eastern Peridotite/Eagle East, and Western Peridotites/Eagle with 650 m apart. The peridotites are weathered and composed of feldspathic peridotite (at the top), melatroctolite, olivine melagabbro, and feldspathic pyroxenite. The melatroctolite unit contains 30–45% olivine and forms a sandwich structure characterized by an olivine melagabbro (10–35% olivine) core wrapped in melatroctolite (Fig. 4.5).

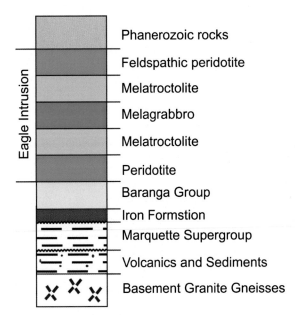

FIGURE 4.5 Geological succession of the Eagle intrusive within Archean basement, Paleoproterozoic Marquette Range Supergroup, Mesoproterozoic Baraga Group, and Phanerozoic cover sequence.

4.2.3.3 Mineralization

The peridotite, troctolite, olivine gabbro, and pyroxenite rocks extend east–west for 480 m, have a maximum width of ~100 m, and dip of near vertical. The Intrusion appears like a funnel and its width narrows to ~10 m at 340 m (90 mRL (meter reduced level)) below the surface. The sulfide mineralization includes pyrrhotite, pentlandite, chalcopyrite, and cubanite. Massive Ni–Cu grades are higher at 6.11% Ni and 4.15% Cu+Co, Au, Pt–Pd than those of the disseminated type. Eagle is one of the richest-Ni grade orebodies in the world. The parental magma had an Ni-rich high-magnesium component, and Ni enrichment is related to the early attainment of sulfide saturation. Four distinct types of sulfide mineralization have been recognized (Ding et al., 2011).

1. Disseminated sulfides: occur as blebs in feldspathic peridotite, melatroctolite, olivine melagabbro, and feldspathic pyroxenite, and are usually uneconomic.
2. Semimassive sulfides: are separated by a zone of massive sulfides forming two separate zones of the Lower and Upper Semimassive units, bordered by disseminated low-sulfide with typical sharp and abrupt contacts (Fig. 4.6). It contains 35–50% sulfides with ~2.11% Ni, 2.19% Cu, 0.06% Co, 0.27 g/t Au, 0.48 g/t Pt, and 0.30 g/t Pd.
3. Massive sulfides: large and coherent with simultaneous depositional layering of up to 100% sulfides grading ~6.11% Ni, 4.15% Cu, 0.17% Co, 0.32 g/t Au, 1.14 g/t Pt, and 0.77 g/t Pd. The massive sulfide zone contains rich IPGE, PPGE (Pt, Pd and RH), and unfractioned PGE.
4. Sulfide veins: Massive sulfides often extend 10–20 m out into adjacent country rocks forming rich mineralization.

4.2.3.4 Reserve Base and Mining

Eagle is extensively explored by close-space drilling at 30-m interval to a depth greater than

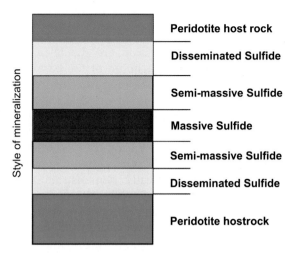

FIGURE 4.6 Style of mineralization at the Eagle intrusive with the massive sulfide zone in the core sandwiched between semimassive and disseminated sulfides to make it a complete package of mineral concentration.

340 m, and outlined an economically rich orebody. Exploratory drilling at Eagle East identified Ni–Cu-rich massive, net-textured and disseminated sulfide. Reserves and resources of the Eagle Mine, as in 2013, is estimated at 5.33 Million tonnes (5.9 million tons) with 3.05% Ni, 2.51% Cu, 0.29 g/t Au, 0.70 g/t Pt, 0.47 g/t Pd, and 0.08% Co.

Eagle is a decline-accessed shallow underground mine, utilizing bench-and-fill stoping in an updip primary–secondary sequence. Backfilling is undertaken using cemented and uncemented rock fill. Two ventilation shafts are in place, equipped with emergency outlets. Commercial concentrate produced in the fourth quarter of 2014, and target mine production of 2000 tonnes per day ore continues from 2015.

4.2.4 Sonju Lake Intrusion

The Sonju Lake intrusion is located within the Beaver Bay Complex of northeast Minnesota, and formed during Midcontinental rifting (1100 Ma). The sheet-like tholeiitic layered mafic intrusion with thickness up to 1200 m forms

part of the Duluth Complex, and outcrops over 9 sq. km (Fig. 4.3). The host units comprise marginal troctolite, a thin dunite layer, and a thick sequence of troctolite, gabbro, Fe–Ti oxide gabbro, and apatite-bearing olivine diorite to olivine monzodiorite indicating a process of continuous fractionation with no influx of fresh magma.

Exploration drilling during 2002 intersected an 85-m thick precious metal zone within homogeneous well-foliated iron rich oxidized gabbro. Sulfide mineralization is strongly zoned, and composed of copper–gold–palladium, and platinum. Stratiform PGEs are distinctly similar to the reef-type PGE mineralization at the Skaergaard tholeiitic intrusion.

This sulfide occurrence suggests an orthomagmatic process with mineralization being related to fractional segregation of sulfide melt from silicate magma. Lundstrom and Gajos (2014) opined that the layered intrusions form by a top-down process of sill injection, and react as thermal migration-zone refining. The reefs form as a moving sulfide band passes downward through a mineral-melt mush at a particular temperature of sulfide saturation. Li et al. (2007) observed that Pt and Pd tellurides, antimonides, and arsenides may have formed during both magmatic crystallization and subsolidus hydrothermal alteration at the Sonju Lake Intrusion.

4.2.5 Bovine Igneous Complex

4.2.5.1 The Intrusion

The Bovine Igneous Complex is located 8 km southeast of L'Anse town, southern margin of Lake Superior, Michigan (Fig. 4.3). It occurs within the Marquette–Baraga dike swarm related to the Midcontinental Rift system, and emplaced during the early magmatic stage with similarities to Eagle. The complex is small and layered, extends over 1200 m E–W, and is 450 m wide. A small satellite intrusion (Little Bovine) is located 100 m west of the main Bovine. The main intrusion is a funnel- to bowl-shaped body with 650-m thickness at the bottom.

4.2.5.2 *Geological Setting*

The host rocks are sedimentary rocks of the Marquette–Baraga Supergroup comprising slates, iron formations, quartzites, and older cherty carbonate. The main Bovine intrusion is composed of a large gabbroic inner core with two thin rims in succession. The inner core (upper unit) is composed of fine-grained gabbro with thickness varying between 30 and 150 m. The inner rim (middle unit) is composed of feldspathic olivine clinopyroxenite with thickness between 30 and 150 m. The outer rim (basal unit) comprises medium-grained feldspathic wehrlite/olivine melagabbro with thickness between 50 and 200 m (Fig. 4.7). Little Bovine is a keel-shaped intrusion, and exclusively composed of feldspathic wehrlite/olivine melagabbro with maximum thickness up to 50 m.

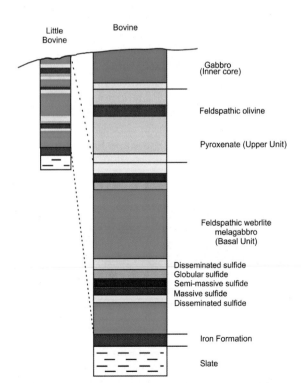

FIGURE 4.7 Simplified stratigraphic succession and nickel–copper mineralization in the Bovine and Little Bovine Complexes, located on the western margin of Lake Superior.

4.2.5.3 *Mineralization*

The most favorable host rocks for rich Ni–Cu sulfide mineralization are the feldspathic wehrlite/olivine melagabbro of the basal unit. The feldspathic olivine clinopyroxenite of middle unit hosts low-grade mineralized zones. The gabbroic inner core seldom contains small disseminated mineralization. The sulfide minerals include pyrrhotite, chalcopyrite, pentlandite with traces of cubanite, and pyrite. Copper and nickel minerals typically demonstrate strong positive correlation. Cu–Ni sulfides occur in both intrusions, with higher enrichment in Little Bovine, being composed entirely of the basal unit. Main Bovine and Little Bovine are formed by multiple magma pulses, saturation of several several sulfides, and distinct sulfur isotope compositions, and suggests that sulfide saturation was promoted via addition of country rock-derived sulfur.

Donoghue et al. (2014) classified the mineralization into four principal types based on sulfide enrichment and texture:

1. Disseminated sulfide: Low-grade disseminated type occurs in all the three mafic–ultramafic units in variable proportion, and contains up to 3% S.
2. Globular sulfide: Globular-type sulfide occurs as small blebs in contact with silicate and oxide minerals. Disseminated and globular-textured sulfide mineralization occurs at the interfaces between all three mafic–ultramafic units containing 3–11% S.
3. Semimassive sulfide: Semimassive sulfide includes globular, net-textured, and fill within the interstices between silicate grains. Sulfides usually occur on either side of the massive zones containing 11–30% S.
4. Massive sulfide: Massive sulfide intervals occur at the contact between the feldspathic wehrlite/olivine melagabbro basal unit and the footwall metasedimentary rocks, and near the basal contacts of feldspathic olivine clinopyroxenite and core gabbro units containing 30–35% S.

4.2.6 La Perouse Intrusion

The La Perouse Intrusion (Brady Glacier Ni–Cu deposit) is located at an average elevation of 1006 m (3300 ft) on Brady Glacier Mountain, Alaska. M/s Fremont Exploration Company conducted a helicopter exploration survey in 1958 over the Brady Glacier region, drilled 32 core holes as follow up, and discovered the La Perouse Ni–Cu sulfide mineralization in mafic–ultramafic intrusive host rocks.

The intrusion is completely covered by ice, except a few exposures, and extends over 1600 m along the northeast–southwest direction. La Perouse is the largest of the four, layered mafic–ultramafic plutons in the Fairweather Range in Glacier Bay National Park. The complex is composed of interlayered olivine gabbro and norite with an exposed thickness of ~1829 m (6000 ft). Thin layers of ultramafic rocks are most abundant near the base. Layering and other similar features suggest that the mafic–ultramafic rocks are formed primarily by cumulus processes (30 Ma). Host rocks include magnesian augite and bronzite, plagioclase, and olivine with accessory chromite, ilmenite, magnetite, and graphite.

The La Perouse Ni–Cu deposit consists of stratigraphically continuous disseminated-sulfide zones, locally more than 120-m thick containing up to 10% sulfide minerals. Massive sulfide zones, with thicknesses up to 3 m, occur locally, typically near contacts of gabbroic and ultramafic cumulates with thickness exceeding 305 m where Ni+Cu grade is ≥0.50%. The deposit contains low concentrations of ~0.18 g/t PGEs. The sulfide minerals include pyrrhotite, pentlandite, chalcopyrite, cubanite, and niccolite ± bornite, mackinawite, and violarite.

Drill-indicated reserves have been estimated at 100 Mt grading 0.5% nickel, and 0.3% copper. PGE average is ~0.18 g/t and exceeds 1 g/t in massive-sulfide units.

4.2.7 Duke Island Complex

The Duke Island ultramafic intrusion is located 45 km (30 mi) south of Ketchikan, in southeast Alaska, discovered and staked by Quaterra Resources Inc. in 2001.

The Duke Island intrusion is a Urals–Alaskan–type complex that hosts sulfide-rich intervals primarily in an olivine clinopyroxenite unit. The Duke complex (110 Ma) consists of two separate, well-exposed zoned ultramafic bodies that join at depth. The Marquis Zone (exposure one) is in the north-central part of the island, near Hall Cove sea inlet. The Raven Zone (exposure two) is in the northwestern part of the island, close to Judd Harbor. The ultramafic complex is characterized by visible graded layering and other cumulate features, especially in the dunite, peridotite, and olivine pyroxenite. It is concentrically zoned outward from a core of dunite and peridotite, surrounded by shells of olivine pyroxenite and hornblende pyroxenite. The Raven Zone is in a plug of gabbro and pyroxenite ~800 m in diameter surrounded by quartz diorite, granodiorite, and quartz monzonite.

Sulfide mineralization consists of disseminated chalcopyrite, pyrrhotite, pentlandite, and chromite in serpentinized clinopyroxenite and olivine pyroxenite. The mineralization comprises massive, net-textured, and disseminated assemblages formed by fractional crystallization and interaction between primary magma and sulfidic–graphitic country rocks. The sulfide mineralization is characterized by strong depletion in Ir, Os, and Ru (IPGEs), and relative enrichment in Pt, Pd, and Cu. Platinum-rich placers in the Urals type have been derived from chromitites and dunites. Relative enrichments in Pd and Cu are the results of fractional crystallization before achieving sulfide saturation (Thakurta et al., 2014).

Quaterra Resources Inc. completed airborne and ground geophysical surveys, geological mapping, soil and rock sampling, and 1820 m of core drilling in 11 holes (eight in Marquis, and three in the Raven Zone) between 2001 and 2006. Copper Ridge Explorations Inc. drilled an additional five holes on the Marquis Zone in 2010. The four holes are barren and one drill hole intersected 4.9 m grading 0.294% Cu, and 0.075% Ni, and 13.4 m with 0.148% Cu and 0.039% Ni.

No deposit of substantial size and grade has been identified so far on Duke Island.

4.3 CANADA

Canada is a leading producer and exporter of zinc, uranium, gold, nickel, PGMs, aluminum, steel, iron ore, coking coal, lead, copper, molybdenum, cobalt, and cadmium. Metallogenic provinces include the magmatic Ni–Cu–PGE deposits of the Sudbury mafic–ultramafic complex, the palladium deposits from the Neo-Archean Intrusive complex in Lac des Iles, the volcanogenic massive-sulfide deposits from the Archean greenstone belts of Quebec and Ontario, and the Proterozoic volcanic belts of Manitoba. Other nickel–copper–PGE deposits include Coldwell, and River Valley intrusions, Ontario, Muskox Layered Intrusion, Nunavut, Abitibi and Raglan, Quebec, and Voisey's Bay, Labrador.

4.3.1 Sudbury Igneous Complex

4.3.1.1 *Location and Discovery History*

The Sudbury Igneous Complex (SIC) contains abundant sulfides, especially near the base, and hosts one of the largest nickel and copper deposits in the world (Cawthorn, 2005b). The Sudbury Basin or Sudbury Structure or Sudbury Nickel Irruptive, is a major geologic structure in the city of Greater Sudbury, Ontario, Canada. The impact of a meteorite resulted in excavation of a transient crater, 60-km diameter ($62\,km \times 30\,km = 1860$ sq. km surface area), and 15-km deep. The final form of the Sudbury Basin (250-km diameter) is the second-largest known impact crater after Vredefort (300-km diameter) in South Africa. Sudbury is known for the largest resources of nickel–copper + PGEs over centuries. Vredefort is at initial search for gold and uranium. The Sudbury Structure is the remnant of a deformed multiring impact basin and hosts a vast amount of sulfide Ni–Cu–PGE mineralization (Doreen, 2008).

Alexander Murray of the Geological Survey of Canada first reported "presence of an immense mass of magnetic trap" containing sulfide minerals at the present-day site of the Murray mine in 1856. Yet the discovery of this significant Ni–Cu resource could be documented only by a blacksmith in 1883 during construction of the first transcontinental Canadian Pacific Railway. Identification of Ni–Cu metals resulted in a prospecting, production, and staking rush since 1886. Growth of the Sudbury mining camp continues with involvement of giant copper and nickel companies. PGE realization came much later.

4.3.1.2 *Regional Setting*

The Sudbury Igneous Complex (SIC) is structurally placed between the Neo-Archean Levack tonalitic granite-gneiss (+2500 Ma) and intrusive quartz monzonite of the Superior Province to the northern and much of the western and eastern outer margins, and the Early Proterozoic (2400 Ma) metasedimentary and metavolcanic Huronian Supergroup rocks of the Southern Province to the southern outer margin. The Huronian Supergroup unconformably overlies Neo-Archean basement and thickens to the south of the southern margin. Rocks of Huronian Supergroup were deposited in an intercontinental rift-basin environment, and is overlain by the Whitewater Group in a succession of the Onaping tuff Formation, the Onwatin Slate Formation, and the Chelmsford sandstone Formation. The core or inner margin of the SIC is occupied by the Whitewater Group of rocks.

The footwall and older basement rocks are extensively brecciated, deformed, and metamorphosed due to the explosion and impact of a remarkably large-intensity meteorite. The megastructures related to this impact created Sudbury and the Footwall Breccia. The Sudbury breccia is composed of fragments of country rocks up to 20 km from the outer margin. The Footwall Breccia is composed of crushed fragments of the country rocks and a mafic–ultramafic intermix and forms a 10–50 m thick layer between the SIC and the footwall Archean basement and designates a marker zone.

4.3.1.3 Geology of Sudbury Structure

The SIC represents multilayered rings of massive structure (Fig. 4.8). The SIC and strata overlying it are exposed in a series of concentric, crudely elliptical rings dipping to the center of the Complex and suggest that the structure is a basin (Naldrett, 2004). The total configuration of igneous complex can be attributed into three components based on size, shape, and nature of distribution. The components are: Main Mass, Sublayers, and Offsets. The Main Mass is the key unit and encompasses the major part of the complex, composed of all four layers of marginal mafic/quartz-rich norite and gabbro, felsic norite, quartz-rich gabbro, and granophyre. Sublayers occur discontinuously around the Main Mass/Complex. Offsets are dike-like bodies emplaced radially to the Main Mass and extend out 20 km with width ranging between 70 and 100 m, eg, the Nickel Offset (Foy Offset), Copper Offset (Copper Cliff), and Worthington Offset (Fig. 4.8).

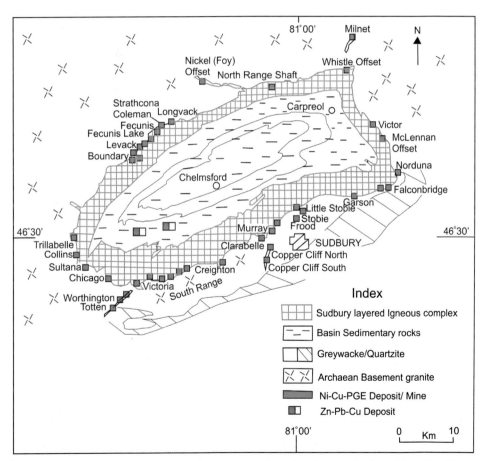

FIGURE 4.8 Simplified geologic map of the Sudbury Igneous Complex showing Ni–Cu–PGE and Zn–Pb–Cu deposits/ mines. Sudbury Camp is ranked as the principal base and PGE metal mining district of Canada *Haldar, S.K., 2013. Mineral Exploration – Principles and Applications. Elsevier Publication, p. 374.*

The outer rims are widely known as North Range and South Range. The structure is postulated to have formed because of a large meteorite (1850 Ma) impact that produced a 150–280 km multiring crater, comprising a 2–5-km thick sheet of andesitic melt. The immiscible sulfide liquid collected into topographic lows, where it differentiated into Cu–PGE-dominated contact deposits by crystallization of a monosulfide solid solution. The residual sulfide liquid migrated into the footwall to form a variety of vein and disseminated types of deposits that underwent remobilization of their metals. The Footwall Breccia hosted remobilized high-grade Ni–Cu–PGE mineralization. The period of mineralization and ore-forming process related to cooling of the SIC can be over 100 to one million years. Many deposits in the South Range have been modified by deformation.

4.3.1.4 Geochemistry

The major and trace element geochemistry of the SIC studied by many researchers eg, Naldrett (2004), Dare et al. (2014), and Mukwakwami et al. (2014). The stratified Sudbury magma separated into two layers: a high-density norite at the base and overlain by a low-density granophyre. The igneous complex was emplaced in a near-surface environment in a continental setting. Naldrett (2004) compared the major elements with those of continental flood basalts. The high SiO_2 (~60%) and K_2O (~1.5%), low CaO (~6%), and low Na_2O/K_2O ratio of quartz-rich norite equated to the Keweenawan and Columbia River flood basalts.

Mukwakwami et al. (2014) identified characteristic features of Ni–Cu–PGE mineralization at the Garson Mine, located in the South Range. The disseminated sulfides in the Main Mass norite and quartz–calcite sulfide veins within a breccia zone present low Pd/Ir ratios, low Rh–Ru–Ir abundances, and limited ranges of Ni content. The Fe/Ni ratios are similar to the undeformed disseminated sulfides and contact-type massive ores at the Creighton Deposit. Garson breccia ore is depleted in Cu–Pt–Pd–Au and enriched in Rh–Ru–Ir, and high Pd/Ir ratios consistent with their mobilization and deposition from hydrothermal fluids. The footwall vein-type mineralization is strongly enriched in Cu–Pd–Pt–Au depleted from the Main Mass breccia ores of monosulfide solid-solution cumulates.

Copper-rich massive sulfides, comprising chalcopyrite, cubanite ± pentlandite, are an important source of Pt and Pd in magmatic Ni–Cu–PGE deposits. It has been generally accepted that the late-magmatic and/or hydrothermal fluids play a significant role in the collection of precious metals in an orebody. Dare et al. (2014) invoked that PGEs in Cu-rich ore have a magmatic origin and concentrate in a small volume of late-stage S-bearing melt trapped between intermediate solid solutions during crystallization of the magma. A sequence of PGEs, followed by accessory tellurides and sulfides, crystallized from this late-stage melt and formed composite grains.

4.3.1.5 Mineralization

Ni–Cu–PGE mineralization is divided into three main types: (1) Contact/Marginal (50%), (2) Offset (25%), and (3) Disseminated and Footwall Breccia (25%), containing Ni–Cu–PGEs in various proportions. Footwall and Breccia types show enrichment and locally very high PGE contents.

Contact deposits are typically located at the interface between the SIC and Neo-Archean or Paleoproterozoic basement rocks. Massive pyrrhotite-rich, Ni–Cu–PGE sulfide ores occur within the contact megabreccia zone. The basal igneous Sublayer norite ± mafic–ultramafic inclusion hosts disseminated sulfides. A series of contact deposits of the North Range are located primarily along a 10-km stretch of the contact extending from the Longvack Mine in the east to Boundary Mine in the west (Fig. 4.8). The Strathcona deposit in the North Range is a type example of contact mineralization occurring at the contact between the Sublayer norite, the Footwall Breccia, and

FIGURE 4.9 Nickel-bearing massive pyrrhotite and patches (pale green) of pentlandite are intergrown from trench cutting the Trill "offset inclusion quartz dyke," Wallbridge Mine, Sudbury Camp. The other nickel minerals, niccolite, millerite, and gold occur as the microlevel intergrown type. The mineralized dike contains 1.2% Ni, 1.0% Cu, and ~7 g/t Pt+Pd+Au. *Dr. Tom Evans.*

FIGURE 4.10 Sperrylite crystal on weathered chalcopyrite mat from the Broken Hammer Ni–Cu–PGE Deposit, Wallbridge Mine, Sudbury Camp, North Range, Ontario. The crystal is approximately 8 mm across. The host rock is quartz diorite Offset Dike. *Dr. Tom Evans.*

Archean gneiss, and extends into the country rocks. The Murray, the Creighton, and a series of other mines are type examples of marginal mineralization from the South Range.

Offset-type deposits are hosted within radial or concentric quartz diorite offset dikes that may extend more than 20 km from the SIC. The economic mineralization is located at the Nickel Offset dike, and the Milnet deposit at Whistle Offset dike at the North Range, the Kelly Lake deposit on the Copper Cliff Offset dike, the Totten deposit on the Worthington Offset dike, South Range, and Ni–PGE mineralization on the Trill Offset dike (Fig. 4.8). PGE–Cu–Ni content increases at distance from the SIC. Massive to semimassive vein-type sulfides typically occur in steeply plunging orebodies along the length of the offsets. The deposits are dominated by pyrrhotite with less abundant pentlandite and chalcopyrite.

The metallic minerals can be placed in three groups: Fe–Ni-sulfide, Fe–Cu-sulfide, and platinum group minerals. Fe–Ni-sulfide minerals include pyrrhotite (Fig. 1.11), pentlandite (Fig. 4.9), and pyrite (Fig. 1.9). The ferromagnetic pyrrhotite is the most abundant host for nickel sulfides in the Sudbury Complex. Fe–Cu-sulfide minerals include abundant and ubiquitous

chalcopyrite (Fig. 1.9), minor cubanite, and rare bornite. Common platinum group minerals are sperrylite, ($PtAs_2$) (Fig. 4.10), michenerite (PdBiTe) (Fig. 1.9), moncheite, $PtTe_2$, insizwaite $PtBi_2$, sudburite (PdSb), froodite ($PdBi_2$), kotulskite (PdTe), niggliite (PtSn), and merenskeyite ($PdTe_2$).

It is observed that sulfide mineralization in all deposits of the Sudbury Camp contribute an average nickel content with minor variation. Copper, platinum, and palladium coexist with average high metal content characteristically in the Offset and Footwall copper deposits.

4.3.1.6 Exploration

The initial economic recognition of the Sudbury Mining District started with the discovery of large Cu–Ni sulfide resources. Exploration and mining for new deposits continued in the complex and endorsed to identify a polymetallic Ni–Cu–PGE program. The current exploration expenditures are mainly focused on the footwall-hosted Cu–Ni– PGE systems containing high Cu and PGE grades. This has been rewarded by the discovery of many new deposits like Victor and Levack Footwall. New exploration models for low-sulfide, high-PGE polymetallic deposit styles of mineralization are being refined for both North and South Ranges. Base metal (zinc–lead) deposits are regularly discovered in the Sudbury camp with evolving deposit models, exploration techniques by introducing deep drilling (Bremner et al., 1996), and genetic models.

The knowledge-base 3-D exploration modeling coupled with application of borehole Electromagnetic (EM) geophysics and interpretation of exploration outcome has resulted in identification of deep-seated targets for Ni–Cu–PGE systems. The followup single-collar multiple-wedge orientation deep drilling of +2000-m depth paid dividends to many success stories in the past decades and on into the 21st century. The geophysical program employing bore hole UTEM survey in old and new holes is credited with the discovery of the Victor contact Ni–Cu–PGE (1970 and 1980s), the Fraser–Morgan and McCreedy East-153 zone (1990s), the Kelly Lake Ni–Cu–PGE in Copper Cliff Offset (1995), the Totten Depth Ni–Cu–PGE (!997), the Ni Rim South (2001), and Podolsky (2002). The other significant discovery is the Levack Footwall deposit in 2005 that intersected at a depth of ~1033 m at 26.2% Cu, 3.0% Ni, and 14.45 g/t Pt + Pd + Au.

4.3.1.7 Mineral Resources

The total mineral resources are estimated at +1500 Mt at 1% Ni, 1% Cu, and 1 g/t Pd + Pt. The figure includes past production, reserve, and resources (Doreen, 2008). The district also hosts significant polymetallic resources comprising U–Ni–Cu–PGE–Au, and Zn–Pb–Cu deposits. The total metal-making episode established Sudbury Camp as the major base and PGE metal exploration and Mining District of Canada.

4.3.2 Lac des Iles Palladium Complex

4.3.2.1 Location and Discovery

Lac des Iles layered mafic intrusion hosts the largest palladium orebody in Canada. The intrusive complex is located at a distance of 85 km northwest of the community of Thunder Bay, in northwestern Ontario. The State Department of Mines mapped the area and discovered a large layered gabbro intrusive complex at Lac des Iles by aeromagnetic survey in 1958. Subsequently airborne, ground Magnetic, and Electro-Magnetic surveys were conducted, followed by five drill holes to test four anomalies. Drilling and sampling continued and identified eight subparallel N–S-trending mineralized zones by 1963.

4.3.2.2 Geology

Lac des Iles Intrusive Complex has intruded into Early Archean granite, gneisse, and greenstone country rocks of the Wabigoon Subprovince of Superior Province. The mafic–ultramafic intrusion extends over 300 km from Rainy Lake to Lake Nipigon. The Las des Iles complex is the largest of a series of mafic–ultramafic intrusions, and collectively they define a circular pattern of 30 km in diameter. The Lac des Iles complex has been dated at 2738 ± 27 Ma by samarium–neodymium (Sm–Nd) dating on pyroxenite. The intrusions constitute a large layered ultramafic body centered on Lac des Iles, and the body is divided into a North and a South Ultramafic Complex, composed of pyroxenite, peridotite, wehrlite, websterite, gabbronorite, and hornblendite. Mafic gabbroic cycles, located in the extreme south of the complex, comprise hornblende-gabbro, magnetite-gabbronorite–pyroxenite–anorthosite, vari-textured gabbro (pegmatoidal), gabbro breccia, and felsic rocks with sharp contacts (Fig. 4.11).

4.3.2.3 Mineralization

The PGE-bearing Ni–Cu mineralization is hosted by both mafic and ultramafic rocks. However, the economic deposits occur only within the

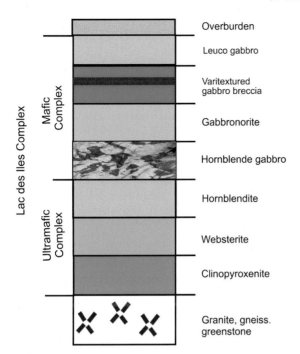

Overburden

Leuco gabbro

Varitextured gabbro breccia

Gabbronorite

Hornblende gabbro

Hornblendite

Websterite

Clinopyroxenite

Granite, gneiss. greenstone

Lac des Iles Complex

Mafic Complex

Ultramafic Complex

FIGURE 4.11 Stratigraphic succession of the mafic–ultramafic intrusive complex at the Lac des Iles Deposit showing the preferred host rock of vari-textured gabbro breccia for rich platinum group metals (red band).

gabbroic pegmatite and gabbro-breccia zones. The major mineralization occurred in the southern part of the complex, in a concentrically zoned elliptical mafic intrusive package (Mine Block) measuring 3-km long and 1.5-km wide with an average thickness of 500 m. Mineralization was enriched in the magmatic breccia, varitextured gabbro, and pyroxenite. PGEs also occur with sulfide-poor, varitextured to pegmatitic gabbro throughout the Roby Zone. It represents coarse-grained disseminated chalcopyrite, pentlandite, and iron sulfide with Pd, Pt, Au, Cu, and Ni.

The ore minerals include pyrrhotite, pentlandite, chalcopyrite, pyrite, and rare PGE minerals: vysotskite (Pd,Ni)S, braggite (Pt,Pd,Ni) S, kotulskite $[Pd(Te,Bi)_{1-2}]$, isomertite $(Pd_{11}S-b_2As_2)$, merenskite $[(Pd, Pt)(TeBi)_2]$, sperrylite $(PtAs_2)$, and moncheite $[(Pt,Pd)(Te,Bi)_2]$. The two major mineralized zones are delineated as Roby

and Offset. The other small zones include: High-Grade, Twilight, Baker, Moore, Creek, Cowboy, Outlaw, Sheriff, and Southern North occurring around Roby. The orebodies show remarkably high Pd/Pt ratios, averaging 7 in Roby and Twilight and even higher up to 14 in locally High-Grade Zone, in contrast to Pd/Pt ratios between 0.5 and 3 in most Pt–Pd-dominated deposits (Barnes and Gomwe, 2011a).

The Roby surface zone is a bulk-mineable deposit with north-to-south length of 950 m, width of 815 m, and depth over 1000 m (including Twilight, Powerhouse, and Moore). It represents multiple styles of mineralization like high grade, breccia, and disseminated. The zone appears to be terminated down dip at Offset Fault.

The Offset Zone, containing high-grade PGEs, is located below the Offset Fault structure. It is displaced downward and shifted to the west by about 300 m. The Offset orebody splits into three horizons to subzones: High Grade Subzone, Mid Subzone, and Footwall Subzone. The Offset Zone contains palladium mineralization with unusually high Pd/Pt and Pd/Ir ratios in rocks that range from relatively unaltered norite to amphibolite, and chlorite–actinolite–talc schist (Boudreau et al., 2014).

4.3.2.4 Exploration, Reserve Base, and Mining

The area has been recurrently explored since 1963–1964 by mapping, geophysical surveys, overburden stripping, long-trench cutting, and 1882 holes drilled totaling 616,283 m (~383 miles; McCracken et al., 2013). The sampling had been unique with data verification, and Quality Control/Quality Assurance (QC/QA) analysis. Many significant exploration targets exist around the Lac des Iles mine, a number of potential palladium-rich deposits discovered, and exploration potential will be further complemented by existing surface and underground infrastructure and the excess capacity of existing mills.

The total Measured and Indicated reserves, as of March 31, 2012, stand at 71.47 Mt grading 1.98 g/t Pd, 0.20 g/t Pt, 0.14 g/t Au, 0.07% Ni, and 0.06% Co. The total Inferred Resources have been estimated at 15.04 million tonnes (16.6 million tons) grading 2.67 g/t Pd, 0.20 g/t Pt, 0.17 g/t Au, 0.08% Ni, and 0.07% Cu (McCracken et al., 2013).

The Lac des Iles mine commenced production as an open pit (Roby) in December 1993, and extended to the underground method in 2006 for a winning Roby zone at depth. The mine commenced an expansion program to access the Offset Zone in late 2010. The access changed operations from ramp haulage (2012) to shaft haulage in 2013. These changes enhanced the benefit from increased underground mining rates and decreased operating costs by the utilization of a shaft. This resulted in turning the Lac des Iles mine into a low-cost producer with a rising production profile.

4.3.3 Voisey's Bay Intrusive

4.3.3.1 *Location and Discovery*

Voisey's Bay mafic intrusion is one of the large nickel–copper deposits in the Canadian province of Newfoundland and Labrador, close to Voisey's Bay and the Labrador Sea in the northeast. The deposit was discovered in the hills along the western shore of Potato Island in September 1993 by Albert Chislett and Chris Verbiski of Archean Inc., a prospecting firm appointed by Diamond Fields Resources Inc. It was routine stream-sediment sampling looking for diamonds. The team identified a blue-stained gossan on top of the hill, 45 km southwest of Nain village.

4.3.3.2 *Geology*

The regional geology of the area is vastly covered by the Nain Plutonic Province in the east and the Churchill gneissic complex in the west. The Nain granitoid basement of Archean age is composed of Makhavinekh Lake and Voisey's Bay granites. The overlying Churchill complex

of reworked Archean and Paleoproterozoic units includes Enderbitic Orthogenesis and the Tasiuyak paragneiss.

The Voisey's Bay Mesoproterozoic mafic intrusion of 1334 Ma age was emplaced into older granitic and gneissic terrain cutting across in an east–west trend between 1350 and 1290 Ma (Fig. 4.12). The sequence of rocks from bottom to top and east to west are Feeder Troctolite, Variable-textured Troctolite, Normal Troctolite, and Olivine Gabbro (Fig. 4.13).

The Feeder Troctolite is the Western Deeps Intrusion, exposed at the surface, and composed of leucotroctolite and/or olivine gabbro containing plagioclase ± olivine cumulate. Varitextured and normal troctolite, and olivine gabbro occur as the Eastern Deeps Intrusions, and exposed to surface. Varied-textured Troctolite contains large pegmatitic plagioclase laths ± olivine, and inclusions of gneiss with spotty sulfide. The Normal Troctolite is medium-grained, uniformly textured, and contains randomly oriented plagioclase laths and olivine cumulus ± orthopyroxene, and hornblende as overgrowth, often containing bands of ilmenite. The Olivine Gabbro is a layered sequence and occurs in the Eastern Deeps in direct contact with older granites. It is composed of elliptical olivine, and euhedral plagioclase cumulate. Reaction rims and serpentinization of olivine are rare. Bulle and Layne (2015a) studied compositional variations in olivine from the economic Voisey's Bay intrusion (bimodal and primitive) and from the primarily barren Pants Lake intrusion (homogeneous and evolved). The results provide potential as a regional-scale mineralogical fertility indicator for mafic intrusions with comparable types of sulfide mineralization.

Basal/Feeder/Magmatic Breccia is a distinct unit invariably enriched with massive sulfides. Ni–Cu–Co sulfide mineralization is spatially and genetically related to a Magmatic Breccia zone associated with an extensive Feeder Dike. The dike system initiates the first entry point of magmatic emplacement into a larger intrusion of Eastern

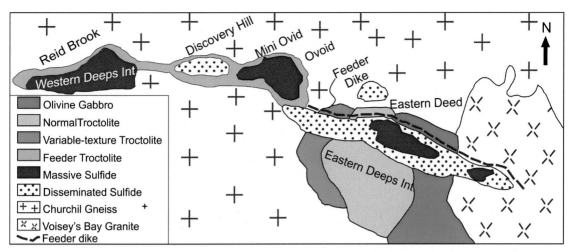

FIGURE 4.12 Simplified geologic map of Voisey's Bay intrusion displaying massive and disseminated mineralized zones *After Bulle, F., Layne, G.D., 2015a. Trace element variations in olivine from the eastern deeps intrusion at Voisey's Bay, Labrador, as a Monitor of assimilation and sulfide saturation processes. Economic Geology 110, 713–731.*

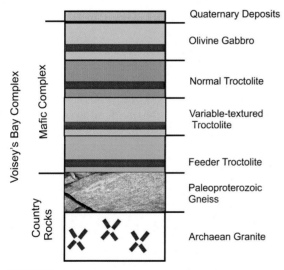

FIGURE 4.13 Stratigraphic succession of Voisey's Bay Ni–Cu Complex. The four mafic sequences host sulfide mineralization in variable proportions.

troctolite with serpentinization of olivine and sericitization of plagioclase. The degree of reaction between troctolitic magma and inclusions of Tasiuyak gneisses gradually increases from the Reid Brook Zone in the west through Discovery Hill Zone to the Far Eastern Deeps.

4.3.3.3 Mineralization

Voisey's Bay mafic intrusion is separated into two distinct segments: Eastern Deeps Intrusion and Western Deep Intrusion with respect to the magma chamber, rock composition, and sulfide concentrations.

The Eastern Deeps is exposed to the surface with uniformly low dip at 25° to the south covering largest length, represents the upper magma chamber, and is composed of vari-textured troctolite, normal troctolite, and olivine gabbro. The deep-seated gabbroic–troctolitic source magma identified by a feeder sheet/dike exposed to the surface. The mineralized blocks are delineated from east to west (Fig. 4.13). The Ovoid is 600-m long, 350-m wide, and a 350 by 110-m deep lens of massive sulfide underlain by a feeder sheet, and overlying Basal Breccia. Mini Ovoid is a separate body located between Ovoid and Discovery Hill.

Deeps. The massive sulfide and breccia-bearing inner basal margin of this upper intrusion is enveloped by disseminated mineralization (Bulle and Layne, 2015b). Breccia zones include fragments of gneisses, ultramafics comprising melatroctolite, wehrlite, and dunite, and sulfide-poor irregular

The Western Deeps Intrusion is exposed to the surface as the Reid Brook with a feeder sheet that opens out into a deeper-level intrusion. Mafic-to-ultramafic intrusive systems necessarily involve conduits through which magma flow enroute to shallow levels of the crust. The near-vertical and horizontal conduit systems (Feeder dikes) are key sites for accumulation of immiscible sulfide liquids and localize large sulfide Ni–Cu–(PGE) mineralization as evidenced at Voisey's Bay (Ripley and Li, 2011a). Localization of magmatic sulfide mineralization is strongly reinforced by interaction between intrusive magmas and preexisting wall-rock structure (Saumur et al., 2015). Isotopic and geochemical studies suggest crustal contamination of the primary magma as an activator for sulfur saturation and the formation of the deposit. Multiple S isotopes identified a bacterial sulfate reduction biosignature in the Tasiuyak gneiss in the footwall to Voisey's Bay (Hiebert et al., 2013).

Sulfide minerals are pyrrhotite, troilite, pentlandite, chalcopyrite, cubanite, and magnetite. Mineralization is grouped into four different types ubiquitously distributed in most of the mineralized envelope (Naldrett, 2004).

1. Massive sulfides: Ores are in abundance in all deposits containing >85% sulfides of coarse-grained pyrrhotite, pentlandite, chalcopyrite, and cubanite with average metal content that varies between 2.50 and 4.10% Ni, 0.85 and 2.50% Cu, 0.18% Co, 33 and 37% S, and 200 and 400 ppb Pt + Pd.
2. Disseminated sulfides: Occurs in the lower part of vari-textured troctolite from the east to the west end decreasing upward into irregular patches up to 30 cm in diameter of sulfide intergrown with coarse silicates (plagioclase and olivine). Average metal content is low in general.
3. Leopard-textured sulfides: Characterized by black spots of augite and olivine crystals in a yellow matrix composed of pyrrhotite, pentlandite, and chalcopyrite in the Eastern Deeps, the Mini-Ovoid, Discovery Hill,

and Reid Brook. Average metal content varies between 1.0 and 1.4% Ni, 0.7 and 1.3% Cu, 0.5% Co, 12 and 16% S, and 40 and 90 ppb Pt + Pd.
4. Sulfides hosted in Basal Brecciated Sequence: Occur as small massive lenses, and blotches of sulfide varying between 0.6–1.2% Ni, 0.3–1.1% Cu, 0.05% Co, 5–15% S, and 60–140 ppb Pt + Pd.

The low content of PGE in all types of mineralization at the Voisey's Bay intrusion is due to melts generated in an anorogenic setting having unusually low concentrations of platinum–palladium to form economic PGE deposits. Nevertheless, the nickel–copper sulfide saturation is suitable for forming major Ni–Cu–Co sulfide deposits (Lightfoot et al., 2011a).

4.3.3.4 Reserve Base and Mining

Systematic drilling started in January 1995, outlined 31 Mt of near-surface reserves grading 2.83% Ni, 1.68% Cu, and 0.12% Co, and named as "Ovoid" and "Mini-Ovoid." Exploratory drilling continued, and discovered Eastern Deeps in October 1996 (Naldrett, 2004). Current ore reserves plus indicated and inferred resources are estimated at 141 Mt averaging 1.60% Ni, 1.25% Cu, and 0.10% Co (Naldrett and Li, 2007).

Mining rights belong to Toronto-headquartered Vale INCO Company. Vale has operated an open-pit mine and concentrator plant at 6000 tonnes/day capacity since 2005. Vale produces two types of concentrate: Ni–Co–Cu bulk concentrate, and copper concentrate.

4.3.4 Coldwell Complex

The Coldwell Complex is a large subcircular gabbro–syenite intrusion, located on the north shore of Lake Superior in northwestern Ontario. The complex has a 25-km diameter covering ~580 sq. km surface area. The Marathon PGE–Cu–Ni deposit is part of the large magmatic system of Coldwell Complex and located ~10 km north of Marathon town.

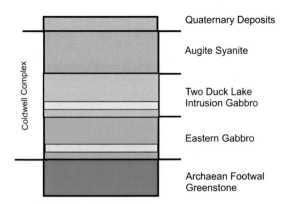

Quaternary Deposits

Augite Syanite

Two Duck Lake
Intrusion Gabbro

Eastern Gabbro

Archaean Footwal
Greenstone

FIGURE 4.14   General geological succession of the Eastern half of the Coldwell Alkaline Intrusive, showing concentration of PGE and copper (yellow bands) in the Eastern Gabbro and Two Duck Lake Gabbro.

The Coldwell mafic–alkaline complex intruded into and bisected the much older Archean Schreiber–Hemlo Greenstone Belt along the northern edge of the North American Midcontinental Rift. The Coldwell complex originated in the Midcontinental Rift system (1100 Ma). The complex was emplaced as three nested intrusive centers. The Center I intrusion includes Eastern gabbro, Western gabbro, amphibole–quartz syenite, iron-rich augite syenite, monzodiorite, and mafic volcanic and subvolcanic rocks. Center II includes amphibole–nepheline syenite and alkaline gabbro. Center III includes quartz syenite and amphibole–quartz syenite (Fig. 4.14).

The Eastern Gabbro occurs as an arcuate-shaped composite intrusion around the eastern and northern margins of the Coldwell Complex. Fine-Grained Series (basal) is fine-grained olivine melagabbro overlain by the Layered Series that occurs stratigraphically ~250 m above the main mineralized zone. The Eastern Gabbro extends over 33 km, thickness up to 2 km, and forms part of a very large magmatic system. It is considered the oldest intrusive phase and interpreted to have formed by at least three discrete magmatic intrusions within a ring dike associated with ongoing collapse. The Two Duck Lake

gabbro is a late intrusive phase of the Eastern Gabbro Suite.

Copper sulfide- and PGE-associated gabbroic and ultramafic intrusions form the youngest intrusive group in the Eastern Gabbro Suite and exhibit a characteristic lithogeochemical signature. The series is dominated by the Two Duck Lake gabbro, layered troctolite sill, and frequent small melatroctolite–olivine clinopyroxenite sill/pod-shaped bodies crosscutting the Fine-Grained Gabbro Series (Good et al., 2015). Mineralization is predominantly the disseminated type forming thick and continuous shallow-dipping lenses at 20°–30°→west parallel to the footwall contact. The orebodies are delineated as Footwall, Main, Hanging-wall zones, and the West Horizon. Sulfides located in the Main zone, including the feeder channel, continue down dip to over 550 m. The mineralization includes chalcopyrite and pyrrhotite, minor pentlandite, and bornite with extreme enrichment of PGMs (25–50 g/t) with respect to Cu and Ni (<0.02%).

Exploration for copper and nickel in the Coldwell Complex was initiated in the 1920s and continued until the 1940s. Marathon PGM Corp. completed advanced exploration and diamond drilling between 2004 and 2009, totaling 65,000 m in 320 holes. The mine plan schedule suggests for a primary pit and satellite pits that will feed a 22,000 tonne/day mill with total estimated proven and probable mineral reserve of more than 91 Mt. Project is planned to develop an open-pit mine to extract copper, nickel, and PGE ore.

4.3.5 Thompson Nickel Belt

The Thompson Nickel Belt is a Neo-Archean and Early Proterozoic geological region located in north central Manitoba known to host a large Ni–Cu–PGE deposit. The Thompson group of mines is located 740 km north of Winnipeg and is named after John F. Thompson, chairman of International Nickel when a significant orebody was discovered in 1956. Thompson nickel belt is second largest komatiite-associated Ni–Cu–PGE camp in the world, and second largest Ni–Cu–PGE mining camp in Canada after Sudbury.

The regional geology that hosts the Thompson nickel belt forms a 10–35-km wide terrain of variably reworked autochthonous Archean basement orthogneisses and migmatite granulite (3100–2600 Ma), and Early Proterozoic cover rocks along the northwestern margin of the Superior Craton. The Ni-Belt extends over 300 km in the NNE–SSW direction. The Thompson orebody extends over 6 km and down-dip depth over 1200 m. The mineralization is associated with inconsistently serpentinized ultramafic intrusions (1900 Ma), lensoid or tabular in shape, that include dunite and pyroxenite hosted by clastic and chemical sediments of the Ospwagan Group.

This mineralization is associated with komatiite, characteristically stratabound, locally stratiform, that exhibits a strong tectono-metamorphic overprint, and occurs in various forms and types. Matthews et al. (2011) described mineralization in five distinct categories. Type-I occurs in basal stratiform with disseminated/net-textured/massive sulfides. Type-II is internally stratabound disseminated sulfides hosted by komatiitic dunite intrusions. Type-Iva presents Ni-rich sulfide layers. Type IVb is hydrothermal-type mineralization. Type-V is tectonically displaced breccia sulfides hosted by adjacent Pipe Formation sulfide facies iron formations and metapelites.

The relict igneous textures in type II ores, basal stratigraphic positions of type I ores, and high Ni/Cu, low Pd/Ir, and high S/Se ratios of types I and II ores indicate that sulfide mineralization is derived by interaction between komatiitic magmas and sulfides incorporated from enclosing iron formations at relatively low magma/sulfide ratios (Matthews et al., 2011).

Systematic exploration was initiated in 1946, and discovered two low-grade nickel deposits at two opposite ends, Moak in the northeast and Manibridge in the southwest in 1955. International Nickel Corporation Ltd carried out over 72 km of exploration diamond drilling leading to discovery of the nickel-rich, large-volume main Thompson orebody in early 1956. This discovery led to development of the current mining camp. Mine production started in 1961.

The premining resource is estimated at 150.3 Mt grading 2.32% Ni, 0.16% Cu, 0.046% Co, 0.10 g/t Pt, 0.54 g/t Pd, 0.046 g/t Rh, 0.072 g/t Ru, 0.033 g/t Ir, and 0.041 g/t Os (Naldrett, 2004).

4.3.6 River Valley Intrusion

The River Valley intrusion is located at the easternmost extended part of the East Bull Lake intrusive suite that exists along an east-northeast trend, over a distance of ~250 km, roughly run on the present site of the Sudbury Igneous Complex. The River Valley Ni–PGE Project is located in the southern part of the intrusive complex, ~60 km east of Sudbury. The project is road accessible to Canada's premier Ni–Cu–PGM mining and smelting district with exceptional infrastructure, and community support for mining activities. River Valley intrusion was identified in early 1990 by researchers as part of a nickel search in the Huronian–Nipissing Magmatic Belt.

The East Bull Lake Intrusive Suite (2480 Ma) comprises a series of three major and several smaller mafic igneous plutons. These intrusions are eroded remnants of layered sills that formed from tholeiitic, plagioclase-rich magmas emplaced during a Paleoproterozoic rifting event. The River Valley intrusion is shallow dipping, layered, and ~900 m thick. The rock types are leucogabbro-norite and leucogabbro with gabbros and anorthosite. The River Valley intrusion is either in thrust contact with quartzite of Huronian Mississagi Formation or in contact with mafic and felsic metavolcanic rocks of the lower Huronian Supergroup with the nature of the contacts unknown. The River Valley intrusion is made of basal ~100 m of unlayered units, underlain by Archean footwall gneiss, and migmatite, and overlain by ~800 m of layered cumulates containing autoliths of gabbroic rocks and inclusion-bearing footwall gneiss, and amphibolite, all within a gabbroic matrix. The Basal unit is distinctly marked upward by a footwall breccia, a boundary unit, a breccia unit, and an inclusion-bearing unit. The layered units comprised upward by olivine gabbro, gabbronorite, and leucogabbro-norite.

PGE-rich magmatic Ni–Cu mineralization occurs ubiquitously throughout the Basal unit encompassing inclusions, breccia, and matrix as blebby and disseminated sulfides. The PGE–Ni–Cu mineralization was identified in associated fertile host rock along the margin of the intrusion. The best potential structure for economic accumulations is in the Breccia Zone that includes the main mineralized zone at ~20 m of intrusive contact with country rocks. This contact-type magmatic sulfide zone extends for over 9 km of prospective strike length and hosts currently defined resources. The Main Zone of breccia-hosted PGM mineralization averages 20–50 m in thickness, continues to depths of +200 m, and is open along strike and down dip. The River Valley intrusion appears to have a much more complex history regarding the timing of sulfur saturation. The critical ore genesis processes might have occurred much earlier, prior to emplacement (Holwell and Keays, 2014).

Pacific North West Capital Corp and Anglo American Platinum Corporation Limited carried out eight diamond-drilling programs between 1999 and 2008, that included 13,500 m of resource drilling (2006) focused on Dana South and North Zones. Pacific Capital Corp further completed 140 line km of 3-D Induced Polarization geophysical survey and 15,500 m of drilling in 2011.

Measured + Indicated reserves at River Valley has been estimated at 91.34 Mt grading 0.58 g/t Pd, 0.22 g/t Pt, 0.02 g/t Rh, 0.04 g/t Au, 0.34 g/t Ag, 0.06% Cu, 0.02% Ni, and 0.002% Co. Inferred resources stand at 36 Mt at 0.36 g/t Pd, 0.14 g/t Pt, 0.014 g/t Rh, 0.03 g/t Au, 0.11 g/t Au, 0.06% Cu, 0.03% Ni, and 0.002% Co (Pacific North West Capital Corp).

4.3.7 Raglan Ni–Cu–PGE Deposit

The Raglan Ni–Cu–PGE sulfide deposits in the Cape Smith Belt are located in the Nunavik area of New Québec region on Ungava Peninsula, northern Canada. The area is located 100 km south of Deception Bay. Low-grade mineralization at the western end of the Cape Smith Belt was discovered in 1898 by the Geological survey of Canada.

The deposits occur within a series of 50–200 m thick mafic–ultramafic complexes that outcrop discontinuously along the contact between the Proterozoic Chukotat and Povungnituk groups in the east-central part of the Cape Smith Belt. The country rocks, host rocks, and ores have been regionally metamorphosed. The ultramafic complex comprises two main facies: conduit-facies assemblages composed of peridotite, and channelized-sheet assemblages composed of peridotite and massive gabbro. Raglan mineralization consists of sulfide lenses associated with Proterozoic ultramafic flows that lie along the contact of tholeiitic basalts, Povungnituk Group sediments, and Chukotat Group of komatiitic basalts (2004–1920 Ma) within the Labrador Trough. William et al. (2011) proposed through mathematical modeling that an initially 10-m thick komatiitic basalt magma flowed turbulently near the vent and thermomechanically eroded unconsolidated pelitic sediment during emplacement to cause the degree of contamination at distances of ~30–60 km downstream from the source.

Nine peridotite flows contain significant sulfide mineralization over a distance of 55 km identified that include the Donaldson, Boundary, West Boundary, 13–15, 5–8, Kaatinniq, 2–3, East Lake, and Cross Lake deposits. The minerals are pentlandite and pyrrhotite, chalcopyrite, magnetite, and pyrite with associated Co and PGEs. Local enrichment of PGE in the South Raglan Trend is due to hydrothermal alteration. PGE grades in the vicinity of carbonate alteration attain levels of potential economic significance of the Ni–Cu–PGE sulfide assemblage (Liu et al., 2016).

Systematic exploration started from the 1930s, and high-grade mineralization at "Deception Creek" (Katinnig) was discovered in 1956 and eventually to develop "New Quebec Raglan Nickel Mines." Naldrett (2004) reported the reserve base at Raglan at 24.7 Mt grading 2.72% Ni, 0.70% Cu, 0.054% Co, and 3.76 g/t of total PGEs. M/s Falconbridge estimated proven and probable reserves at 11.5 Mt at 2.94% Ni, 0.77% Cu, and 0.07% Co, as of June, 2009. Measured and indicated resources totaled a further 16.4 Mt at 3.18% Ni, 0.90% Cu, and 0.07% Co. Inferred resources stand at 14 Mt at 2.9% Ni, 0.90% Cu, and 0.1% Co (www.mining-technology.com/projects/raglan/).

The Raglan mine is a large underground operation stretched over 72 km in the east–west direction. The onsite mining–mineral processing is operated by Toronto-based Xstrata Nickel (Falconbridge Ltd.). Concentrate is transported by road to port at Deception Bay, shipped to Quebec City, and transported by rail to Falconbridge's smelter at Sudbury. Smelted matte is returned to Quebec City and shipped to company's Nikkelwerk refinery in Norway for final nickel production. The mine production started from 1997 with a mine life of 25 years at capacity of 1.1–1.3 Mt ore/year from three underground mines and two open-pit operations.

4.3.8 Abitibi Ni–Cu–PGE Complex

The Abitibi Complex forms a part of the Abitibi–Wawa terrane and represents one of the best exposed, largest, and most richly mineralized Archean (2800–2600 Ma) greenstone belts in the world. It spans the Ontario–Quebec border. The belt is composed of volcanic and mafic–ultramafic intrusions, granitoids, and early mid-Precambrian sediments. The mafic–ultramafic–komatiite hosted significant Ni–Cu–PGE deposits (Dumont, Alexo, Bird River, and Shebandowan). The Abitibi greenstone belt has a long history of exploration and mining of Ni–Cu–PGEs, with several periods of extensive exploration and discovery, including a major renewal in recent decades (Houlé and Lesher, 2011a).

The Abitibi greenstone belt represents a conformable sequence and includes four distinct mafic–ultramafic lithotectonic domains overlain by arc-related igneous rocks. Mafic komatiite domains are accreted Archean deformed oceanic-plateau material. Komatiites occur intermittently throughout the Superior Province with best exposures at Alexo, Pyke Hill, and Spinifex Ridge. Komatiites are identified as komatiitic–basalt sequences, and komatiite–rhyolite–dacite–andesite sequences that host almost entire Ni–Cu–PGE endowments. The younger calcalkalic sequence represents extension-related volcanism erupted through deformed plateau material, as a consequence of ridge subduction in a regime. Houlé et al. (2011b) observed that the host komatiite flow thermomechanically eroded the footwall andesite at the contact between the komatiite and the andesite.

Komatiite-associated Ni–Cu–PGE mineralization occurs as the basal stratiform and internally disseminated type. The basal stratiform type is the most common type and widely distributed. The stratabound disseminated type is sparsely spread within the belt. The host rocks include undifferentiated olivine mesocumulate units as lava channels, subvolcanic sills, or feeder dikes. Mineralization broadly occurs as massive to semimassive, and net-textured to disseminated within olivine pyroxenite to pyroxenite. The sulfide minerals include nickel-bearing pyrrhotite, pentlandite, pyrite, chalcopyrite, millerite, magnetite, and ferrochromite. The average grade of nickel in the Abitibi greenstone belt ranges between 0.60% and 3.20% Ni.

4.3.9 Dumont Sill

Dumont sill is a differentiated ultramafic to mafic intrusion forming a part of the Abitibi–Témiscamingue region of Québec. It is located 25 km northwest of Amos town, 60 km northeast of Rouyn-Noranda city, and ~70 km northwest of Val-d'Or.

The Dumont sill is a zoned intrusive body within the Abitibi Greenstone Belt. It extends over 7 km along the NW–SE direction and dips 60°–70° toward the northeast. The mafic–ultramafic sill is differentiated into a lower ultramafic zone, composed of lower peridotite and lower dunite, overlain by a mafic zone with clinopyroxenite, gabbro, and quartz gabbro subzones. The sequence is followed by an upper dunite and upper peridotite, overlain by metavolcanics (gabbro). A complete sequence of the mafic–ultramafic intrusion and preferential concentration of magmatic nickel sulfide is given in Fig. 4.15.

The upper dunite unit contains several layers enriched in primary Ni-sulfide minerals adding considerable value to the deposit. Half of the nickel resource consists of awaruite (Ni_3Fe), heazlewoodite (Ni_3S_2), and minor pentlandite-bearing serpentinized dunites with formerly contained Ni in the silicate minerals. Serpentinization causes remobilization of nickel from olivine to enrich cumulus sulfides in layers containing primary accumulations of magmatic sulfides. In layers deficient in primary sulfides, serpentinization causes formation of awaruite by reduction of Ni originally hosted by olivine. The early stages of serpentinization were marked by the generation of large volumes of Fe-rich serpentine containing abundant ferric iron and lesser amounts of brucite rich in ferrous iron (Sciortino et al., 2015).

The area has been systematically drilled at close interval. Proved reserves at the Dumont deposit are estimated at 179.6 Mt grading 0.32% nickel ± PGE, and mine life is estimated to be 33 years. The Probable reserves stand at 1 billion tonnes (1.1 billion tons) grading 0.26% nickel (June 2013, Royal Nickel Corporation).

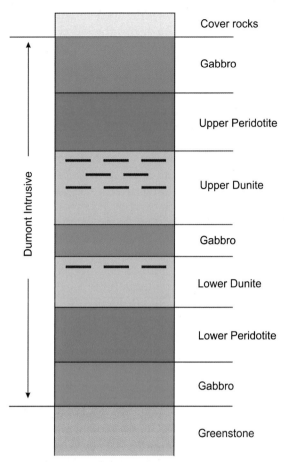

FIGURE 4.15 Complete sequence of a mafic–ultramafic intrusion showing preferential concentration of magmatic nickel sulfides as in the Upper Dumont Sill.

A large feasibility open pit has been developed covering the central widely mineralized zones. The open-pit mining at the rate of 100,000 t/d will be the fourth largest nickel sulfide mine in the world once in full operation. The mine is owned by the Royal Nickel Corporation of Canada.

4.3.10 Bird River Sill

The Bird River mafic–ultramafic layered-intrusive sill complex is located in the southern part of the Bird River greenstone domain, about

120 km northeast of Winnipeg in southeastern Manitoba. Since the 1920s, the sill was known to contain Ni, Cu, and Cr mineralization. The mafic–ultramafic intrusion has hosted two Ni–Cu mines, the Dumbarton mine (1969–1974), and the Maskwa West mine (1974–1976). The presence of PGEs was discovered by P. Theyer in the early 1980s.

Bird River Supercrustal greenstone belt extends for 150 km from Lac du Bonnet (Manitoba) in the west to Separation Lake (Ontario) in the east. The supracrustal rocks are divided into six formations composed of volcaniclastic sedimentary rocks and mafic lavas, pillowed basalt–tuff, rhyolite flows, pyroclastic breccia, andesite, dacite, conglomerate/sandstone, graywacke–mudstone turbidites, interbedded conglomerates, and iron formations.

The Bird River Sill intruded within the older Bird River greenstone belt. The sill extends over 22 km in length and up to 800-m thickness in the NNE–SSW direction from Bird Lake in the east and Poplar Bay in the west. The Bird River intrusion is composed of a differentiated, layered, mafic–ultramafic sill separated by a 240 m transition series. The Lower Ultramafic package (<200 m) comprises subzones named Contact, Megadendritic, Layered, Massive Peridotite, and Chromitiferous. Upper Mafic package comprises Lower Gabbro/Xenolith/Leucotonalite/and Upper Gabbro. It is reported to be of 2743 ± 52.2 Ma U–Pb age (Scoates and Jon Scoates, 2013). The sill is divided into a number of tectonic fragments, mineralization blocks, or mines from east to west: Bird Lake, the Dumbarton Ni–Cu mine, the Maskwa West Ni–Cu mine, the Ore Fault Ni–Cu project, Page, Peterson, Chrome, and National Ledin projects/blocks.

The Bird River intrusion hosts significant mineralization of chromium–nickel–copper, and locally anomalous concentrations of platinum and palladium. Chromitite occurs as stratiform layers with significant lateral continuity and irregularities at the local level. The shifting is mainly due to synmagmatic faulting, occurs in up to six main intervals over a thickness of ~60 m in the Lower Ultramafic part of the intrusion. Ni–Cu sulfide mineralization is hosted in the ultramafic rocks at the base of, or just below, the sill (Maskwa–Dumbarton mines).

Mustang Minerals drilled 6938 m in 38 boreholes at the Makwa Deposit during 2008–2009. Probable category reserves have been estimated at 9.855 Mt grading 0.54% Ni, 0.11% Cu, 0.02% Co, and 0.434 g/t Pd + Pt. The reserves of other blocks are reported as: Dumbarton mine (1.539 Mt at 0.81 Ni and 0.30% Cu), and Maskwa West mine (0.366 Mt at 1.16% Ni and 0.20% Cu).

4.3.11 Muskox Layered Intrusion

The Muskox layered mafic intrusion outcrops in the northwestern corner of the Canadian Shield. The pluton crosses the Arctic Circle at a point about 145 km east of Port Radium on Great Bear Lake and ~90 km south of Kugluktuk village. The intrusion was discovered by H. Vuori of Canadian Nickel Company in 1956.

The basement rocks are composed of gneiss and metasediments trending NNW–SSE with steep dip. The basement rocks are unconformably overlain by a package of sandstone, dolomite, and basaltic flows dipping to the north. The Muskox complex intruded into the basement rocks following a deep feeder channel and is exposed as a swarm of dikes near the surface. The intrusive structure extends over 120 km in the NNW–SSE direction dipping inward at 23° to 57° SW. The internal structure of the Muskox intrusion is divided into four major units:

1. The feeder unit composed of bronzite gabbro and picrite–norite zones parallel to the nearly vertical footwalls.
2. The marginal zones occur parallel to the walls of the intrusion and grade inward from the bronzite gabbro at the contact through picrite and feldspathic peridotite, to peridotite, and locally dunite.

3. The central layered series is 2.60 km thick with 38 main layers of dunite, peridotite, pyroxenites, and gabbros. The layers are nearly flatlying, discordant to the marginal zones. The layer thickness varies between 3 and ~550 m.

4. The upper border group is ~60 m thick and is characterized by an upward gradation from granophyre-bearing gabbro to granophyre.

The Muskox Ni–Cu–PGE deposit is a Greenfield exploration project. Drilling along the marginal zone near the eastern boundary of the intrusion established massive sulfides with an average grade of 1.2% Ni and 2.6% Cu. Prize Mining drilled two holes on the rim and intersected 4.7% Ni, 10.6% Cu, 2.2 g/t Pt, and 11.1 g/t Pd over 5.4 m and 3.22% Ni, 7.52% Cu, 2.2 g/t Pt, and 17.5 g/t Pd over 5.5 m.

4.3.12 Giant Mascot Ni–Cu–PGE Deposit

Giant Mascot Ni–Cu–PGE deposit/mine is located east of Harrison Lake, and ~12 km north of Hope in southwestern British Columbia.

The country rocks hosting the Ni–Cu–PGE mineralization and the associated ultramafic intrusive suite are Upper Triassic "Settler Schists," and the Late Cretaceous "Spuzzum pluton." The "Settler Schist" is an assemblage of pelitic and quartz feldspathic schist and micaceous quartzite. The "Spuzzum pluton" is composed of an upward succession of quartz diorite, hornblende gabbronorite, and gabbronorite–diorite. The Giant Mascot ultramafic suite (93 Ma) is intrusive into the "Spuzzum pluton" (95 Ma). The host ultramafic intrusive suite occurs as an elliptical plug of ~3-km length and ~2-km width, and predominantly consists of remarkably fresh dunite in the core, with an outer sequence of peridotite, pyroxenite, hornblende pyroxenite, and hornblendite.

The sulfide mineralization is mostly hosted in olivine-rich rocks, including dunite, peridotite, olivine-bearing pyroxenite, and hornblende pyroxenite. Ni–Cu–PGE-bearing thin parallel lenses are exposed at the surface, and occur at near-vertical dip down to a depth of +600 m. The orebodies are open at depth indicating additional ore at greater depth.

Preproduction reserves were estimated at ~4.2 Mt of ore grading 0.77% Ni and 0.34% Cu with minor Co, Ag, and Au, and Ni/Cu = 2.3 (Manor et al., 2016). The deposit was under production by underground mining method between 1958 and 1974.

4.4 GREENLAND

Greenland is a phenomenally massive island and autonomous Danish territory. The state company Nunamineral and Greenland Minerals and Energy are in the forefront for the prospecting and development of mineralbased industries. The minerals of interest include iron ore, uranium, aluminum, zinc, nickel, copper, PGEs, titanium, and Rare Earth Elements (REE). Greenland is a democratic, promining country with a transparent regulatory system, competitive mining tax, and no land-claim issues.

4.4.1 Skaergaard Au–PGE Intrusion

Several mafic, layered intrusions are exposed in Eastern Greenland and the Skaergaard Intrusion is the simplest and smallest. The Skaergaard intrusion is located within the Kangerlussuaq region, and is part of an igneous province of the extensive North Atlantic Tertiary. The intrusion was discovered by Lawrence Wager during 1961 on a routine British Arctic Air Route Expedition. The area is located 420 km west of Iceland and the nearest towns are Safjordur, Iceland (430 km east), Tasiilaq, Greenland (390 km), and Scoresbysund, Greenland (460 km).

The regional geology of the Skaergaard area is covered by Archean basement gneisses that host the intrusive and the extrusive magmatic complex composed of a chain of Tertiary gabbro and syenite intrusions along the eastern coast. The Skaergaard intrusion was formed by emplacement of titanium-rich massive tholeiitic magma about 55 million years ago, commensurate with the opening of the North Atlantic Ocean. The intrusive body represents a single magmatic pulse that crystallized from the bottom upward and top downward through a holistic process of magmatic differentiation, fractional crystallization, and development of layering. This has resulted in exceptionally well-developed characteristic cumulate layering defined by variations in abundance of crystallized olivine, pyroxene, plagioclase, and magnetite. The intrusive body includes gabbro, ferrodiorite, anorthosite, and granophyre.

Mineralization outcrops at the surface for a length of ~10 km, and extends to a depth of more than 1.1 km. Mineralization consists of flatlying reefs dipping at 20° toward the south referred to as the Triple Group: two gold reefs and one palladium reef. The palladium reef, hosted by the Skaergaard intrusion, is a rarer type of PGE deposit. It is characterized by a Pd–Au–Cu-dominant, and Ni–Pt-poor, sulfide assemblage present in the upper parts of the host intrusion. Such deposits are considered to form by continued fractional crystallization of the magma, with crystallization of magnetite playing a distinct role in late-stage sulfur saturation (Holwell et al., 2011, 2014). The broad geological succession of Skaergaard intrusion is grouped as in Fig. 4.16:

1. Upper Border Series under the roof.
2. Layered Series accumulated up from the floor of the intrusion.
3. Marginal Border Series along the walls.

Over 35,000 m of diamond drilling has been completed in the Skaergaard to delineate indicated and inferred mineral resources. The reserve and resources have been estimated as 203 Mt grading 0.88 g/t Au, 1.33 g/t Pd, 0.11 g/t Pt, and reported in accordance with Joint Ore Reserves Committee (JORC) Code (2012) guidelines. Resources include a combined total of 5.7 million ounces of gold, 8.7 million ounces of

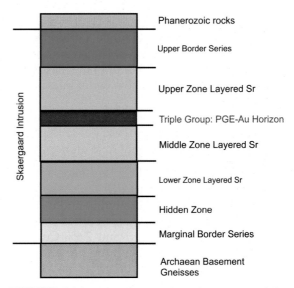

FIGURE 4.16 Generalized geological succession of the Skaergaard intrusion in regional Archean Basement gneiss showing magmatic intrusion and mineral-bearing horizon (yellow).

palladium, and 0.79 million ounces of platinum confined within three reefs (H0, H3, and H5) of the Triple Group, the major host location for all noble metals.

4.4.2 Maniitsoq Nickel Complex

The Maniitsoq nickel deposit is located next to the east of Maniitsoq town and Napasoq on the southwest coast of Greenland, and is centered ~125 km north of the capital city of Nuuk. The pack ice-free environment and mild coastal climate is suitable for mining and shipping of concentrate around the year.

Geological mapping, prospecting, and shallow diamond drilling between 1959 and 1973 delineated a 75-km long and 15-km wide belt of nickel–copper–cobalt–PGE-bearing norite intrusions, referred to as the Greenland Norite Belt. Diamond drilling intersected a zone of 9.85 m grading 2.67% Ni and 0.60% Cu at Imiak Hill. The advance exploration techniques, including Helicopter-borne

Time Domain Electro-Magnetic (TDEM), Borehole TDEM, Surface EM, and gravity survey, continued in the area by many exploration companies: Cominco, Falconbridge, and others. An advance deep-drilling program was conducted between 2012 and 2014 totaling 14,590 m in 73 holes at three areas: Imiak Hill (Mikissoq), Spotty Hill, and Fossilik II intersecting significant semimassive to massive Ni–Cu–Co–PGE sulfide mineralization.

Regional geology of Maniitsoq area is represented by strongly deformed and metamorphosed Meso-Archean tonalitic-gneisses of the North Atlantic Craton, amphibolite, and metasedimentary supracrustal rocks. The mafic–ultramafic intrusions (3000 Ma) intruded the gneissic country rocks and are concentrated in the Greenland Norite Belt. These intrusions range in size from small dikes and plugs to elongated bodies covering up to 8 sq. km (Fossilik intrusion). The mafic–ultramafic complexes include leuconorites, norites, pyroxenites, and peridotites with a distinct signature of surface gossan.

Sulfide mineralization consists of disseminations, blebs, net-textures, semimassive and massive sulfides in the form of sulfide breccia, veins, and stringers. The primary sulfide phases are pyrrhotite, pentlandite, chalcopyrite, and pyrite. Pentlandite and chalcopyrite are the primary nickel and copper minerals, respectively. The mineralization is open at depth indicating future enhancement of reserves and resources.

The general trend of drill-hole intersections indicates 2.89 m at 2.24% Ni, 0.63% Cu ± PGE (Fossilik), 9.85 m at 2.67% Ni, 0.60% Cu ± PGE (Imiak Hill), and 4.95 m at 1.97% Ni, 0.43% Cu ± PGE (Quagssuk).

CHAPTER

5

Deposits of South America

5.1 BACKGROUND

South America is a continent with the majority of its surface area in the Southern Hemisphere and a small portion in the Northern Hemisphere. The continent has the fourth largest surface area after Asia, Africa, and North America. The continent includes 12 sovereign states (Argentina, Bolivia, Brazil, Chile, Colombia, Ecuador, Guyana, Paraguay, Peru, Suriname, Uruguay, and Venezuela), two nonsovereign areas (French Guiana and the Falkland Islands), and the Republic of Trinidad and Tobago, the twin-island country lying just off the northeastern coast of Venezuela.

South American economies have traditionally relied on a foundation of forestry, agriculture, fishing, mining, tourism, and nonexportable manufactures. The major mineral resources of South America are gold, silver, copper, nickel, platinum group of metals, iron ore, tin, coal, and petroleum.

5.2 MINERAL DEPOSITS

The mineral occurrences are exceptionally localized on the South American continent: few countries have well-balanced mineral resources (Argentina, Brazil, Chile, and Colombia), and others are practically devoid of mineral resources (Paraguay and Uruguay). South America contributes about 20% of the world's iron ore reserve and resources, mainly from Brazil. Chile alone contributes about a third of the world copper production and the second largest reserves of molybdenum. Peru is the largest silver producer and the second-ranked producer of copper.

South America, in general, is moderately gifted with platinum group of elements (PGEs), nickel, and cobalt, and deficient in chromium resources. These metals occur as low grade in mafic–ultramafic intrusions, laterites, and as a value-added byproduct of porphyry copper–gold deposits (Alumbrera porphyry $Cu \pm Mo$, Au, and Pd, Pt). The mineral occurrences exist in eastern Brazil, the central and northern Argentine Andes, Colombia, Cuba, and Venezuela. South America is a producer of nickel and cobalt from Brazil and Colombia.

5.3 ARGENTINA

The Republic of Argentina, the second largest country on the continent, is located in southeastern South America. The economy is based on agriculture, oil and gas, copper–gold–silver mining, and others including a small recovery of PGEs from copper–gold ore. The region of Argentina adjacent to the Andes is one of the most metal-rich areas in the world. The Bajo de la Alumbrera mine, owned by Minera Alumbrera Limited, produces copper and gold + PGEs.

5.3.1 Alumbrera Copper–Gold–Platinum Deposit

5.3.1.1 Location

Bajo de la Alumbrera is a classic porphyry-type copper–gold deposit with occurrence of PGEs. The deposit was discovered in 1949, and commercial production commenced in February 1998. The deposit is located 1100 km northwest of Buenos Aires at an average altitude of 2500 m in Catamarca Province, Argentina. Alumbrera mining, owned and operated by Minera Alumbrera Limited, is one of the largest and lowest-cost gold and copper production mines in the world.

5.3.1.2 Geology

The host rock is dacite–quartz-feldspar porphyry intruded into volcanic andesite, andesite breccia, conglomerate, sandstone, and tuff. The basement rocks are granite, schist, and gneiss. The length and width of surface alteration over the deposit are 2.5 and 1.75 km, respectively, with an overall surface coverage of 3.4 sq. km. The orebody extends over a strike length of 1.1 km, width of 0.6 km emplacement depth of 2.5 km, and emplaced ~7.5 Ma. The main ore mineral is chalcopyrite. The weathering and alteration at the surface, supergene enrichment, and stockwork structure are common features at depth.

5.3.1.3 Reserve Base

The reserve base stands at 338 Mt of proven category grading 0.4% copper, 0.39 g/t gold, 2.50 g/t silver, and 130 g/t molybdenum (2009). The reserve base includes another 10 Mt of probable ore grading 0.33% copper, 0.3 g/t gold, 2.5 g/t silver, and 150 g/t molybdenum. The possible resources stand at 450 Mt grading 0.53% Cu, 0.64 g/t Au, 2.50 g/t Ag, and 130 g/t Mo.

5.3.1.4 Mining

The conventional open-pit mine covers about 3.24 sq. km area. Prestripping involved removal of 25 Mt of weathered cover. The initial strip ratio was 1.4:1. The concentrator was designed to treat 80,000–85,000 t/d, yielding 700,000 t/y of copper–gold concentrate. Design recovery rates were 91% for copper and 70% for gold. The molybdenum plant was commissioned in 2008 and has produced 450 t of molybdenum.

Alumbrera produced 28 Mt of ore, grading an average of 0.50% Cu and 0.55 g/t Au during 2008. The operation treated 37 Mt of ore during the same year and produced 156,893 t of copper in concentrates, and 60,484 oz of gold in doré (semipure alloy of gold and silver). The Alumbrera mine produced approximately 25 Mt copper ore in 2010.

5.3.2 Chromitites of Western Ophiolite Belt, Pampean Ranges, Córdoba

5.3.2.1 Location

The ophiolitic chromitite bodies are considered potential hosts for PGEs formed as traces enclosed in chromitites during magmatic differential crystallization, segregation, and precipitation. Two large Neoproterozoic ophiolitic belts are located in Eastern Pampean Ranges of Córdoba, Argentinian Central Andes. The ophiolitic ultramafic bodies are known as Los Congos and Los Guanacos (Fig. 5.1).

5.3.2.2 Geology

The basement rock of the Eastern Pampean orogeny comprises metasedimentary and metaigneous rocks, consisting of gneisses and schist intruded by granitoid suites of Early Cambrian to Devonian age. These units were metamorphosed under amphibolite facies. Hypersthene-bearing gneisses and amphibolite occur as migmatites with intercalations of cordierite and garnet.

Neoproterozoic ophiolitic mafic–ultramafic rocks are aligned in two distinct belts (Los Congos and Los Guanacos) with an NNW–SSE trend. The belts cropped out as stretched lenses in tectonic contacts with the surrounding

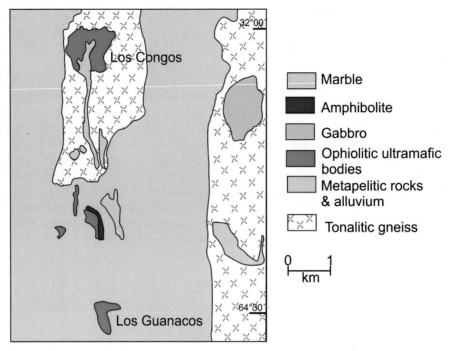

FIGURE 5.1   Geologic map of Los Congos and Los Guanacos ophiolitic ultramafic bodies in the Argentinian Central Andes. Chromitites are the potential host for PGMs. *After Proenza, J.A., Zaccarini, F., Escayola, M., Cábana, C., Schalamuk, A., Garuti, G., 2008. Composition and textures of chromite and platinum-group minerals in chromitites of the western ophiolitic belt from Pampean Ranges of Córdoba, Argentina. Ore Geology Reviews 33 33–48, Elsevier. www.sciencedirect.com.*

migmatites and gneisses. The ophiolitic belts are composed of a metamorphosed suite of harzburgites and peridotites, intruded by dolerite dikes and pyroxenites. These ultramafic rocks are considered a mantle ophiolitic sequence represented by the porphyroclastic harzburgite tectonites. The mafic–ultramafic rocks were emplaced into the upper crust in Neoproterozoic times during a collisional event between the Pampia terrine and the Rio de la Plata craton (Proenza et al., 2008).

5.3.2.3 Los Congos

The Los Congos ultramafic body is located to the north of Santa Rosa River and is one of the largest in the region with a surface spread of 1300-m long, and 700-m wide. The country rocks are mainly tonalitic gneisses with distinct occurrences of marbles to the south (Fig. 5.1).

The ultramafic body is strongly serpentinized (harzburgite) in association with lizardite (serpentine), antigorite, tremolite, brucite, talc, and magnetite. The ultramafic body of the Los Congos is considered a portion of the mantle sequence of an ophiolite (Proenza et al., 2008). The Los Congos area was known for active mining sites during the 1940s from Los Congos, Sol de Mayo, and Los Coquitos mines. The chromitites occur predominantly as podiform or small pods about 10-m long and 3-m wide, surrounded by variably thick dunite envelopes. The typical feature of pyroxenite dikes crosscutting the chromitite bodies is common.

5.3.2.4 Los Guanacos

Los Guanacos spreads over a surface area of 1500-m long, and 600-m wide. The ultramafic suite is located in tectonic contact with metapelitic rocks (Fig. 5.1). The ultramafic rocks are predominantly depleted harzburgites accompanied by minor dunite and saturated peridotites. Major and trace element characteristics indicate a residual origin, although they are strongly metasomatized. The mineral phases that occur in the serpentinized harzburgites are olivine, anthophyllite, antigorite, lizardite, brucite, talc, and carbonates. The serpentinized impregnated peridotites are composed of olivine, tremolite, and antigorite. The pyroxenite dikes of variable dimensions often crosscut the ultramafic body, host dunite, and chromitites. The ultramafic rocks of the Los Guanacos body have been considered to form part of a metamorphosed Moho Transition Zone from an ophiolitic sequence (Proenza et al., 2008).

A number of small and lenticular podiform chromitites occurs in dunite and harzburgite host rocks, and mined between 1940 and 1945. The largest single chromitite body at Ume Pay mine extends over tens of meters at less than 5 m thickness, and has produced over 50,000 tonnes (~55,000 tons) of chromite ore.

5.3.2.5 *Mineralization*

Electron microprobe analysis of massive chromites from Los Guanacos and Los Congos indicate invariably low SiO_2 (0.2 wt%), low Fe_2O_3 (<8.0 wt%), and high MgO contents (>15 wt%), that are characteristics of ophiolitic primary chromite (Proenza et al., 2008).

Platinum group minerals identified include native osmium, laurite (RuS_2), erlichmanite (OsS_2), irarsite [(Ir, Ru, Rh, Pt)AsS], platinum, and a number of inadequately identified phases such as an oxide or hydroxide of Ru, Pt, and Ir-Ru, Pt telluride, Ir-Ru-As-Se, and Ir-Ru-Ti compounds. Native osmium was the only platinum group metal which remained unaltered. The other platinum group metals underwent mineralogical reworking during metamorphism. Proenza et al. (2008) suggested that the Ru-Os-Ir platinum group of metals (PGMs) in the Los Guanacos and Los

Congos chromitites were modified in situ, producing redistribution of these PGEs on a small scale. The presence of rare Pt and PGE-As-Se minerals was possibly related to remobilization of Pt, As, and Se by fluids during the alteration processes.

5.4 BRAZIL

The Federative Republic of Brazil is the fifth largest country, the seventh largest economy in the world, and the largest in South America. Brazil is endowed with rich mineral reserves and has been partly exploited including gold, iron ore, tin, copper, nickel–cobalt, and bauxite. Brazil is the fifth largest producer of gold in the world. The Carajás open-pit mine is the largest iron ore mine in the world with an estimated reserves and resources of 7.2 billion tonnes (7.9 billion tons) of iron ore containing gold, manganese, copper, and nickel. Brazil is one of the major producers of nickel and cobalt, with most production in the form of ferro-nickel alloy. A series of magmatic Ni–Cu–PGE deposits exist along the eastern coastal region that include from north to south: Limoeiro, Caraiba, Santa Rita, Boa Vista, Niquelândia, Americano do Brasil, and Fortaleza (Fig. 1.20). The deposits are of different size and shape, sulfide or lateritic, and offer excellent potential target area. There is currently no production of PGMs in Brazil. Extensive exploration is continuing in many projects.

5.4.1 Pedra Branca Mafic Ultramafic Complex

The Pedra Branca deposit represents one of the best-undeveloped platinum–palladium projects in the Western Hemisphere. The prospect being explored covers a large mafic layered-intrusive complex that hosts several chromite horizons similar to those found in the Bushveld Complex.

5.4.1.1 Location

Pedra Branca mafic–ultramafic intrusive complex is located in the central region of Ceará State, northeast Brazil, and at a distance of 50 km from the Antas copper–gold deposit. Pedra Branca and Antas are located in the world class Carajás Mineral Province. The Carajás region is regarded as one of the prospective mineral provinces for copper–gold and iron ore resources. Pedra Branca PGM project consists of 57 exploration concessions totaling ~700 sq. km surface area. The land package is mostly contiguous farmland with good access and no environmental limitations. The area benefits from excellent infrastructure within a district experiencing substantial investment from Vale's development of "Serra Sul," soon to become one of largest iron ore mines.

5.4.1.2 Geology

The Pedra Branca area belongs to Borborema Structural Province and represents the older crystalline basement of a common tectonomagmatic orogeny of Upper Proterozoic age. The complex is part of the Tróia Median Massif in the Santa Quitéria structural block, and comprises five mafic–ultramafic bodies, all mineralized with chromite layers (Fleet et al., 1993). The country rocks in the southern part are mainly gneisses with intercalations of quartzite, marbles, amphibolite, schist, and calc-silicates. The central part of the massif, which includes the Pedra Branca Complex, is represented by reworked Archean rocks. The mafic–ultramafic complex is interpreted as Archean in age. The geology of the intrusive complex is metagabbros, metapyroxenites, pillowed metabasalts, amphibolite, quartzite, metasediments (with sulfides and manganese), marbles, and graphitic schist.

5.4.1.3 Mineralization

Chromite grains in chromitite contain lamellar inclusions of chlorite oriented parallel to (1 1 1). The chromite grains have a core of aluminous chromite and a broad margin of ferrian chromite (ferritchromite). The chlorite is a Cr-bearing IIb clinochlore, and the inclusions occur preferentially within the ferritchromite. The orientation relationship is equivalent to the orientation of chlorite–olivine. Chloritization of the chromite was broadly contemporaneous with the alteration to ferritchromite (Fleet et al., 1993). Platinum-group mineralization occurs in stratigraphic layering of the ultramafic (iron–magnesium-rich) body. The PGEs are associated with chromite. The PGEs are also more closely related with minor sulfide concentrations.

5.4.1.4 Exploration and Resources

The Joint Venture between Pedra Branca do Mineração and Anglo Platinum at Pedra Branca project completed 318 core-drill holes. Pedro Branca PGE is in the advance stage of exploration in 2015 by Anglo Platinum Group. Anglo Platinum plans to complete up to 6000 m of core drilling on several magnetic geophysical targets identified because of a regional airborne magnetic survey conducted in 2013.

The Pedra Branca deposit has an estimated potential of 1 million ounces of platinum–palladium metals. The inferred resources of Pedra Branca copper–gold has been estimated at 46 Mt grading 1.20% Cu and 0.33 g/t Au, as of June 2013.

5.4.2 Limoeiro Ni–Cu–PGE

5.4.2.1 Location and Discovery

Limoeiro nickel–copper–PGE sulfide mineralization is located in Borborema Province, northeastern Brazil. The deposit was discovered by Votorantim Metais Ltd in 2009 after extensive and systematic soil geochemistry, and an airborne magnetic and gamma geophysical survey program looking for base metals. The ultramafic intrusion and Ni–Cu sulfide mineralization crop out in the eastern portion of the tube-like intrusive (Bofe target block). The surface signature is characterized by a unique multicolored gossan cover of the weathered outcrop. The gossan is enriched with nickel, copper, and PGMs.

5.4.2.2 Geological Setting

The regional country rock is part of the Neoproterozoic Braciliano orogenic system and composed of high-grade paragneisses and schist. The basement is overlain by the Vertentes complex and composed of a metavolcanic-sedimentary sequence metamorphosed under the granulite facies.

The magmatic sulfide deposit is hosted by an intrusive ultramafic suite of irregular configuration affected by partial tectonic deformation and high-grade metamorphism. The tube-like ultramafic body extends over 5 km in the east–west direction, exposed only at the eastern end (Fig. 5.2), and lies subhorizontally at a maximum depth of 500 m. A series of ductile and brittle faults affected the ultramafic complex by horizontal and vertical shifts, resulting in four distinct blocks from east to west as Bofe, Picarra, Retiro, and Parnazo (Fig. 5.3). The ultramafic rocks include orthopyroxenite and harzburgite. The primary magmatic structure of the intrusion hosting the Limoeiro deposit is remarkably well preserved. The tabular ultramafic conduit consists of three main magma influxes involving fractionation, sulfide saturation, and segregation. The sequences of the ultramafic complex from the top to the bottom are Lower Chromium, Upper and Lower sequences, each one consisting of a core of harzburgite enveloped by orthopyroxenite, with an irregular and discontinuous outer shell of amphibolite (Silva et al., 2013). The low-Cr sequence has the lower grades of MgO (25%), and Cr (441 g/t). The Upper and Lower sequences contain 29% MgO, and 1933 g/t Cr.

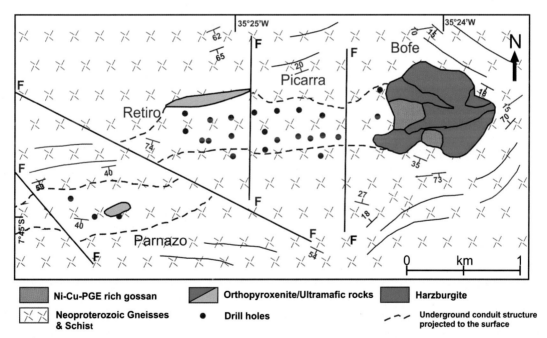

FIGURE 5.2 Simplified surface geology map of the Limoeiro nickel–copper–PGE-sulfide mineralization, projecting a new exploration and mining opportunity in eastern Brazil. The ductile and brittle faults moved the mineralized blocks both horizontally and vertically. *After Silva, J.M., Filho, C.F.F., Giustina, M.R.S.D., 2013. The Limoeiro deposit: Ni–Cu–PGE sulfide mineralization hosted within an ultramafic tubular magma conduit in the Borborema Province, Northeastern Brazil. Economic Geology 108 1753–1771.*

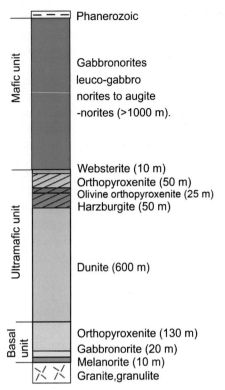

FIGURE 5.3 Simplified stratigraphic column of the mafic–ultramafic sequence at Santa Rita deposit showing nickel–copper–cobalt–PGE mineralization.

5.4.2.3 *Mineralization*

The magmatic system resulted from two major pulses of identical composition. The sulfide saturation and segregation is restricted to the second magma pulse in the Upper sequence. Ni–Cu–PGE sulfide mineralization is largely concordant with the Upper sequence. Mineralization often occurs as disseminated sulfides in the upper horizon, and massive sulfide layers at the bottom of the orebody. The orebodies are formed by tectonic slicing of the originally continuous tube-like structure. The length and thickness of individual orebodies are up to 1 km and 150 m, respectively.

The common ore minerals are pyrrhotite, pentlandite, chalcopyrite, pyrite, and PGMs. The massive ores contain homogeneous

Pt–Ni–Bi-bearing merenskyite ($PtTe_2$) enclosed primarily by pyrrhotite as an exsolution from monosulfide solid solution. The massive base-metal sulfides also host palladium. Sperrylite ($PtAs_2$) is the second most abundant platinum mineral in the disseminated sulfide ores.

5.4.2.4 *Exploration Opportunity*

Since discovery, Votorantim Metais Ltd explored the deposit by multidisciplinary techniques including +27,000 m of diamond drilling. Small conduit-type mafic–ultramafic intrusions are a common geologic setting for magmatic Ni–Cu sulfide±PGE deposits. This preferred association is widely recognized as a key exploration guideline. The discovery of the Limoeiro deposit has opened a new window in northeast Brazil for exploration opportunities in this large orogenic belt.

5.4.3 Caraiba Cu–Ni Deposit

The Caraiba copper–nickel deposit is located in the Curaçá River Valley, in the Bahia region of eastern Brazil. The country rocks of Curaçá Valley are high-grade migmatitic gneisses, granulite, and metasediments. The mafic–ultramafic bodies form part of the Caraiba Group. The mafic igneous complex comprises multiple intrusions of dikes, veins, and igneous breccia of norites and hyperthenites, with minor proportions of amphibolized gabbronorite and peridotite xenoliths transported by the magma from deeper levels in the lithosphere. The host rocks are polydeformed and metamorphosed.

Caraiba is the largest Brazilian copper deposit under exploitation and consists of unusually high copper and low nickel. The mineralization mostly of disseminated and remobilized bornite and chalcopyrite is hosted in early Proterozoic norite and hypersthenite.

The Caraiba deposit was operated initially as an open pit and subsequently changed to underground mining. Mineração Caraíba owns the underground Caraíba mine, with installed capacity of 1.20 Mt/y of copper–nickel sulfide ore. Mineração Caraíba plans to reschedule throughput at its Pilar underground copper sulfide mine in Bahia state to a capacity of 900,000 t/y in 2016.

5.4.4 Santa Rita Ni–Cu–Co–PGE Deposit

5.4.4.1 Location and Discovery

Santa Rita is a large Ni–Cu–PGE deposit currently exploited in Brazil. The deposit is located about 260 km southwest of Salvador, the capital of Bahia State, and 8 km south-southeast of the town of Ipiaú, in Brazil.

The presence of mafic–ultramafic layered intrusive complexes within the Sao Francisco craton in Bahia was reported in 1976 based on regional aeromagnetic surveys by the Brazilian Geological Survey. Various mining agencies conducted follow-up detail exploration programs including regional mapping, ground geophysics, stream-sediment and soil sampling, and diamond drilling, and identified a nickel–copper–cobalt–PGE geochemical anomaly near Fazenda Mirabela (Santa Rita deposit) in the late 1990s.

5.4.4.2 Geological Setting

The Archean to Paleoproterozoic basement rocks within the São Francisco craton represent granite, granodiorite, and migmatite crust, greenstone belts, and charnockite plutons.

The Fazenda Mirabela and Fazenda Palestina mafic–ultramafic layered intrusions were emplaced into the core of antiformal structures. The mafic–ultramafic complexes were intruded into the supracrustal succession and orthogneisses, typically deformed charnockites at the granulite facies. The metasediments of this sequence are sulfide poor. Both the Fazenda Mirabela and Fazenda Palestina intrusions are undeformed, and show no evidence of the penetrative deformation that has affected the surrounding rock.

5.4.4.3 Host Rocks

The Santa Rita nickel–copper–cobalt–PGE deposit is exclusively hosted by the Fazenda Mirabela intrusion (Barnes et al., 2011b). The typical mafic–ultramafic cumulate rocks of the layered-intrusive body outcrop as an ellipsoidal configuration with its longer axis trending NNE–SSW over 4.2 km. The shorter axis extends over 2.5 km. The funnel- or boat-shaped body covers an approximate area of 7 sq. km. The layered-intrusive complex occurs in three sequences from bottom to top as: a Basal unit in the west, an Ultramafic unit in the center, and a Mafic unit in the east.

The Basal unit is composed of melanorite, gabbronorite, and orthopyroxenite with a thickness of 160 m and devoid of any mineralization. The Ultramafic unit comprises fine-grained dunite and harzburgite, with a lower (western) ultramafic zone, varying from dunite to harzburgite, olivine orthopyroxenite, orthopyroxenite, and a thin cap of websterite with a total thickness of ~735 m. The olivine-bearing cumulates are completely serpentinized at the surface, with the preservation of relic texture. The layered ultramafic unit hosts the sulfide mineralization in the top layers and laterite nickel in the dunite layer. The Mafic unit lies to the east of the mafic–ultramafic layered complex, is ellipsoidal in exposed shape, and occupies two-thirds of the surface area of the ntrusion. The Mafic unit is entirely composed of nonmagnetic homogeneous coarse-grained gabbronorites, leucogabbronorites to augite norites having thickness over 1000 m without any mineralization (Fig. 5.3).

5.4.4.4 Mineralization

The Santa Rita deposit comprises disseminated nickel and copper sulfide mineralization with cobalt and PGE credits. The deposit is up to 140-m thick at places with an average thickness of 40 m over a continuous strike length of more than 2 km and a depth of ~1 km. The maximum thickness and continuity of the rich orebody are located at the north and south ends of the deposit.

The mineralization occurs as a continuous stratiform layer containing magmatic Fe–Ni–Cu–Co sulfides, confined entirely within a thick stratigraphic envelope from the top of the dunite upward to the base of the Mafic unit. The disseminated-sulfide mineralization occurs within a sequence of unaltered harzburgite and orthopyroxenite. This orthopyroxene-dominated sequence lies between the dominantly olivine-rich cumulates that comprise the lower half of the intrusion, and the gabbroic cumulates that make up the upper half. It varies from a single 50–60-m thick, well-defined layer near the northern margin, to a more complex interval with fluctuating sulfide abundance over an interval of 100–180 m toward the south-central portion of the intrusion. Grades and thicknesses increase, and the continuity of the mineralization improves, toward the axis of maximum thickness of the ore zone. The mineralization tends to occur over a wider zone as multiple layers of fluctuating sulfide abundance toward the southern margin. The mineralization resembles a zone of complex intercalation of pyroxenite and dolerite with minor semimassive sulfide veinlets. The sulfides exhibit composite polymineral blebs composed of a magmatic sulfide assemblage of enriched pentlandite, chalcopyrite, pyrrhotite, and ubiquitous pyrite. Pyrite is often forms as an intergrowth with pentlandite.

5.4.4.5 Reserves and Resources

The Santa Rita ore zone is an unusual example of high-tonnage magmatic Ni–Cu–Co–PGE-sulfide mineralization developed as a stratiform layer within the main cumulus sequence of a layered intrusion.

The Joint Ore Reserves Committee (JORC) compliant reserve base has been estimated at +160 Mt at 0.52% Ni, 0.13% Cu, 0.015% Co, and

Pt + Pd credit. The reserve is under the proved-and-probable category, and will be outlined for open-pit mining at a strip ratio of 5:1. The inferred resources of ~40 Mt at 0.79% Ni, 0.23% Cu, Co, and PGEs is planned for underground mining.

5.4.4.6 Mining

Mirabela acquired the Santa Rita deposit in 2004, commissioned the mine and plant in 2009, and commercial production started from January 2010. The operation consisted of a low-cost, drill-and-blast, load-and-hall open-pit mining process. The conventional plant produces a high-quality nickel–copper–cobalt–PGEs concentrate with metal content of approximately 20% and an average nickel content of ~13–15%. The concentrate is sold in bulk to overseas customers via the port at Ilheus, Bahia, Brazil.

The future plan of Mirabela is to operate the zones of thicker and continuous ore in the north and south ends of the deposit. The first phase involves mining in the north pit (Pit 6). The first phase will provide feed to the plant for a 6-month period at an average grade of 0.50% nickel. The second phase (Pit 8) in the south end will provide feed for 3.5 years with the ore quantity for the two phases sufficient for a 4-year production plan at an average strip ratio of 2.35:1.

The open-pit operation will ultimately change over to underground mining.

5.4.5 Niquelândia Lateritic Nickel Deposit

5.4.5.1 Location

The Niquelândia Lateritic Nickel Deposit is located 200 km north of Brasilia, the Federal Capital of Brazil, in Goiás State. Laterite weathering of ultramafic rocks formed important nickel reserves in this area.

5.4.5.2 Geology

The mafic–ultramafic complex is about 40 km long and 20 km wide, extends in the NNE–SSW

TABLE 5.1 Stratigraphic Sequence and Lithology of the Igneous Complex at Niquelândia Containing a Nickel-Mining Opportunity

Stratigraphy	Age	Lithology
Upper sequence	Proterozoic	Upper amphibolite
		Gabbro and anorthosite
Lower sequence		Gabbro
		Gabbro and norite
		Interlayered pyroxenite and peridotite (Ni–Cu–Co host rock)
		Massive dunite and harzburgite (Ni–Cu–Co host rock)
		Interlayered gabbro, pyroxenite, peridotite, and mylonite
Basement	Archean	Gneiss, quartzite, and amphibolite

direction from the Niquelândia village. The igneous complex of Archean–Proterozoic age rests over the Archean basement metamorphic rocks comprising gneiss, quartzite, and amphibolite. The dips of the mafic–ultramafic intrusive layers range between 40 and 60 degree toward the NNW. The complex consists of a lower and an upper sequence. The lower sequence is in direct tectonic contact with the basement rocks defined by fault boundary on all sides of the complex. The stratigraphic sequence and rock types are given in Table 5.1.

5.4.5.3 Mineralization

The soil geochemical survey at the Niquelândia area outlined a nickel anomaly covering 3.2-km long, and 150–450-m wide. The soil-sample values range between 0.15% and 1.36% nickel. This anomaly coincides with copper soil-sample values ranging between 0.025% and 0.21% copper. This coexistence of nickel and copper content in soil anomalies suggests a possible nickel-bearing sulfide source in the lateritic host rock. The potential stratabound nickel–copper–cobalt-sulfide mineralization occurs at the lower contact of the ultramafic sequence of massive dunite and harzburgite, and interlayered pyroxenite and peridotite (Table 5.2). The mineralization consists of disseminated pyrrhotite,

TABLE 5.2 Laterite Profile at Loma de Niquel, Venezuela

Description	% Ni	g/t Co	Thickness (m)
Ferruginous overburden	0.69–0.82	300–500	1
Ferruginous laterite	0.96–1.49	1000–3400	2
Argillaceous saprolite	2.24	600	~4
Coarse saprolite	2.43	400	

pentlandite, chalcopyrite, and the associated precious group of metals.

Two typical lateritic-weathering profiles (Jacuba and Angiquinho) formed over parental pyroxenite and dunite at the lower intrusive ultramafic sequence of Niquelândia supergene nickel–copper deposits. The Jacuba and Angiquinho weathering profiles are located within parental interlayered pyroxenite and massive dunite, respectively. The pyroxenes are replaced by goethite and kaolinite through a series of transitional Ni-bearing phyllosilicates from the base to the top of the profiles. The mineralogy and chemical composition of these clay minerals depends on the degree of fracturing and serpentinization of the pyroxenite and location of the pyroxenite with respect to neighboring dunite. Smectite and pimelite $[Ni_3Si_4O_{10}(OH)_2.4H_2O]$ pseudomorphs after pyroxene are exceptionally Ni rich, and in fact, are the most Ni-enriched clay minerals now known in lateritic weathering within the Jacuba profile (Colin et al., 1990).

5.4.5.4 Reserve Base

The 20.36 sq. km Niquelândia property is a highly prospective geological environment hosting several pastproducing and producing nickel deposits. The reserve and resource base of laterite nickel deposit was initially estimated at 60 Mt at 1.45% Ni. The property is shared by International Nickel Co Vale (INV), Canada, Anglo American, and Votorantim Metals. The Codemin property reportedly contains reserves of 3.7 Mt grading 1.33% nickel, and resources of 6.9 million tonnes (~7.6 million tons) grading 1.27% nickel.

5.4.5.5 Mining

The lateritic nickel deposit is widespread and most suitable for a low-cost open-pit mining operation. The block of INV is contiguous to the Codemin mine of Anglo American and the Niquelândia mine of Votorantim Metals, to the north and south, respectively. Anglo American's Codemin nickel mining and processing unit produced 9800 tonnes (~10,800 tons) of nickel metal in 2006. Votorantim Metals is the largest producer of electrolytic nickel in Latin America, exporting nearly 70% of its product. The Votorantim property produces about 26,000 tonnes (~28,600 tons) nickel metal and 1200 tonnes (~1320 tons) of cobalt per annum.

5.4.6 Ipueira–Medrado Chromite Deposit

5.4.6.1 Location

Ipueira–Medrado Sill is part of the 70-km long and 20-km wide, north–south-trending swarm of the mafic–ultramafic Jacurici Complex, situated near the northern margin of the São Francisco craton, Brazil.

5.4.6.2 Geology

The mafic–ultramafic host rocks are composed of layered complexes and intruded into the basement Archean gneissic blocks. The mafic–ultramafic host rocks contain abundant stratiform chromite bodies of Paleoproterozoic age. The intrusive complex comprises orthopyroxene-rich sequences with highly magnesian bulk composition, and fractionation trends consistent with komatiitic parent magmas.

The Ipueira–Medrado Sill crops out as a synform structure 7-km long with intrusive thickness of 300 m. The Sill is subdivided into three geographical segments: Ipueira I, II, and Medrado. The mafic–ultramafic sequence is similarly subdivided into three zones based on rock composition: the Marginal Zone, the Ultramafic Zone, and the Mafic Zone. The Marginal, Basal, or Contact Zone is 5–20-m thick and composed of sheared gabbroic rocks. The Ultramafic Zone is up to 250-m thick, and further subdivided into the Lower Ultramafic Unit, the Main Chromitite Layer, and the Upper Ultramafic Unit. The Lower Ultramafic Unit is 100–180-m thick and composed of dunite interlayered with harzburgite and rare chromitite seams up to 0.5-m thick. The Main Chromitite Layer is 5–8-m thick with highly massive chromite layers of variable thickness and textures of individual seams. The Upper Ultramafic Unit is up to 50-m thick, and composed of harzburgite interlayered with dunite and thin chromite seams. The Mafic Zone is up to 40-m thick, and composed of norite with variable proportions of orthopyroxene and plagioclase (Marques et al., 2002). The Ipueira–Medrado chromitite contains low concentrations (<350 ppb) of PGMs (Lord et al., 2004).

5.4.6.3 Reserves and Resources

The estimated chromite reserves and resources in the Jacurici mafic–ultramafic Complex stand at +30 million tonnes (~33 million tons), probably ranking it as the largest chromite deposit in Brazil. The estimated chromite reserves at Ipueira–Medrado Sill Complex amount to +4.5 Mt grading 30–40% Cr_2O_3. The chromite ore from the

Ipueira–Medrado deposits is mined by underground methods following continuous single main layer averaging 5–8m in thickness. The thickness of other seams varies between 0.3 and 1.1 m.

5.4.7 Campo Formoso Chromite Deposit

The Campo Formoso chromite deposit is located about 50 km south of the Ipueira–Medrado Sill, Bahia State, Brazil.

The complex consists of a tabular and arch-shaped body of ultramafic rocks about 40-km long and 100–1100-m wide. It hosts the most prominent chromite deposits of Bahia State, Brazil. The deposit is hosted by a stratiform ultramafic intrusive, and forms part of the Jacurici Complex of Paleoproterozoic age. The ultramafic complex is intruded between the Campo Formoso older granite and the Jacobina sedimentary sequence. The peridotite host rock was completely serpentinized, and affected by multiple cycles of hydrothermal influxes forming emerald in the roof zone. The hydrothermally invaded chromitite sequences are extremely rare in containing unusual minerals like monazite, apatite, galena, bismuthinite, and antimony.

Seven chromitite layers of +10-m thickness have been identified in the southwest limb of the intrusion. The complex has nine mines, located along the northern border of the Serra da Jacobina mountain range. The Coitezeiro mine is one of the largest units. The average metal content for lumpy ore is 38% Cr_2O_3. The other types, e.g., friable, ribbon, and disseminated, contain 18% Cr_2O_3. The open-pit mining method is adopted in the Campo Formoso region. The highest metal grade reported is 57% Cr_2O_3.

5.4.8 Americano do Brasil Ni–Cu Deposit

5.4.8.1 Location and Discovery

The Americano do Brasil Ni–Cu Deposit is located in the Tocantins Province in central Brazil. The complex was discovered in 1969 during a regional geologic mapping program, and the sulfide deposits were identified during regional exploration in the early 1970s.

5.4.8.2 Geology

The Complex is part of a cluster of coeval synorogenic mafic–ultramafic layered intrusions emplaced during the Brasiliano/Pan-African Orogenic Cycle (Neoproterozoic, ~600 Ma). The Neoproterozoic mafic–ultramafic complex intruded into the basement rocks of Archean greenstone belts and tonalite–granodiorite associations, known as the Goiás Massif. The Massif also includes the other three large layered intrusions, namely, Barro Alto, Niquelândia, and Cana Brava. The medium-sized Americano do Brasil consists of interlayered dunite, peridotite, websterite, and gabbronorite. The intrusive was formed by crystallization of tholeiitic high-MgO parental magmas.

5.4.8.3 Mineralization

There are three distinctively different nickel–copper-sulfide orebodies at Americano do Brasil deposit. The orebodies are S1 and S2 located in the Northern Sequence, and G2 in the Southern Sequence.

The S1 orebody is the largest by size in the complex, but the lowest in metal content, and Ni:Cu ratio. The orebody is hosted by websterite and gabbronorite in a more fractionated sequence of the mafic–ultramafic complex. It occurs as a cluster of several irregular and discontinuous concentrations of disseminated Ni–Cu sulfides. The individual lenses are irregular and unevenly cylindrical shape, about 50–100 m in diameter extending for 500 m along strike. The Cu/Cu + Ni ratios are higher ranging between 0.5 and 0.8 and exceedingly depleted in PGEs. The S1 orebody has been interpreted to have formed from a later event of sulfide segregation in the magma chamber, possibly following the event that originated the G2 orebody.

S2 is a steep-dipping EW-trending tabular orebody that extends for 1.2 km and 250 m in depth. The orebody occurs as stratiform massive sulfide (0.5–4 m thick) hosted by partially serpentinized dunite and peridotite in the interior of a layered intrusion. The orebody resulted from sulfides accumulated at the transient base of the magma chamber following a new influx of parental magma. The orebody is in sharp contact between massive sulfides and dunite–wehrlite in the hanging wall, and a gradational contact from massive sulfide to net-textured sulfide and barren dunite–wehrlite in the footwall. The S2 orebody is rich in nickel and copper contents, and the highest Ni/Cu ratio. The mineralization consists of rich pyrrhotite, pentlandite, and chalcopyrite (Ni–Cu–Pt–Pd rich). This association occurs as thin veins 0.5–5-cm thick, small pods of 5–20-cm thick, and chalcopyrite-rich ore within the pyrrhotite and pentlandite-rich mineralization.

The G2 orebody occurs as an irregular and unevenly cylindrical shape, consisting mainly of net-textured sulfides. The G2 orebody is hosted by peridotite and pyroxenite, and located stratigraphically below the S1 orebody. The S2 and G2 orebodies are characterized by low Cu/Cu + Ni ratios below 0.4 (Silva et al., 2011).

5.4.8.4 Reserve Base and Mining

Systematic drilling between 1975 and 1978 outlined a small Ni–Cu-sulfide deposit. The reserve base has been estimated at 3.1 Mt grading 1.12% Ni, and 1.02% Cu. Prometalica Ltd acquired the mining rights in 2004 and designed for underground sublevel stoping, and matching conventional sulfide-flotation process. Production of Ni yielded a concentrate of about 6000 t/y started in 2006.

5.4.9 Barro Alto Nickel Deposit

5.4.9.1 Location and Discovery

The Barro Alto nickel deposit is located in the state of Goias, Brazil, about 170 km from existing Codemin nickel operation. The deposit

was discovered in 1960, explored since 1970, and Anglo American acquired mineral rights in 2002.

5.4.9.2 Geology

Barro Alto mineralization occurs primarily in saprolite overlain by laterites. It extends for 35 km in an arc from southwest to northeast. The deposit is in the Barro Alto mafic–ultramafic complex forming part of the Precambrian Shield. The complex is composed of a sequence of serpentinized dunites and pyroxenites, enveloped by gabbros. The mineralization corresponds to the surficial weathered portions of the serpentinites. There are three types of saprolitic ore. The West-type ore tends to have higher nickel grades and silica/magnesia ratios than the East-type and Plain-type ores.

5.4.9.3 Reserve and Mining

The total mineral resources are 1.27 Mt grading 1.48% Ni. The project consists of an open-pit mine and a ferronickel plant. The shallow pits extend over 35 km in length and width of less than 2 km. No blasting is required to access the relatively soft laterite/saprolite ore with a maximum working bench height of 2.5 m. The open-pit project was approved in 2006 and the first metal was produced in 2011 with a rated capacity of 36,000 t/y of nickel metal over a mine life of 25 years.

5.5 COLOMBIA

The Republic of Colombia is situated in northwest South America. The country is well endowed with minerals that include coal, nickel, platinum, gold, silver, and gemstones. The platinum resources are located in the Pacific coastal lowlands of western Colombia. The platinum–gold operations historically continued from the small-scale placer deposits in the Choco region, western Colombia.

5.5.1 Platinum

The element platinum was first reported from the Choco region of western Colombia in 1557. Colombia, at that regime, was endowed with the primary source of precious metals for the Spanish Empire. The country remains the sixth largest producer of platinum in the world and produced 28,359 ounces of platinum in the fiscal year 2011 from small-scale operations in the Choco region. The Choco region has been a historical producer of platinum since the 16th century. The San Juan River valley in the Choco region is one of the richest platinum regions in the world. The mining is carried out mostly by small mining cooperatives or individual prospectors. Large sections of the rainforest in the Choco are being destroyed by illegal miners searching for and extracting gold and platinum.

5.5.1.1 Condoto Mafic–Ultramafic Complex

5.5.1.1.1 LOCATION

The group of small-scale placer platinum–gold operations, namely, Novita, Martinez, and Condoto are located in the historical platinum- and gold-producing Choco region, western Colombia, 85 km south of Quibdo, the capital of Choco, 200 km southwest of the city of Medellin, and 80 km east of the Pacific Ocean. Panned-sample concentrate indicated average values ranging between +1 g/t and 13.50 g/t Pt and Au, with the maximum value of 709.52 g/t Pt + Au.

5.5.1.1.2 CONDOTO MINE

The Condoto mine is located in the Choco region. The area is historically known for potential hard-rock source for alluvial platinum and gold. Extensive mapping, stream-sediment sampling, and drilling in the early 1990s identified a series of zoned mafic–ultramafic intrusions as a possible source of alluvial precious metals. The Condoto mafic–ultramafic Complex is one such potential source and holds 140 sq. km of concessions in Choco.

The Condoto area is part of an island arc setting of Tertiary age (20 Ma). The oceanic crust of the Nazca plate to the west subducted and accreted under the Cretaceous Western Cordillera oceanic crust in the foreground to the northern end of the Andes. The Atrato San Juan Basin is located to the west of the town of Condoto. The valley has been formed by deposition of platinum- and gold-rich Tertiary clastic sediments in the forearc of the island arc system.

The emplacement of magmatic intrusions into the shallow crust includes: Alto Condoto Ultramafic Zoned Intrusive Complex, and Vira–Vira and El Paso high-magnesium basalt with peridotite lenses. Both the intrusive complexes are genetically related with the same source of parent magma and former intrusive complex representing early fractionation of Alto Condoto in the Tertiary–Lower Miocene (20–23 Ma). The inner core of Alto Condoto zoned ultramafic complex is composed of dunite, containing recorded enrichment in PGMs and gold. The Alto Condoto complex covers an area of ~50 sq. km with a maximum length of 9 km and maximum width of 6 km. The successive zones/layers of the inner core are composed of wehrlite, olivine clinopyroxenite, hornblende-magnetite clinopyroxenite, a gabbro–dioritic injection zone, and high-grade and low-grade metamorphic rocks.

The host peridotites of the two complexes are extremely enriched in PGMs and gold.

5.5.1.1.3 CHOCO PLACER DEPOSIT

The Choco metallogenic belt, in western Colombia, is a placer platinum–gold concentration formed during the Eocene period, and extends over 200 km in a north–south trend. The deposits produce the majority of significant platinum output. The Choco region is also known as a top gold-producing province. The majority of this production is from small-scale operations.

5.5.2 Nickel

Colombia has significant lateritic nickel reserves, and is the second largest nickel producer in South America. The major nickel mining/smelting operation in Colombia is from the Cerro Matoso deposit.

5.5.2.1 Cerro Matoso Nickel Deposit

The Cerro Matoso deposit is located near the town of Monlelibano, Córdoba, northwest Colombia owned by Broken Hill Proprietary (BHP) Billiton and listed on the Australian Security Exchange. The operation combines a lateritic nickel ore deposit/mining from 1980 with a low-cost ferronickel smelter producing high-purity and low-carbon granules from 1982 (second largest in the world), and an additional second line of production of high-purity nickel metal from 2001.

5.5.2.1.1 GEOLOGY

The deposit is hosted by a peridotitic protolith, exposed in the form of an elongated low-lying hill. The surface-weathering profile is significant and variable both vertically and laterally, forming 10 distinct lithostratigraphic units. The weathering profiles, from bottom to top, change from weakly serpentinized peridotitic protolith to saprolitized peridotite, green saprolite (main ore horizon), tachylite, black saprolite, yellow laterite, red laterite, and finally capped by a magnetic/nonmagnetic ferricrete. The thickness of the units is extremely variable, with the major horizons having maximum thicknesses of tens of meters. The area presents abundant microfault- and joint-related silicate veins in the form of stockworks in the lower part of the sequence.

5.5.2.1.2 RESERVES

Cerro Matoso represents one of the largest nickel reserves in Colombia having estimated ore reserves of 108 Mt of ore grading 0.57% Ni, contains 615,000 tonnes (~678,000 tons) of nickel metal. Cerro Matoso has an estimated reserve life of 42 years based on current production levels. Significant opportunities to expand the resource exist in the area.

The mine production is exclusively by low-cost open pit. The ore is transported to smelter to produce ferronickel and high-purity metal. It produces an average of 52,000 tonnes (~57,000 tons) of nickel/year, which places this mine in the second place of worldwide nickel producers. The Cerro Matoso ferronickel operation is rapidly becoming a world-class operation. After the current expansion program at the mine, Cerro Matoso will be one of the world's largest ferronickel producers.

5.6 VENEZUELA

The Federal Presidential Republic of Venezuela is a country on the northern coast of South America. Venezuela is a major producer and exporter of minerals, notably bauxite, coal, gold, nickel, and iron ore. The Loma de Niquel deposit is the sole nickel-producing mine in Venezuela, and it would produce almost 1% of the world's nickel output at full capacity. The state controls most of the country's vast mineral reserves.

5.6.1 Loma de Niquel Deposit

5.6.1.1 Location

The Loma de Niquel (Loma de Hierro) deposit is located in Miranda and Aragua States of Venezuela, about 80 km southwest of Caracas (Fig. 1.20). The infrastructures available close to the deposit are electric power, natural gas, paved roads, and within 200 km of the ocean port of Puerto Cabello.

5.6.1.2 Geology

The country rock is represented by sedimentary–volcanic terrain of Mesozoic age. The host rock is the Loma de Hierro ophiolite unit, composed of basalt, gabbro, peridotite, serpentines, and serpentinized harzburgite of Upper Jurassic to Lower Cretaceous age. The peridotite body is 21-km long, 0.70–5.0-km wide, and trends in the ENE–WSW direction. The host rock has undergone prominent weathering during the Tertiary period with concentration of nickel–cobalt mineralization. The nickeliferous minerals include kerolite, limonite, montmorillonite, serpentine, talc, and garnierite.

The Ni–Co-bearing Loma de Hierro lateritic deposit occurs as the blanket type on a hill top and at the gentle slope. The lateritic deposit covers a surface area of 6.2 sq. km. The length, width, and laterite thickness are 8 km, 1 km, and 10 m, respectively. The thickness of mineralization is continues up to 7 m from top. The laterite profile is described in Table 5.2.

5.6.1.3 *Reserve and Resources*

The ore reserves (Proved + Probable) at 0.80% Ni cutoff has been estimated at 32.40 million tonnes (~35.7 million tons) grading 1.43% Ni following the JORC standard. The resources stand at 15.60 million tonnes (~17.2 million tons) grading 1.52% Ni.

5.6.1.4 *Mining and Smelting*

The mining at Loma de Hierro project is by the conventional low-cost open-pit method with matching ferronickel smelting. Loma de Niquel was previously run by Anglo American and since 2012 has been managed by a government subsidiary, Petroleos de Venezuela exporting ferronickel to countries such as the United States, Belgium, India, South Korea, and China. The trend of nickel production from Venezuela stands as 13,200 tonnes (~14,550 tons) in 2009, 14,300 tonnes (~15,800 tons) in 2010, 14,000 tonnes (~15,400 tons) in 2011, 8100 tonnes (~8900 tons) in 2012, and 0 tonnes in 2013.

6.1 BACKGROUND

Asia is the largest, most populous continent with 49 countries that follow a multilingual system and coexist under multireligious faiths. The continent has the second largest nominal Gross Domestic Product next to Europe and the highest Purchasing Power Parity. The continent is endowed with rich mineral resources covering metallic and nonmetallic minerals, coal, natural gas, and petroleum.

6.2 AFGHANISTAN

Afghanistan is a landlocked country located within South and Central Asia. A new era in the study of geology and mineral resources began by financial and technical support from Germany, Italy, France, the former Soviet Union, and the United Nations (UN). The mineral resources identified include gold, copper, lithium, uranium, iron ore, cobalt, zinc and chromite.

Geologic mapping, sampling, and limited shallow diamond drilling identified many chromite deposits in the early 1990s located near Jurgati in Parwan Province, Werek in Logar Province, and Sperkay and Shandal in Paktia Province.

The Logar Valley chromite deposits are largest and located in Muhammed Agha district, 35 km south of Kabul. The host ophiolite complex outcrops as an elongated body over 65 km in a northwest–southeast direction, and up to 45-km wide. The external contacts of the ophiolite complex are mostly tectonic faults and thrusts. The ophiolite includes an ultramafic sequence ~2800-m thick. The basal part of ~2400 m is composed of dunite and harzburgite, overlain by a 200-m thick pyroxenite with intercalated dunite at its base. The ultramafic sequence continues up into a thin unit of troctolite, pyroxenite, and gabbronorite 50-m thick. The chromite bodies occur predominantly in harzburgite within small dunite pods.

The chromite deposits occur in two groups, ~10 km apart, on the eastern side of Logar Valley. The northern cluster is within 5 km to the northwest of Muhammed Agha, and the Southern cluster occurs ~10 km south of Muhammad Agha. Chromite lodes occur as massive lenses, pods, and irregular masses. The largest two lenses are dimensioned as 97×10 m, and 65×5 m. The grade varies between 31 and 37% Cr_2O_3, and 8.2 and 9.6% Fe. PGEs concentrations are low with maxima of 6.5 ppb Pt and 5.5 ppb Pd.

6.3 CHINA

The People's Republic of China is one of the fastest growing economies in the world. China has abundant nickel and cobalt resources. China is the largest nickel producer in Asia having more than 120 nickel–copper±PGE deposits/mines, with the Jinchuan group producing the maximum share.

6.3.1 Jinchuan Ultramafic Intrusion

6.3.1.1 Location and Discovery

Jinchuan is the single largest Ni–Cu–PGE-sulfide deposit in the world and stands third with a 515 Mt reserve base, after Noril'sk (1309 Mt) in Russia, and Sudbury (1648 Mt) in Canada. The deposit is located in Jinchuan District, Gansu Province, Northwest China. The Yongchang nickel deposit was discovered in 1958 at the foot of Longshou Mountain in middle Hexi Corridor, and renamed Jinchuan Nonferrous Metals Corporation in 1961. Jinchuan Group Limited was formally established in 2001 with development of a large mining–beneficiation–smelting–refining and industrial city complex.

6.3.1.2 Geology

The Jinchuan ultramafic intrusion of Proterozoic age occurs at the margin of the Sino–Korean massif. The magma was emplaced into late Archean metasedimentary rocks consisting of gneisses, chlorite–quartz schists, banded marbles, and migmatites. The intrusion extends over 6.3 km in an NW–SE-oriented lens, 20–525-m wide with more than 1100 m of vertical depth. Jinchuan intrusions dip steeply to the southwest caused by the extensive movement of a regional thrust fault. Phase-equilibrium analysis using whole-rock and mineral chemical data confirms that the parental magma originated from high-Mg basaltic composition, and the primary magma may have contained MgO up to 18.5 wt.% (Li et al., 2011). The intrusion consists of five ultramafic units that include lherzolite, dunite, plagioclase lherzolite, olivine websterite, and websterite in order of decreasing abundance. Dunite is usually sulfide bearing with 8–30% Ni–Cu sulfides (Naldrett, 2004). The sequence of ultramafic units is subdivided into Upper, Lower, and Transition Layers.

The Upper layer unit consists of dunite, lherzolite, and pyroxenite, and is largely free of mineralization or weakly disseminated sulfides at the base of the unit. The Lower layer unit consists of coarse-grained dunites (37–40 wt% MgO) and lherzolites (28–35 wt% MgO), and comprises the oxidized No. 24 orebody. Mineralization is primarily of disseminated, net-textured, and minor massive sulfides, forming the third largest orebody in Jinchuan. The disseminated ores have remarkably high-Cu/Pd ratios (24,200–85,600), and low-PGE tenors (Chen et al., 2013).

The central part of the intrusion is characterized by a concentric distribution of rock types with a core of sulfide-bearing dunite enveloped by lherzolite. The intrusions and the deep-seated magma chambers comprise a complicated magma plumbing system. Normal faults played a significant role in the formation of this magma plumbing system and provided pathways for magmas (Song et al., 2012). A simplified geologic map (Fig. 6.1) and generalized sequence of an ultramafic intrusion and its preferred concentration of mineralization at the Jinchuan Intrusion are given in Fig. 6.2.

The superlarge magmatic sulfide deposits occur in magma conduits as the open magma system provided a perfect environment for extensive concentration of immiscible sulfide melts along deep regional faults (Song et al., 2011). PGEs and S–Hf–Sr–Nd isotope data from ongoing lower-level mining indicate an increase of PGE tenors in bulk sulfide ores eastward, except for two fault-offset ore zones in the western segment. The two fault-offset ore zones show depletions in iridium-series platinum group of elements (**IPGE:** Ir, Ru, and Os) by a more evolved parental magma. The eastward increase of PGE content in the rest of the deposit is by the upgrading of preexisting

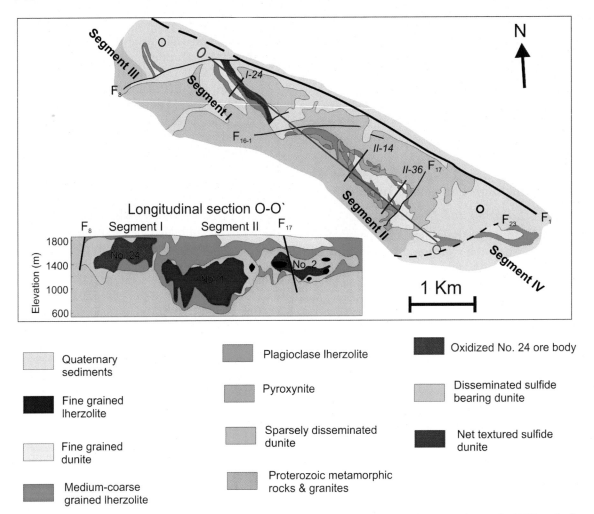

FIGURE 6.1 Geological map of Jinchuan ultramafic intrusion showing a superlarge single Ni–Cu–PGE orebody extending over 6 km.

After Song, X., Wang, Y., Chen, L., 2011. Magmatic Ni-Cu-(PGE) deposits in magma plumbing systems: Features, formation and exploration. Geoscience Frontiers 2(3), 375–384. Production and hosting by Elsevier.

sulfide liquid in a subhorizontal conduit by a new surge of magma moving through the conduit from west to east (Duan et al., 2015).

The deposit is divided into four segments by E to W- to NE to SW-trending strike-slip faults displacing the ultramafic body up to a km in depth. Each segment represents a mine block having approximate lateral extent, and are referred to as Mines III (~500 m), I (~1500 m), II (~3000 m), and IV (1100 m). Mines I and II

intrusions are exposed at the surface hosting rich mineralization in the dunite. Mines III and IV are covered by alluvium.

6.3.1.3 Mineralization

Mineralization is relatively rich in copper with average Ni/Cu ratio of 1.76. The average PGEs and gold concentrations are moderately high at 1 g/t in the sulfide ore (Naldrett, 2004). Sulfide-orebodies are situated in the middle and

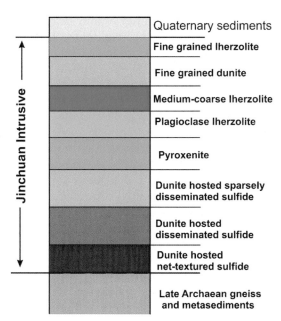

Quaternary sediments

Fine grained lherzolite

Fine grained dunite

Medium-coarse lherzolite

Plagioclase lherzolite

Pyroxenite

Dunite hosted sparsely disseminated sulfide

Dunite hosted disseminated sulfide

Dunite hosted net-textured sulfide

Late Archaean gneiss and metasediments

Jinchuan Intrusive

FIGURE 6.2 Generalized sequence of ultramafic intrusion at Jinchuan Complex showing preferred concentration of sulfide mineralization in dunite host rock at the basal layer. The individual unit discontinues along strike.

lower regions of the intrusion, with the major orebodies of Mines I and II accounting for more than 90% of the Ni–Cu reserves in the deposit. Ore minerals include pyrrhotite, pentlandite, chalcopyrite, cubanite, mackinawite, and pyrite. Sulfide minerals generally increase in abundance toward the base. Nickel sulfides are the dominant ores and occur in the form of disseminated to net-textured, with sulfide contents ranging between 1 and 40% by volume. The occurrence of massive sulfide ore is rare.

6.3.1.4 Reserve Base

Reserve base has been estimated at 515 Mt grading 1.18% Ni, 0.63% Cu, 0.019% Co, 0.13 g/t Pt, 0.10 g/t Pd, 0.005 g/t Rh, 0.010 g/t Ru, 0.010 g/t Ir, and 0.011 g/t Os, making a total PGE content of 0.26 g/t (Naldrett, 2004). The annual capacity of production includes 150,000 tonnes (~165,000 tons) of nickel, 600,000 tonnes (~661,000 tons) of copper, and 10,000 tonnes (~11,000 tons) of cobalt and PGE + gold.

6.3.2 Huangshandong Mafic–Ultramafic Intrusion

The Huangshandong mafic–ultramafic complex is the largest among six intrusions hosting significant Ni–Cu sulfide mineralization in the eastern Tianshan orogenic belt. The deposit is located at the southern margin of the Central Asian Orogenic Belt along the Huangshan–Kanguer fault in north Xinjiang, northwest China. Huangshandong intrusion extends over 3.5 km trending NNE–SSW, with width of 1.2 km at central bulge, and a vertical depth of ~1 km.

The regional rocks of the eastern Tianshan terrane consist of Precambrian crystalline basement and calc-alkaline volcanics, overlain by passive continental margin sedimentary rocks. The mafic–ultramafic complex (270 Ma) intrudes the Carboniferous Gandun Formation. The intrusion is composed of a massive unit at the base in the western part of the complex, and a weakly deformed layered unit in the east. The massive unit is composed of gabbronorite, inconsistently altered to serpentine, talc, tremolite, and actinolite. The layered unit is composed of troctolite, chrome-bearing lherzolite, olivine gabbro, hornblende gabbro, gabbro–diorite, and diorite (Fig. 6.3). the host intrusion consists of a massive gabbronorite unit and a layered sequence composed of two to three ultramafic layers overlain by gabbroic rocks with visible modal layering. Significant sulfide mineralization is associated with massive gabbro and the ultramafic rocks (Sun et al., 2013).

More than 15 large orebodies are hosted by the gabbronorite of the massive unit. These bodies are 200–500-m long, 30–200-m wide, and 8–9-m thick dipping south at steep angles of 50–70 degree. There are + five sulfide zones concentrated at the contact of layered units associated with the transitional zones. These sill-like lens-shaped orebodies are 1–2.5-km long, 200–1000-m wide, and 10–30-m thick. Ore minerals include pyrrhotite, pentlandite, and chalcopyrite forming disseminated to

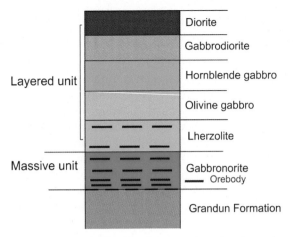

FIGURE 6.3 Geological success of mafic–ultramafic intrusion with preferential concentration of mineralization predominantly in basal massive gabbronorite and upper and lower contacts of layered-lherzolite unit, at the Huangshandong Intrusion, northwestern China.

net-textured or massive to semimassive sulfides. The oxide-free sulfide mineralization has low-PGE concentrations. The oxide-rich sulfide mineralization contains 1–5% magmatic magnetite with exsolution lamellae of ilmenite and spinel. The oxide-rich sulfide mineralization has high-PGE concentrations. Gao et al. (2013) proposed that the host intrusion and associated sulfide mineralization are derived from high-Mg basaltic magmas from a mantle source that was previously modified by a subducted oceanic slab. Sulfide saturation is activated by crustal contamination and sulfide segregation from PGE-depleted magma produced by early sulfide removal at depth (Deng et al., 2014).

Reserve base has been estimated at ~135 Mt of sulfide ores grading 0.52 wt% Ni and 0.27 wt% Cu. The total PGE content in Huangshandong rocks and sulfide ores is low ranging between 0.25 and 99 ppb.

6.3.3 Poyi Ultramafic Intrusion

Poyi ultramafic intrusion is located in western part of Beishan Rift forming part of Pobei complex (278 Ma) located within Beishan fold belt in the Tarim Craton. The fold belt belongs to Central Asian mega orogenic terrain hosting significant economic Ni–Cu sulfide deposits. Regional geology is represented by Archean–Paleoproterozoic basement of granite, quartz diorite, gneisses, and schists, and the Mesoproterozoic–Early Neoproterozoic successions of Tarim Craton. Poyi intrusion is a pipe-like body, extends over 2.2 km in an NNE–SSW direction, average 0.6-km wide and a down-depth continuity of +1.6 km. The Poyi ultramafic complex is composed of dunite, lherzolite, and olivine websterite cumulates in the core and an outer layer of olivine gabbro and gabbro. Sulfide mineralization is dominated by concordant layers/zones/lenses of disseminated sulfides with thicknesses varying between 10 and 80 m within steeply dipping layers of lherzolite. The significantly higher contents of PGEs in sulfide separates versus whole rocks from the Poyi intrusion indicate that immiscible sulfide liquids were the primary collectors of PGEs (Xia et al., 2013).

6.3.4 Kalatongke Ni–Cu–PGE Sulfide Deposit

Kalatongke (Karatungk) mafic–ultramafic intrusion is part of the Central Asian Orogenic Belt and located in the Eastern Junggar terrane, Northern Xinjiang, northwest China. The deposit was discovered in the 1970s.

A number of small, elongated mafic–ultramafic intrusions have been emplaced into the regional geological setting of Lower Carboniferous volcanic tuffs and shales in the Kalatongke area. The mafic–ultramafic intrusions of the Paleozoic Era (283 ± 3 Ma) are composed of diabase, norite, gabbro, and diorite. The economic Ni–Cu–PGE mineralization is hosted by three of the lens-shaped intrusions (Y1, Y2, and Y3) that extend over 5 km in the NNW–SSE direction with a gap inbetween. The orebodies continue up to 600 m in vertical depth.

Y1 orebody, exposed to the surface at the NW end, is 700-m long, up to 300-m wide at the center and dips toward the NE. The Y1 intrusion is primarily composed of a sequence from base to top of diabase, norite and diorite. The unexposed Y2 and Y3 intrusions lie to the southeast of Y1 intrusion and occur 150 m below the surface. The Y2 orebody, with a gap of ~500 m to the SE, is ~2-km long and averages 500-m wide. The Y2 intrusion is composed of norite to diorite from the base to the top. The Y3 orebody, at the SE end, is 1.5-km long and over 500-m wide. The Y3 intrusion is dominated by diorite host rock with a small proportion of norite at the base. There exist a few small and parallel intrusions in the northeast of these three (Y1, Y2, and Y3) mineralized intrusions.

Mineralization is characterized by concentration of high-sulfide zones within the Y1 intrusion and exclusively hosted by the norite unit at the base. The mineralization occurs as massive sulfides in the inner core zone, enclosed by an intermediate densely disseminated sulfides zone, and fenced by a sparsely disseminated sulfides in the outer zone. The mineralization in Y2 and Y3 intrusions are relatively poor and occurs as layered disseminated sulfides settled at the basal norite zone. The three intrusions are connected at depth and belong to a magma conduit with the root located in the vicinity of Y1.

The deposit has been in systematic exploration since 1970. The mineral reserves and resources of the Kalatongke Ni–Cu–PGE-sulfide deposit has been estimated at 33 Mt sulfide ore grading 0.8 wt% Ni and 1.3 wt% Cu. The PGE contents of the disseminated ores (14–69 ppb Pt and 78–162 ppb Pd) are lower than those of the massive ores (120–505 ppb Pt and 30–827 ppb Pd) (Song et al., 2009).

6.3.5 Jinbaoshan Ultramafic Intrusion

The Jinbaoshan mafic–ultramafic Intrusion is located in Midu County, Yunnan Province, at the western margin of the Yangtze block, southwest China, and known for hosting the largest sulfide-poor magmatic Pt–Pd deposit in China. The mafic–ultramafic bodies are associated with late Permian Emeishan continental flood basalts. The Jinbaoshan intrusion is a sill/sheet-like subhorizontal body, about 5-km long, ~1-km wide, and 25–175-m in vertical thickness. Host rock is exclusively wehrlite and minor gabbro, hornblendite, and pyroxenite. The mafic rocks are formed by an early intrusive phase. The ultramafic rocks are composed of a thick wehrlite unit in the core and thin pyroxenite units at the margins (Fig. 6.4). The wehrlites represent residual assemblages formed by passing magma in a dynamic conduit (Tao et al., 2007). PGE minerals include sperrylite, moncheite, rustenburgite, and tetraferroplatinum.

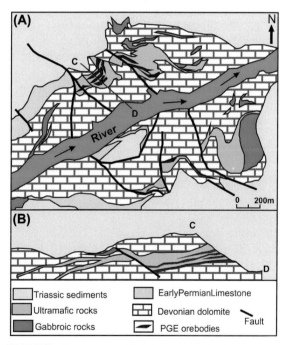

FIGURE 6.4 (A) Surface geologic map and (B) cross section showing the sill-like mafic–ultramafic intrusion with layers of PGE orebodies at Jinbaoshan. *After Song, X., Wang, Y., Chen, L., 2011. Magmatic Ni-Cu-(PGE) deposits in magma plumbing systems: features, formation and exploration. Geoscience Frontiers 2 (3), 375–384. Production and hosting by Elsevier.*

6.3.6 Sichuan Layered Intrusions

Mafic–ultramafic layered intrusions of the Sichuan Province are located in the Pan–Xi rift and Danba areas of southwest China. The widespread magmatic intrusions are emplaced into the western margin of the Yangtze Platform composed of late-Archean to late-Proterozoic basement and stable sedimentary sequences. The Yangtze Platform was reactivated by the Hercynian–Indosinian orogeny and magmatism between 350 and 250 Ma. The mafic–ultramafic intrusions host a number of economic to subeconomic deposits containing chalcopyrite, pyrrhotite, and pyrite ± PGE. The individual deposits are:

1. Panzhihua V–Ti–Fe–PGE deposits: The PGE-bearing intrusion is located in the southernmost part of the Pan–Xi rift zone at the Panzhihua community. The deposit is 19-km long in NE, 2-km wide, and +2 km of thickness, composed of a Marginal hornblende/olivine gabbro zone, Lower and Middle zones of rhythmic layered melagabbro, and Upper zones of leucogabbro, peridotite, dunite, and anorthosite at the top. A superlarge Fe–Ti–V oxide + PGE deposit with 1333 Mt reserves (Zhou et al., 2005), the deposit supports a large open-pit mine and is a major producer of these metals.

2. Xinjie V–Ti–Fe–PGE deposits: The Xinjie V–Ti–Fe–PGE deposits are located ~50 km northeast of Panzhihua and 5 km north of Miyi. It extends over 7 km in the NW–SE direction, is 1-km wide, and composed of three cumulate cycles of peridotite–pyroxenite, Fe-gabbro–pyroxenite–peridotite, and pyroxenite–quartz diorite. The primary ore minerals are disseminated pyrrhotite, millerite, neckelite, chalcopyrite, and PGEs at the base and characterized by magmatically layered Cu–Ni–PGE mineralization.

3. Lufangqing PGE-bearing mafic Intrusion.

The fault-related hydrothermally altered Cu–Ni–PGE mineralization is located at Lufangqing village, 2 km south/southern extension of the Panzhihua V–Ti–Fe–PGE deposits. The deposit extends over 1.5-km along N–S strike and 1-km wide. The mineralized mafic–ultramafic complex intruded into the Proterozoic amphibolite schist, metagabbro, and marble in the east, and Triassic conglomerate, sandstone, and shale in the west. The intrusive complex is composed of peridotite, gabbro, and pyroxenite. Two Cu–Ni–PGE mineralized zones have been outlined at the eastern and western contacts between mafic–ultramafic intrusion and the Precambrian marble. The ore minerals are chalcopyrite, bornite, pyrrhotite, and pyrite. The orebody in the eastern zone extends over 1500 m and is 2–9-m wide. The trench-sample grades vary between 0.55 and 0.90% Cu, <1.73% Ni, and between 0.1 and 1.73 g/t PGEs. The western mineralized zone is relatively small at 50-m long and 2–4-m wide (Yao et al., 2001).

4. Badong PGE-bearing Intrusion: The mafic–ultramafic intrusion is located at ~5 km south of Dechang County in the northern part of the Pan–Xi metallogenic rift belt. The mafic–ultramafic body intruded into Proterozoic metamorphic rocks of the Kangding Group and comprises peridotite, gabbro, and pyroxenite hosting fault-related Ni–Cu–PGE mineralization.

5. Dayanzi PGE-bearing Intrusion: The mafic–ultramafic intrusion is located at ~3 km north of Huili County in the south of the Pan–Xi metallogenic belt. The Cu–Ni–PGE mineralized zone at the Qinshuihe intrusion is 2-km long and 5–64-m wide, with mineralized zones of 1200-m long and 1–9-m wide grading 0.29–0.44% Cu, 0.25–0.60% Ni, and 0.43–1.71 g/t Pt + Pd (Yao et al., 2001).

6. Yangliuping Cu–Ni–PGE deposit: The deposit is located 30 km south of Danba. Song et al. (2003), proposed that fractional

crystallization of the Yangliuping magma accompanied by the introduction of S and CO_2 from the wall rocks caused the magma to become S-saturated leading to the segregation of magmatic sulfides enriched in Ni–Cu–Co–PGE mineralization. The mafic–ultramafic sills acted as conduits.

7. Baimazhai Ni–Cu–PGE: The intrusion is located within "Emeishan Large Igneous Province" in the Jinping region. Mineralization is massive, net-textured, and disseminated. Ore minerals are magmatic pyrrhotite (85%), with subordinate amounts of pentlandite (10%), and chalcopyrite (5%). The massive ores contain high Ni (1.6–4.2 wt%), Cu (0.4–6.5 wt%), and total low PGE (ΣPGE) contents (85–524 ppb) (Wang and Zhou, 2006).

8. Hongqiling Intrusion: The host rocks of the mafic–ultramafic complex include pyroxenite, olivine websterite, lherzolite, gabbro, and leucogabbro (216 ± 5 Ma). The Cu–Ni magmatic sulfides are predominantly pyrrhotite, pentlandite, and chalcopyrite.

9. Giant Xiarihamu Ni–Co deposit: The deposit is located in East Kunlun orogenic belt, Northern Tibet. The orebody has been explored by close-spaced diamond drilling. The reserves and resources have been estimated at ~157 Mt) sulfide ores with average grades of 0.65% Ni, 0.14% Cu, and 0.013% Co (Song et al., 2016). The PGE content is extremely low at ~200 ppb.

6.4 INDIA

India is a vast South Asian Country with diverse terrain, and the third largest economy by Purchasing Power Parity (PPP) in the world. India is a pluralistic, multilingual, and multiethnic society. India has become one of the fastest-growing major economies and a newly industrialized country. With an open mineral policy, Mines and Minerals encourages 100% Foreign Direct Investment in the Indian economy. India is the second-largest producer of chromium (~4.0 Mt ore). Indian chromite reserves stand at 54 Mt of reserve and 149 Mt of resource totaling 203 Mt (Indian Bureau of Mines, 2013). The single lateritic-nickel ore associated with Sukinda could not make an economic breakthrough due to metallurgical problems. The possibility of PGE reserves at Nausahi and Sukinda mafic/ultramafic complexes is becoming significant. Future resources are expected from Sitampundi, Namakkal district of Tamil Nadu, Hanumalapura block, Devangere district, Karnataka, and many ophiolite rocks. The geographic distribution of chromium, nickel, and PGE deposits (see Fig. 6.5) with significant characteristic features is tabulated at Table 6.1. The individual deposits will be discussed in the same order.

6.4.1 Andhra Pradesh

Chromite deposits in Andhra Pradesh are confined to the Precambrian Eastern Ghats Granulite Belt and Southern Granulite terrain. The deposits are stratiform and of small size. The in situ resources of chromite have been estimated at 187,000 tonnes (~206,000 tons). No production has occurred since 1991–92, except a small quantity from Kondapalle.

The Kondapalle deposit is located near Kondapalle village, 20 km from Vijayawada town, East Godavari district. Chromite-bearing ultramafic intrusions occur as sheet-like bodies along the foliation planes of the country rocks in ~100 sq. km areas in the Kondapalle area. The host ultramafic zone is composed of serpentinite, talc–tremolite schist, enstatite chromitite, pyroxenite, websterite, and dunite. The two orebodies, 400 m apart, occur as N–S trending en echelon lenses and pods aligned parallel to regional trend of charrnockite (Fig. 6.6). These lenses are concentrated in steeply dipping pyroxenite and norite bands within charrnockite. Two types of mineralization are recognized: One formed earlier than olivine, and the other

FIGURE 6.5   Map of India showing existing chromite deposits, and potential areas for PGE exploration targets.

type (the main orebody) is contemporaneous with primary ultramafic rocks. The ore minerals are chromite, pyrrhotite, pentlandite, magnetite, hematite, and sphalerite. The chromite orebodies are small in size, and erratic in distribution with grades of 38–50% Cr_2O_3.

The Lingampet chromite bodies are interlayered with talc–tremolite–chlorite schist derived from layered ultramafic–mafic intrusive parent magmatic rocks within a granitoid and gneissic complex. The mafic–ultramafic layers strike NW–SE with steep (70–80 degree) NE dip, and comprise dunite, pyroxenite, websterite, anorthosite, norite, and gabbro. Chromite occurs as podiform lenses within the layered ultramafic sequence with length varying between 10 and 80 m and width between 2 and 13 m. The ore reserve has been estimated at 160,000 tonnes (~176,000 tons) at 27–39% Cr_2O_3.

The Jannaram chromite deposit is located in the Khammam district. The mafic–ultramafic intrusive bodies are emplaced into Archean granitoid and gneisses of the Peninsular Gneissic Complex. The host rocks are composed of anorthosite, gabbro/leucogabbro, pyroxenite, websterite, and dunite. The area is soil covered with in situ podiform-chromite lenses in layered ultramafic units of dunite, pyroxenite, and websterite. Size of the chromite lenses varies between a few centimeters and up to 12.5 × 2 m across. The reserve has been estimated at 600 tonnes (~660 tons) of in situ ore, and ~500 tonnes (~550 tons) of float ore making a

TABLE 6.1   Distribution of Chromite–Nickel–PGE Deposits in India

State	District	Deposits	Characteristic Features
Andhra Pradesh	East Godavari, Krishna, Khammam	Kondapalle Dendukuru Gangineni Lingampet Jannaram	Precambrian Eastern Ghats granulite belt and southern granulite terrain. Disseminated ores N–S trending en échelon lenses.
Jharkhand	Singhbhum	Jojuhatu and Roro Dalma mafic volcanics[a]	Archean supracrustal greenstone belts. Occurs in small pockets, lenses and bands.
Karnataka	Hassan Davangere Mysore	Nuggihalli schist belt[a] Hanumalapur complex[a] Sindhuvalli belt	Archean supracrustal greenstone belts. N–S trending band and lenses associated with ultramafics within Dharwad craton/schist.
Maharashtra	Bhandara Sindhudurg	Pauni, Kankauli,[a] Kankauli[a] Sateli[a]	Archean supracrustal greenstone belts. Occurs in altered ultramafic rocks intruded along synclinal folds in Sakoli metasediments.
Orissa	Cuttack & Dhenkanal Keonjhar	Sukinda[b] Katpal[a] Boula–Nausahi[a]	Archean supracrustal greenstone belts. Cr–Ni as crystalline, stratiform, and lenses in folded limonitic ultramafic rocks along deep marginal fracture. Bands and lenses within the serpentine, peridotite, pyroxenite, gabbro, vanadiferous magnetite anorthosite, resembling well-known stratiform complexes of the world.
Tamil Nadu	Namakkal Salem	Sittampundi complex[a]	Lens-shape bodies in anorthosite within charnokite country rock.
Jammu and Kashmir	Kargil	Dras	Tertiary ophiolite sequence in orogenic belt. Associated with serpentine and dunite within Cretaceous Dras volcanics.
Manipur Nagaland	Ukhrul Manipur	Sirohi peak, Nepal basti.	Tertiary ophiolite sequence in orogenic belt. Sporadic occurrences in peridotite and serpentinite.
Andaman Nicobar island	South Andaman Middle Andaman, North Andaman.	Chiriatapu–Bedanabad, Rutland–"Jones" point hillock, Kalighat	Ophiolite suit comprising altered dunite, harzburgite, lherzolite, and peridotite assemblage hosting small pods and layers of rich chromite.

[a] PGM environment.
[b] Nickel-bearing limonite.

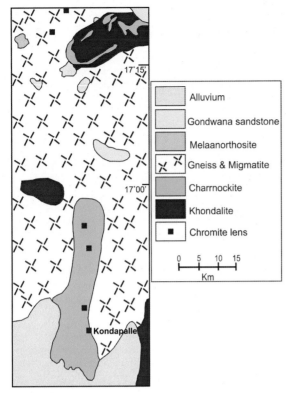

FIGURE 6.6 Geological map of Kondapalle group of chromite deposits in charrnockite rocks, Andhra Pradesh, India.

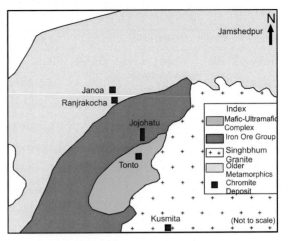

FIGURE 6.7 Map of Archean supracrustal greenstone belt hosting Jojohatu–Roroburu chromite belt emplaced into the Iron Ore Group (IOG) of rocks, Jharkhand, India.

total of 1100 tonnes (~1210 tons) grading 39.26% Cr_2O_3, 13.53% SiO_2, 17.87% FeO, and 10% Al_2O_3 [Indian Bureau of Mines (IBM, 2013)].

6.4.2 Jharkhand

Jharkhand is known for its rich resources of iron ore, bauxite, copper, uranium, and chromite with possibilities of locating PGEs in vast mafic–ultramafic intrusions. Chromite mining of small deposits was well known during the early 1960s at Paschim and Sighbhum districts (Fig. 6.7). The deposits are small, scattered, and grade is usually medium at 30–40% Cr_2O_3.

6.4.2.1 Jojohatu Chromite

The Jojohatu chromite deposits are located about 25 km west of Chaibasa town and

Rail-Station on the South Eastern Railway. Chaibasa is connected to Tatanagar Steel City at a distance of 80 km by rail and road. The Jojohatu ultramafic body intruded into the Archean Iron Ore Group (IOG) in two patches with a cumulative length of about 8 km in the north–south direction and width of ~3 km. IOG comprises shale, slate, phyllite, quartzite, dolomitic limestone, and altered basic lava. The ultramafic intrusion is composed of dunite, peridotite, and pyroxenite, largely serpentinized and chromiferous. Laterization of ultramafic rocks is a common feature.

This ultramafic belt represents, from north to south, three hills or potential blocks, namely, Kimsiburu, Ramla–Kittaburu, and Roroburu-Chitungbure. The mineralization primarily occurs as the banded type 3–50-m long and 10–15-cm wide in serpentinized pyroxenites, saxonites, and dunites near the margin of the ultramafic units. The chromite deposits are confined to an overfolded shear zone and occur as segregations, lenticular masses, and pods with an average thickness of ~22 cm. Chromite/chromitite is usually interbanded with serpentinite. A group of chromiferous/steatite-bearing ultramafics occurring within and north of the shear zone in

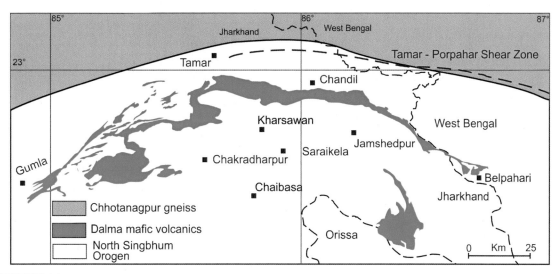

FIGURE 6.8   Geological map of Dalma mafic volcanics—a future exploration target for magmatic Ni–Cu–PGE mineralization, Singhbhum Craton, India.

the Rakha Mines–Chapri area, as well as Ranjra-kocha–Janoa and Bichaburu–Kusumita areas in the Singhbhum district has been reported.

6.4.2.2 *Dalma Mafic Volcanics*

The Singhbhum craton is bestowed with a mineral storehouse that includes large and rich iron ore, copper–gold–nickel, chromium, and uranium deposits producing ore over many decades. A large mafic–ultramafic–felsic pluton and volcanic arcuate package that extends parallel to the Singhbhum Shear Zone situated a few km north of the copper–uranium belt has been demarcated. The Dalma volcanic package starts from Belpahari in the east (West Bengal) to Gumla in the west (Jharkhand) passing through Jamshedpur, Chandil, Chaibasa, Kanaraon, Tamar, and Ranchi (Fig. 6.8). The belt extends over 250 km E–W and has a maximum thickness up to 10 km. The exploratory drilling by the Geological Survey of India at the Belpahari block in the extreme east of the belt in West Bengal crossed gossan overburden, and intersected iron-rich low-sulfide metal grade (1978–80).

Dalma mafic volcanics and felsic plutonic assemblages are composed of mafic–ultramafic

tuffs, basalts, quartzite–conglomerate, high-Mg-lavas, pyroclastic rocks, and granite. The magmatic assemblage (~2800 Ma) was emplaced in the North Singhbhum Orogeny. IOG basalts from Singhbhum craton show low- to moderate-PGE contents marked by 26.23–68.35 ppb of ΣPGE, whereas the Malangtoli basalts display a moderate to high concentration of PGEs (ΣPGE = 43.01–190.43 ppb). The high-MgO Malangtoli basalts exhibit undersaturated sulfide, undepleted PGEs, and are devoid of crustal contamination. The low-MgO Malangtoli basalts are sulfide saturated, PGE depleted and crustally contaminated (Singh et al., 2015).

6.4.3 Karnataka

Karnataka possesses a number of economic chromite deposits mainly in the "Nuggihalli Schist Belt" in the Hasan District. The other chromite deposits in the Mysore district are mostly exhausted or abandoned. Chromite occurrences are found in Chigmgalur and Shimoga districts. The entire chromite-bearing ultramafic/schist belt passing from north to south from Uttar Kannad to Shimoga, Chgmagalur, Chikmagalur,

Hassan, Tumkur, and Mysore districts, is a potential target area for PGE search.

6.4.3.1 Nuggihalli Schist Belt

The Nuggihalli Schist Belt is a narrow (1–2 km) arcuate belt that extends over 60 km from Arsikere town in the north to Kempinkote in the south in the Hassan district, western Dharwad Craton. The east-dipping belt is characterized by interlayered ultramafic–mafic rocks with granitic gneiss in the west and tonalitic gneiss in the east. Rhythmically layered chromite lenses and reefs occur in serpentinized peridotites all through the Nuggihalli schist belt (Fig. 6.9). The rocks are amphibolite-facies derivations of dunite and peridotite with chromite seams, anorthosite, gabbroic anorthosite, and gabbro with titaniferous magnetite bands.

The important producing chromite mines are Belgumba, Mallapura, Gobalihalli, Byrapura, Bhaktarhalli, Tagadur, and Jambur (Fig. 6.10). The chromite mines from Karnataka produced about 1.60 Mt until 2014–15 in the raw state mainly for export purposes. The mining right of the entire Nuggihalli chromite belt is with Mysore Minerals Limited having current chromite reserves of 1.2 Mt of high grade.

The Bhaktarhalli mine is situated close to Byrapur village. The chromite occurs in the form of pods and lenses within altered serpentinite. A large quantity of float/in situ ore has fully been exploited (Fig. 6.11) and abandoned. The grade ranges between 26% and 37% Cr_2O_3. The property is under the Mining Lease of Mysore Minerals Limited.

The Byrapura deposit or mine block is located 15 km from Nuggihalli village, and 7 km west of nonmetal road, Chennarayapatna Taluk, Hassan district. Mineralization occurs as pod, massive sac form, and fissure types and is exposed in open-pit mine benches (Fig. 6.12). The individual sacs vary in size from very small pockets to large-size orebodies yielding 20 to 30,000 tonnes (~33,000 tons) of ore. The main orebody strikes E–W across the general trend of schistosity.

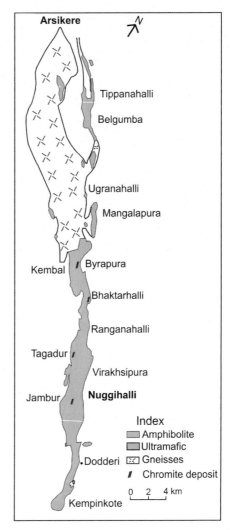

FIGURE 6.9 Geologic map of Nuggihalli Schist Belt extending over 60 km and 1–2 km wide. Mafic–ultramafic complex hosts a number of layered small chromite deposits with possibility of PGE potential, Karnataka, India.

Contacts between the orebody and the enclosing serpentinized peridotite are intensely sheared indicating differential movement. The rose red variety of chromite, Käemmererite [$Mg_5(AlCr)_2Si_3O_{10}(OH)_8$)], is a common associate mineral. A prominent band of anorthositic gabbro with conformable layers of titaniferous magnetite was mapped at the eastern margin

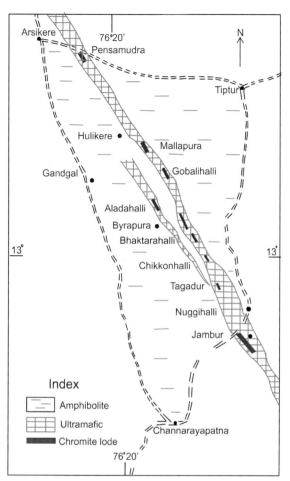

FIGURE 6.10 Elaborated geologic map of Nuggihalli Schist belt showing the active chromite mine/deposit/occurrences under the mining leasehold of Mysore Minerals Limited.

FIGURE 6.11 The Bhaktarhalli chromite–magnesite deposit has fully been exploited by open-pit mine and then abandoned.

FIGURE 6.12 Chromite mineralization occurs as pod, massive sac-form, and fissure type, and exposed in bench of open pit-A, at Byrapur mine. The Byrapur group of mines is the main producer of rich chromite ore (44–48% Cr_2O_3) from Karnataka State.

(Fig. 6.13). Massive bodies of dunite with fresh unaltered olivine are observed at depth. Texturally, the chromite of Byrapur has been classified as massive, layered, disseminated, and altered ores.

6.4.3.2 Hanumalapur Ultramafic Complex

The unusually thick-layered mafic–ultramafic rocks at the southern edge of the Shimoga Schist Belt of western Dharwad Craton were recognized during regional mapping. Smeeth and Iyenger discovered titaniferrous magnetite deposits in the Hanumalapur area in 1916. Mafic–ultramafic intrusives were emplaced in Late Archean Dharwad Super-group (3000–2500 Ma). The mafic–ultramafic rocks of the Hegdale Gudda Formation extends discontinuously over 40 km in the NW–SE direction with an average width of 400 m in the Hanumalapur, Tararekere, Masanikera, Magyanhalli,

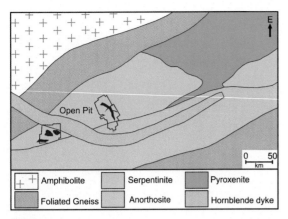

+ + Amphibolite	Serpentinite	Pyroxenite
Foliated Gneiss	Anorthosite	Hornblende dyke

FIGURE 6.13 Surface-geology map of the Byrapura deposit showing two open pits. The current mine production is extending underground sand-fill Cut-and-Fill method from ultimate open-pit bottom.

Magyathahalli, Gangodahalli, Bidaganahalli, Mugalihalli, Hosahallihalu, and Ubrani blocks, Dvangere district (Fig. 6.14). These mafic–ultramafic blocks are enriched with V–Ti–Cu–Cr mineralization.

The area became prominent during the early 1990s with the potential discovery of PGE mineralization in the Hanumalapur mafic–ultramafic offshoot of the Hegdale Gudda Formation by Alapieti et al. (1994, 2002, 2008). The Hanumalapur intrusive extends over 3.5 km, 300-m wide with moderate to steep dip in the east. The central ultramafic constituents are predominantly a complex mix of chlorite, amphibole, iron-rich chromite–ilmenite–Ti-magnetite rock. The fine-grained metamorphosed ultramafic is schistose in nature. PGEs (Pt and Pd) are mainly localized in the fine-grained iron-rich chromite-bearing central ultramafic and part of the eastern chromiferous magnetite band (Alapieti et al., 2008; Devaraju, 2009).

Surface-trench samples showed 0.07–5 g/t PGEs. The systematic exploration is continuing by the Geological Survey of India and eight boreholes have been completed. The borehole-intersected values are ranging between 0.50 and 2 g/t PGEs to a vertical depth of 175 m. An initial resource of 0.294 Mt at 1.79 g/t platinum

group of metals (PGMs) over an average width of 1.43 m was reported. GSI has estimated 0.84 million tonnes (~0.93 million tons) of resources of PGE ore with an average grade of 1.79 g/t at the Hanumalapur area alone.

6.4.3.3 Sindhuvalli Belt

The Sindhuvalli deposit is located at 18 km south on Mysore–Nanjangud road and about 6 km from Sindhuvalli village. Chromite-bearing ultramafic rocks, a distinct but conformable unit within the surrounding basement gneisses, are traceable for nearly 22-km long with an average width of 200 m of the Sargur Group. The mafic–ultramafic complex is layered with compositional variation through harzburgite–dunite, bronzite peridotite to pyroxenite and show effects of subsequent deformation. These are associated with gabbroic rocks that have later been metamorphosed. The layered complex trends N10°W–S10°E to N10°E–S10°W with very steep to vertical dip on either side.

The chromite mineralization occurs either concentrated in layers or as disseminated crystals in dunite and harzburgite. Rhythmic chromite layers vary in thickness between 0.18 and 0.75 m, alternating with thicker serpentinized olivine rock. The mineralization contact is consistently sharp at base and gradational top. A long stretch of narrow (1–10-m wide) subvertical orebody extends over 600 m in an N–S trend. The deposit was mined out by open-pit/underground methods to a depth of 150 m. The mine has produced 136,000 tonnes (~150,000 tons) of high-grade ore containing 48–50% Cr_2O_3 over the entire active period and finally closed (Fig. 6.15). The other deposits are Talur and Dodkatur with grades 45 and 48% Cr_2O_3, respectively.

6.4.4 Maharashtra

Three important chromite-producing belts are found in Maharashtra, namely, Bhandara–Nagpur, Chandrapur, and Sindhudurg. The chromite mineralization occurs as disseminations,

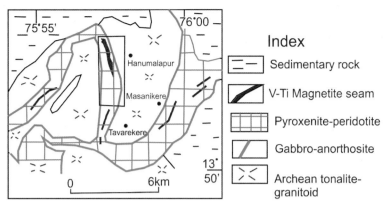

FIGURE 6.14 Geologic map of Hanumalapur layered complex showing V–Ti-magnetite ± PGE mineralization. The area is identified for future exploration target for platinum group of metals.

FIGURE 6.15 The Sindhuvalli deposit was mined by open-pit/underground methods to a depth of 150m, produced 136,000 tonnes (~150,000 tons) of high-grade chromite ore, and finally abandoned.

stringers, veins, segregated pockets, and lenses within altered ultramafic layered complexes. The overall grade varies between 30 and 42% Cr_2O_3.

6.4.4.1 Pauni–Taka Chromite

The Pauni chromite deposit is located within Bhandara–Nagpur ultramafic belt, 2 km NW of Pauni town. The chromiferous ultramafic belt extends over 40 km in an NE–SW trend between Pauni in the east and Kharsingi in

the west, Chandrapur district. The altered ultramafic rocks (dunite and steatitized serpentinite) were emplaced as interstratified layers along synclinal troughs in a tightly folded sequence of the Sakoli Series metasediments. Orebodies show primary magmatic differentiation and occur as concordant pods, lenses, thin layers, stringers, and in disseminated form. The orebodies and associated ultramafic rocks have been folded, faulted, and separated during regional deformation and subsequent granite intrusion. The indicated reserves have been estimated at 487,000 tonnes (~537,000 tons) to a vertical depth of 15 m, and the Cr_2O_3 content in the crude ore ranges between 34 and 52%. The low-grade is elevated to 52% Cr_2O_3 after beneficiation.

The Taka chromite deposit is located near Bhiwapur, Nagpur district. The mineralization is hosted by the Bhandara–Nagpur Ultramafic Belt. Exploration identified six disconnected pods and lenses of chromite lodes comprising several bands. The chromiferous ultramafic zone extends from Chincholi hill in the west to Dhamangaon– Pitechuva village in the east, extending ~5.5 km with 1.4 km width. Total in situ resources has been estimated at 55,580 tonnes (~61,300 tons) [38,900 tonnes (~42,900 tons) recoverable] at 23.50–35.28% Cr_2O_3.

6.4.4.2 Kankavli–Sateli

The chromite deposits of the Sindhudurg ultramafic belt occur in two places—one cluster of three deposits, namely Kankavli–Janoli–Vagda with a large reserve within a distance of 2 km around Kankavli town and the other small orebody at Gosaviwadi. Janoli block is at the NW extension of Kankavli orebody. The Vagda block is situated 1.6 km south of the town.

The area is represented by Archean Dharwad Supergroup comprising biotite schist, banded hematite quartzite, and granite gneisses. The ultramafic rocks associated with the chromite orebodies are intruded into older rocks. The older sequence is unconformably overlain by Kaladgi sediments of Precambrian age and in turn covered by Deccan Traps basalt. The region is extensively covered by laterite and lateritic soil (Fig. 6.16). All chromite lodes are enclosed in ultramafic rocks of dunite, serpentinite schist, talc schist, and tremolite schist. The chromite lodes occur in a few discontinuous outcrops spread over a length of ~525 m. The total resources is expected to be ~146,000 t at +31% Cr_2O_3, 22–28% FeO, and 12% Al_2O_3. Sateli is located at the southern contact with Goa. The layered mafic–ultramafic intrusive rocks within Archean basement can have potential for a PGM-hosting environment.

6.4.4.3 Deccan Traps Basalt

The Deccan Traps flow basalt (65 Ma) is one of the largest volcanic features on Earth, and crops out over 500,000 sq. km of the west-central Indian subcontinent. The trap complex is predominantly composed of multiple layers of tholeiitic flood basalt. The thickness varies from more than 2000 m in the Western Ghats to over 1000 m in eastern part of the province to less than 100 m in some southeastern regions. The basalts progressively overlap the basement from north to south. Most flows are 10–50 m thick, and dip at <0.5°.

PGEs have been analyzed from mafic igneous rocks of the Kutch region, Gujarat, supported by

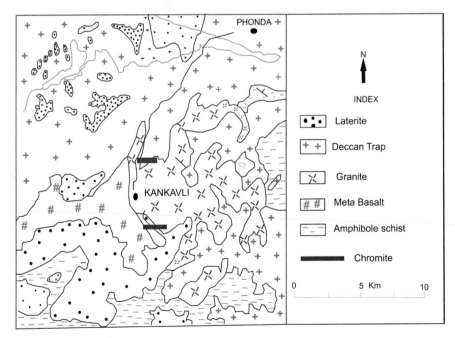

FIGURE 6.16   Surface geologic map of Kankavli group of chromite deposits.

association of Cu, Ni, Cr, and S together with Nd and Sr isotopic compositions. The PGE content is compared to the Deccan basalts of the Western Ghats region that indicated mantle-normalized Ni–PGE–Cu plots, PGE–base-metal ratio plots, and Nd–Sr initial-ratio diagrams (Crocket et al., 2007).

Deccan Trap basalts are comparable with the Siberian Trap at Noril'sk as both igneous provinces are widely considered the product of a mantle-plume event with the presence of large crustal contamination. The Noril'sk Ni–Cu–PGE-sulfide deposits are contemporaneous with the Siberian Trap flood-basalt magmatic event (Keays and Lightfoot, 2010). The S saturation to form magmatic Ni–Cu–PGE-sulfide systems is boosted by volumetrically important crustal contamination. The contaminated southern Deccan Trap lavas did not achieve S saturation, imposing constraints on the potential of the Deccan Trap in southern India to host significant magmatic-sulfide deposits.

The mafic–ultramafic rocks within Deccan Trap in parts of the Thane, Raigad, Satara, Sindhudurg, and Kolhapur districts, Maharashtra, were investigated and values up to 6 ppb Pt and 155 ppb of Pd were detected. Two hundred and fifty core samples covering a width of 2.95 m indicated 13–237 ppb Pt, 81–165 ppb Pd, and 0.1–1.3 ppb Ir. Samples were studied under scanning electron microscopy with energy dispersive X-ray spectroscopy (SEM-EDX-ray). The investigation continued. A vast area of Deccan Traps may hold potential for precious metals.

6.4.5 Orissa

Orissa holds the prime place in chromite reserves (53 Mt) and resources (137 Mt) making a total of 190 Mt. The annual production of +3 Mt accounts for more than 98.6% of India's chromite mining. The major production is contributed from the Sukinda belt with a minor addition from Boula–Nausahi. Nickel occurs as ubiquitous capping on the Sukinda ultramafic

complex and confined to the yellow and red limonitic alteration. Economic significance is yet to be established due to complex metallurgy. PGEs have been identified with numerous samplings within gabbro–norite–anorthosite rocks at Nausahi and Kathpal mine. The favorable milieu of occurrences of PGEs is mafic–ultramafic complexes. Several attempts were made in the last three decades, particularly from 1987 to 1988, by analyzing rocks/soil samples in a grid pattern followed by drilling. The analysis indicated a relatively higher concentration of PGEs in the disseminated variety of chromite of the Boula–Nausahi brecciated gabbro and in chromite and overlying limonite horizons of the Sukinda Complex [Kathpal block of M/s Ferro Alloys Corporation Limited (FACOR Ltd)].

6.4.5.1 Sukinda Mafic–Ultramafic Intrusion

The Sukinda chromite deposits (20°58′ N:85°55E) were a chance discovery in the 1940s. A local tribal villager, employed by Tata Steel at Jamshedpur, picked up an extra heavy stone (float ore) from a village stream out of curiosity, showed it to the company engineers, and made the discovery of the largest chromite belt of India. The Prospecting Division of the Company first brought the occurrence of chromite to light in 1949. The mining lease area was vested with the State Government of Orissa in 1953. Mining started on a small scale using manual open-pit working in December 1960 to meet the requirement of refractory-grade chrome ore. The initial phase of mining was by M/s Misri Lal Jain & Co (Saruabil), Sirajuddin & Co (Sukarangi) in the NE, and Tata Iron and Steel Company Ltd. (TISCO Ltd.) (Bhimtangarh) in the SW of the belt. As of 2016, every inch of the belt is under active mining operation by a dozen companies. The area is located near the triple junction of Jajpur, Keonjhor, and Dhenkanal districts. The nearest railway station is Jaipur–Keonjhor Road (S.E. Rly.), 50 km from the deposits. The highest rainfall is during July and August (~350 mm/month) and June and September (~230 mm/month).

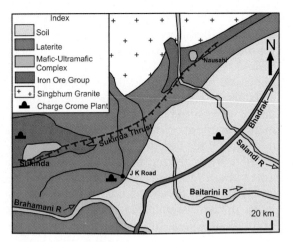

FIGURE 6.17   Simplified geologic map of Sukinda–Boula Nausahi Ultramafic Belt showing the Sukinda Thrust.

The mafic–ultramafic complex emplaced into the sedimentary sequence of Iron Ore Super-group (IOG: >2700 Ma) in Sukinda Valley, Jajput district, that extends in an arcuate belt from the west of Daitari Hill range to Nausahi village, a distance of 50 km and beyond (Fig. 6.17). Sukinda intrusive complex is the most productive ultramafic body for over the last six decades. The intrusion is composed of an earlier chrome nickel-rich phase consisting of montmorillonite rock, limonitized and silicified ultramafites, talc schist and chlorite tremolite schist, and a later chrome nickel-poor orthopyroxenite with marginally developed talc serpentine assemblages (Haldar, 1967). The rocks are found both within the Khondalite–charnokite suite of Brahmani valley and within low-grade metamorphites of the Iron-Ore Stage.

The oldest rock type is quartzite of the Iron Ore Series (Table 6.2), that has been intruded by mafic–ultramafic rocks of two generations into thrust contacts (Sukinda Thrust). The ultramafic rocks have been intruded by younger fresh dolerite dikes. The youngest geological units are high- and low-level laterite and alluvium. The regional trend is NE–SW conforming to Eastern Ghat strike. The quartzites form

two high, subparallel hill ranges, (the Daitari in the north and the Mahagiri in the south), and consist of quartz, with minor sericite, zoisite, zircon, and magnetite. The sericite–quartz association indicates a low grade of metamorphism. The Daitari Range is known for hosting large iron ore reserves. The Mahagiri Range represents massive quartzite devoid of any metallic mineralization.

The NE–SW-trending ultramafic bands extend over 12 km in length with an average width of 2.2 km. The valley portion is occupied by ultramafic rocks composed of rhythmic layers of orthopyroxenite–peridotite–dunite–chromitite–chromite seams, and its weathered variants, such as serpentinite and residual chert with an extensive laterite top. The ultramafic rocks and the associated chromite bodies had undergone relatively simple deformation mostly of postemplacement and postconsolidation (Chakraborty et al., 1980). This phenomenon had resulted in a broadly southwesterly plunging asymmetrical synclinal fold having its southern limb and a part of the hinge zone exposed, but the corresponding northern limb is completely concealed under laterite and alluvium (Haldar, 2011). The Sukinda belongs to the stratified ultramafic complexes of the oxidizing environment. The greater part of the Sukinda ultramafic field is covered by a thick mantle of nickeliferous laterite and soil. The hinge area of the syncline represents the Kansa lateritic nickel block. The Sukinda ultramafic belt is well known for its huge reserves of chromite and nickel ores (Fig. 6.18).

Mineralogical and structural characteristics indicate that the ultramafic rocks of this area are of two different generations. The older ultramafics have been serpentinized, talcified, silicified, and lateritized (limonite) to a considerable extent (Fig. 6.19) and are now represented by silicified ultramafics, serpentinite, talc–antigorite schists, talc–tremolite schists, chlorite–antigorite (bastite) schists, etc. The primary structures have been completely destroyed by later alteration, which is attributed mainly to

TABLE 6.2 Geological Succession of Sukinda Mafic–Ultramafic Belt, Orissa

	Newer Dolerite ≤ 1000 Ma		
	Mayurbhanj Granite, granophyre (1100 Ma)		
	Gabbro Anorthosite; pyroxenite, peridotite (1400 – 1500 Ma)		Basic Intrusives following Dhanjori and Simlipal basins
	Dhanjori Group (1600 – 1700 Ma)	Lavas, interbedded Quartzite, phyllite and basal conglomerate	
	——————— Unconformity ———————		
Orogenic Cycle II	Singhbhum Granite (~ 2700 Ma)	Biotite granodiorite & adamellite	Granite at the end of Iron Ore Orogenic Cycle
	Chromite ultramafic rocks around Nausahi, Sukinda, Saruabil, Katpal, Nilgiri, Manlabhanj, Pasbat etc.	- Peridotite, pyroxenite, talc chlorite tremolite schist - Dunite chromite serpentinite	Intrusive into Iron Ore group and partly into Older Metamorphic
	Iron Ore Group (> 2700 Ma)	Phyllite, metavolcanics, epidiorite, banded chert, Banded Hematite Quartzite with hematite ore, quartzite	
	——————— Unconformity ———————		
Orogenic Cycle I	Older Metamorphic Groups, Amphibolite, hornblende schist, pyroxene granulite (> 3000 Ma)		

autometasomatic processes. The active solution might have come from the ultramafic body itself. The chromite lodes are exclusively confined to these altered rocks. The other younger generation of ultramafic rocks, which is distinctly fresh, is represented by enstatitite, olivine–enstatitite, saxonite, and is completely devoid of any chromite concentration, except a few grains of academic interest.

GSI explored the area by mapping (30,000 sq. km at 1:2000), 1130 cu m of pitting–trenching, and 48,412 m of diamond drilling in 518 holes covering 13-km strike length till 1985. Orissa Mining Corporation Limited drilled 58,600 m in 483 boreholes covering Kalarangi, Sukrangi, and South Kaliapani mine blocks. The entire strike length is leased to various Private- and State-owned mining companies. The companies also conducted surface and underground drilling for delineation of ore for open-pit/underground mining. Most of the open-pit mines will reach ultimate pit bottom in the next couple of years.

Exploration in the belt outlined six persistent/disjointed major chromite bands/seams/lenses running parallel to each other with thickness varying between 10 and 50 m. The major seams are exposed discontinuously over 7 km striking east–west with steep dip. The thickness of the chromite layers is more or less

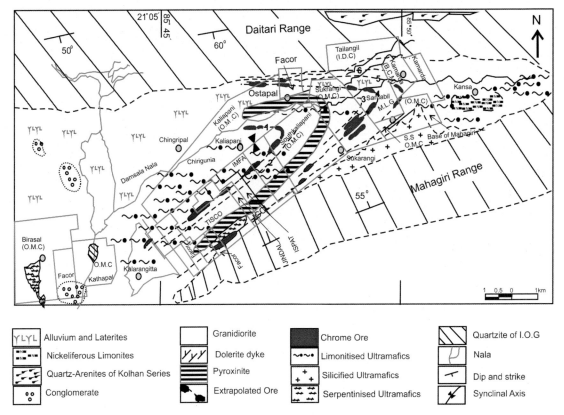

FIGURE 6.18   Geological map of Sukinda Valley ultramafic complex, Orissa, with mining tenements. The exposed chromite bands extend in an NE–SW trend at the eastern limb, take a turn at the synclinal closure, follow the western limb, and finally disappear under weathered rocks. *FACOR, Ferro Alloys Corporation; IOG, Iron Ore Group; OMC, Orissa Mining Corporation; TISCO, Tata Iron and Steel Company Ltd; BCM, B. C. Mohanty; IDC, Industrial Development Corporation; IMFA, India Metals & Ferro Alloys Ltd; ISPAT, Balasore Alloys Ltd; MLG, Misri Lal Jain Group.*

FIGURE 6.19   Older ultramafics have been extensively weathered to form laterite (yellow and red limonite) at the top of the Sukinda intrusion. The nickel-bearing laterite is too complex for current metallurgical economic recovery and has been stacked separately for future technology updates.

uniform for a considerable length, and shows pinching and swelling a number of times. The chromite bodies located in the northeastern part are crystalline, and friable with brown color. The orebodies in the southwest (Kathpal) are massive, hard and gray colored (Haldar, 2011). The chromite seams, from north to south, are designated as:

1. Lower Brown Ore Seam: Dark brown color, mostly loose and friable due to the oxidation of chromite grain boundaries and greater presence of disintegrated silicates to make it loose.
2. Upper Brown Ore Seam.
3. Saruabil–Sukrangi Ore seam: Sickle-shaped tabular gray and brown ore.
4. Kamardah Ore Seam: Covered by a blanket of hard laterite and soil cover, massive to friable podiform chromite deposits.
5. Gray or Mahagiri Ore Seam: Steel-gray color, fine grained, hard, massive and compact, relatively high in Al_2O_3 and MgO, and low in Fe^{3+} and Fe^{2+} indicating earlier crystallization than the brown ore.

All of the mentioned seams (Brown Ore) are typically friable and partly lumpy, except the Gray Ore which is fine to coarse grained, massive, hard, compact, lumpy, and suitable for refractory uses (Fig. 6.20). The ore, in general, shows rhythmic layering within the ultramafic complex by repetition of dunite and chromite members. The area is covered with a thick lateritic profile containing specks of chromite and nickel. This is considered a surface manifestation of mineralization and an excellent exploration guide (Haldar, 2011). Depth continuity has been ascertained more than 250 m from the surface profile and open at depth.

Kathpal chromite bodies are located in the extreme westernmost part of the Sukinda complex and emplaced within the IOG of rocks. The massive gray chromite ore is often structurally disturbed, and occurs as randomly oriented fragments in serpentinized ultramafic rocks (Fig. 6.21) within faulted contact (Fig. 6.22). The host rock preserves the primary layered igneous features. The host ultramafic rocks form irregular xenoliths/pockets within the granite mass. The mineralization displays as lenticular masses

FIGURE 6.20 The Gray Ore Lode-2 from the Sukinda Belt, with subvertical dip at the TISCO mine, is fine to coarse grained, massive, hard, compact, lumpy, and suitable for refractory uses, Sukinda Belt.

FIGURE 6.21 Massive brecciated gray chromite ore is often structurally disturbed, and occurs as randomly oriented fragments in serpentinized ultramafic rocks at the Kathpal underground mine, FACOR Ltd.

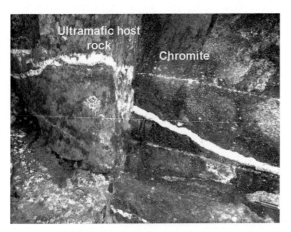

FIGURE 6.22 Massive chromite ore at the Kathpal deposit is structurally disturbed and occurs as a randomly oriented lensoid shape in serpentinized ultramafic rocks within breccia and bounded by faulted contact.

of hard, lumpy, high-grade chromite that are arranged in diverse directions because of intense solid-state reintrusion of the ultramafites before being engulfed by the granite intrusion. The PGE concentration is enriched in the chromite ore compared to the associated rocks (Mohanty and Sen, 2008). Heazlewoodite and nickel-rich pentlandite are the main sulfide minerals, and

occur as intergrowths. Millerite is present as secondary alteration.

6.4.5.1.1 PLATINUM GROUP MINERALS

Sampling in the Sukinda belt during 1982–1983 indicated the presence of PGEs by less than 10–500 ppb. Platinum values in the primary ultramafic rocks vary between <2 and 12 ppb, laterite and limonite overburden between 60 and 290 ppb, chromite ore between <2 and 50 ppb, and chromite horizon between 60 and 400 ppb along with palladium values between <1 and 400 ppb. The stream sediments do not have any specific enrichment of PGE, except higher concentration in the chromite horizon and limonitized overburden. The discovery of PGE mineralization in the Nausahi ultramafic complex prompted GSI to reexamine the possibility of precious metals in the Sukinda ultramafic complex. The existing drill cores of 162 boreholes, drilled at Bhimtangar since 1977, are fully preserved at the GSI core library at Bhimtangar. GSI carried out 2159 m of relogging, 130 cores, 43 bedrock, 36 petrological, and 20 petrochemical sampling work during Field Season 1999–2000.

The area lying between Sukinda and Boula–Nausahi chromite deposits is located on the southeastern fringe of the Singhbhum Craton. Three gravity–magnetic high anomaly peaks were identified near Dantia, Gadabandagoda, and Chitri during a 1998–99 regional geophysical survey by GSI The regional geochemical surveys (soil and stream sediment) also indicated some weak precious metal (Pt, Pd, and Au). GSI has initiated exploratory drilling activity on these anomaly zones from Field Season 1998–99 and have intersected the mafic–ultramafic horizon. The laterite- and alluvium-covered southern continuity of this area up to Bangur Mine in the Nausahi belt has been checked by test drilling and the presence of 0.83 g/t Pt and 0.71 g/t Pd reported.

6.4.5.1.2 NICKEL

A sizable reserve of low- to medium-grade nickel is present in the Sukinda ultramafic complex. The nickel is associated with hydrothermally altered sections and occurs in the oxide/hydroxide state in limonitized ultramafic rocks. Exploration has revealed that nickel ranges from less than 0.7% to 1.9% within the enriched sections and does not show any well-defined geometric configuration due to considerable remobilization of the primary dispersion pattern by interaction of solid-state intrusion and its accompanying hydrothermal alteration. The localized concentrations of Co up to 1% are associated with the silicified fringes of the Ni-rich sections. The in situ nickel-leaching ability of a laboratory stock culture of fungus *Aspergillus Niger* shows maximum leaching of 34% nickel with roasted low-grade chromite overburden material at 2% pulp density, 30°C, and 150 rpm after 28 days incubation, whereas 32% Ni was solubilized by *A. funigatus* under the same conditions (Bohidar et al., 2009).

The resource of nickel ore in the Kansa block has been estimated at 14 Mt at an average grade of ~0.8% Ni. A total of 175 Mt of grade ranging between 0.5 and +0.9% Ni resources has been estimated for the Sukinda Belt.

6.4.5.2 Boula–Nausahi Mafic–Ultramafic Intrusion

The Boula–Nausahi (21°16′:86°20′) Chromite Belt/Group of mines, Keonjhor district, is at a fair-weather road distance of 45 km NNW from Bhadrak Railway station/Town. Bhadrak is 300 km from Kolkata on South Eastern Railways, and 137 km from Bhubaneshwar, the capital of Orissa. The highest elevation is 260 m with valley level at 40 m above mean sea level. The chromite mining in the area dates back to 1942–43.

The Boula–Nausahi Complex is an extension of the Sukinda Thrust Belt occurring as an independent intrusion at a distance of ~50 km ENE of the Sukinda Complex (Fig. 6.20). The gabbro–norite rocks with PGE mineralization are exposed in Bangur quarry at the foot hills of the Boula Range. It covers an area over ~5 sq. km trending NNW–SSE with modest

(55–60°) toward the ENE. The gabbro–norite rocks are intrusive into the ultramafics hosting the chromite lodes in the Boula Nausahi Complex forming an extreme breccia zone (Fig. 2.21). The chromite lodes of the Boula–Nausahi Belt are exclusively confined within ultramafic rocks of peridotite, pyroxenite, and dunite which are intrusive into the IOG of Early Archean age (~3200 Ma). The area was subsequently invaded by Singhbhum granite which was again intruded by a suite of gabbro–anorthosite–granophyre and newer dolerite in succession (Table 6.3 and Fig. 6.23). It is an early magmatic cumulate sequence represented by dunite/serpentinite, harzburgite, enstatite, chromitite, olivine chromite, and websterite. The late magmatic series overlying the ultramafics comprise gabbro–anorthosite, with layers of titaniferous-vanadium–magnetite, norite, dolerite dikes, and granophyre.

GSI drilled 7672 m in 62 boreholes. The ultramafic rock is 4-km long with 800-m wide, thinning out at both ends. The SSW part of the belt is silicified and lateritized. The intrusive complex is an early kinematic multiple intrusion of two different petrographic assemblages:

- Earlier assemblage—chromite (pure in composition)-rich dunite (altered to serpentinite)—chromite suite, olivine–chromite, and chromitite.
- Later assemblage—chromite-poor peridotite (altered to talc serpentinite, talc chlorite schist)/enstatite pyroxenite, harzburgite, websterite, and clinopyroxenite.

The parental magma from which the massive chromitites crystallized was of siliceous high-Mg basaltic or boninitic compositions, similar to the compositions of spatially associated chromite-bearing siliceous high-Mg basalts of the IOG (Mondal et al., 2006). The chromite bodies are confined to a dunite suite of rocks in six stratigraphic levels interlayered with thick layers of dunite usually altered to serpentinite. The thickness of chromite bodies varies

TABLE 6.3   A General Geological Succession Around Boula–Nausahi
Intrusion, Orissa

Alluvium and Soil
Laterite: Alteration of gabbro-anorthosite-ultramafics
~~~~~~~ unconformity ~~~~~~
Newer Dolerite ($\leq$ 1000 Ma)
Granophyre
Dhanjhori Gr (1600 – 1700 Ma): Lavas-gabbro – anorthosite inter-bedded
quartzite,   phyllite and Basal Conglomerate.

~~~~~ Unconformity ~~~~~
Singhbhum granite batholite
Biotite granitoid + pegmatite and aplite veins
Nausahi mafic-ultramafic Intrusion: Chromiferous peridotite-pyroxenite-dunite /
gabbro-anorthosite with Ti-V-magnetite ± PGE / granophyre and granite
Iron Ore Group (~3200 Ma): Massive quartzite, Banded Iron Formation, and
metavolcanics.
~~~~ unconformity ~~~~
Older Metamorphic Group : Schist, Amphibolite, granitised

between 0.8 and 3.9 m of thin rhythmic layers of dunite, chromite dunite, olivine–chromite, and chromite. The layers are continuous and persistent end-to-end and later disrupted by syn- and postemplacement and younger intrusives. The olivine chromitic rock and host dunite exhibit magmatic-layering structure. The Cr/Fe ratio varies between 2.24 and 0.37 from bottom to top.

The Baula–Nausahi Ultramafic Complex, now being mined for chromite, is well exposed in open-pit mine sections showing three cycles of emplacement. The lithologic units comprise an early magmatic-stratified ultramafic suite represented by enstatite, websterite, bronzitite, dunite, and chromitite with chromite concentration and a late magmatic fraction represented by gabbro–anorthosite with layers of Ti–V magnetite concentration, norite, with dolerite dikes, sills, and granophyre of the third cycle. The ultramafic body has a strike extension of 4000 m with a width between 500 and 1000 m. The chromite bands are broadly regular in thickness. Three prominent subparallel chromite seams, trending N–S with a moderate easterly dip, occur within the ultramafic suite. These are known as "Durga (A-B)," "Laxmi (C-D)," and "Shankar (E), Ganga (F)" lodes from west to east. The lodes are demarcated from stratigraphically lower to top levels (Fig. 6.24).

The chromite ore of Durga and Laxmi lodes are stratified, fine to medium grained, loose and soft hosted in altered ultramafic rocks represented by serpentinite and dunites. The chromite of the Shankar–Ganga lodes is dark-brown colored, hard, coarse grained and occurs in a highly brecciated zone at the mafic–ultramafic interface. The chromite ores are often large, massive, sharp-angled, rounded, and ellipsoidal blobs forming alteration rims around in a matrix of gabbro–anorthosite breccia zone (Mohanty, 2009). The altered gabbros are enriched with platinum–palladium (Figs. 6.23 and 6.25).

The Orissa Mining Corporation (OMC) has drilled 4943 m during 1998–2000, and outlined 1.8 Mt of ore around the Bangur mine. The probable reserves have been estimated, to a depth of 30 m from the exposed level in pits and quarries, at 1.37 Mt. The orebodies are continuing to a depth of 120 m, and hence the reserve will be many times more.

The companies adopted both surface open-pit (Figs. 6.26 and 6.27) and underground

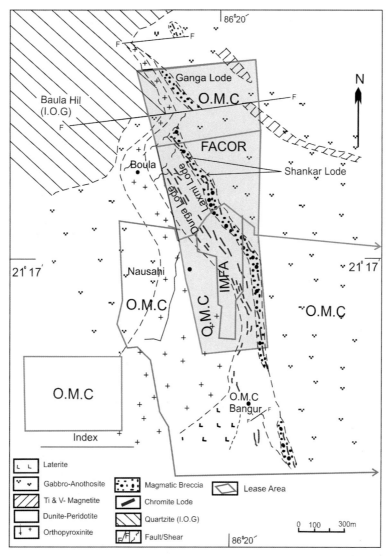

FIGURE 6.23 Geological map of the Boula–Nausahi Complex showing the mafic–ultramafic rocks, magmatic breccia zone, chromite lodes, and lease-hold areas.

(Fig. 6.28) mining methods depending on the location and orebody configuration.

6.4.5.2.1 PLATINUM GROUP OF ELEMENTS

The Early Proterozoic Sukinda–Boula–Nausahi Complex forms a part of the mafic–ultramafic belt in northern Orissa, extending from Nilgiri (21°28′:86°46′) in the east to Barkot in the west passing through the Baula–Nausahi group of mines, Ramachandrapur, the Sukinda group of mines, and Bhuban. In the eastern part of this belt, the igneous complex intruded into the IOG of rocks of Archean age, and in the western part it aligned roughly along the contact

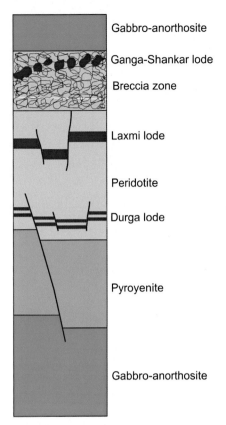

Gabbro-anorthosite

Ganga-Shankar lode

Breccia zone

Laxmi lode

Peridotite

Durga lode

Pyroyenite

Gabbro-anorthosite

FIGURE 6.24   Schematic vertical-column diagram showing succession of mafic–ultramafic rocks and orebodies at
Boula–Nausahi Complex. Pt-Pd-bearing Ganga–Shankar
lode is localized with a breccia zone trapped by a younger
gabbroic intrusion. Large chromite fragments of an earlier
generation with irregular and sharp edges enclosed in the
gabbroic matrix are natural features.

FIGURE 6.25   Chromite ores are often large, massive,
sharp-angled, rounded, and ellipsoidal blobs (center) forming alteration rims (greasy whitish-gray color) around in a
matrix of gabbro–anorthosite breccia zone from the Shankar–Ganga underground mine, Nausahi. The altered gabbros are enriched with platinum–palladium minerals.

FIGURE 6.26   A typical open-pit mine at the Sukinda
Belt, tenement of Tata Iron and Steel Company Ltd., showing the large volume of nickel-bearing limonite cover on the
chromite deposit.

zone between Eastern Ghat granulites and the
IOG of rocks.

The Airborne Mineral Surveys and Exploration (AMSE) wing of the GSI sampled the belt
during 1986–87 and indicated a maximum value
of 2.5 g/t Pt, 3.8 g/t Pd, 0.5 g/t Au, 0.37% Ni, and
0.34% Cu. Subsequently, GSI (1987–88) reported
three anomalous values with total PGE contents
of 244, 334, and 509 ppb.

PGE mineralization at Boula–Nausahi igneous intrusive complex is concentrated in a magmatic–tectonic breccia zone occurring at the
interface between the ultramafic units in the west
and the mafic units in the east (Mohanty et al.,
2008b). The youngest easternmost "Shankar-
Ganga" seam (21°16′35″:86°19′47″) occurs close
to the hanging-wall contact of the ultramafic
body. The magmatic breccia zone stretched over
1500 m, with a width of 30–50 m, and comprised

FIGURE 6.27 Typical open-pit mine in the Sukinda Belt, Tata Iron and Steel Company leasehold, showing the routine activity of loading and dumping of Run-of-Mine ore.

FIGURE 6.28 The underground mine at Boula block of Ferro Alloys Corporation Ltd., showing the main haulage crosscut connected to the production-and-service shaft.

chromite and ferrochromite fragments in a coarse-grained gabbroic matrix. The gabbroic mass has been identified as the most potential for PGE-rich sulfide association (Fig. 2.40). The prominent breccia zone is developed near the interface of the first and second magmatic suites in the eastern part of the complex. The breccia zone is enriched in Cu–Fe–Ni sulfides in the form of chalcopyrite, pyrite, pyrrhotite, and minor pentlandite in the matrix material and

has become the principal site for concentration of PGE, gold, and silver. Overall, Pd dominates over Pt, and Ag has a positive correlation with Pd. These sulfides occur as intercumulus material, disseminations, lenses, and stingers. The PGE mineralogy in sulfide-rich assemblages is dominated by minerals containing Pd, Pt, Sb, Bi, Te, S, and/or As. The enrichment is activated by the process of metasomatism. The interaction of the ultramafic fragments with the evolved fluid-rich mafic magma was key to the formation of the PGE mineralization in the Nausahi complex (Mondal and Zhou, 2010).

The four platinum groups of minerals at the Sukinda–Nausahi Complex include sudburyite, michenerite, palladium–bismuth melonite, and irarsite in decreasing order of abundance. The footwall peridotites in the Ganga pit area show extensive serpentinization with released magnetite. The chromite sulfide assemblage of the matrix part of brecciated zone contains 1.37 g/t Pt, 8.81 g/t Pd, 6.63 g/t Ag, 0.31 g/t Au, 1.00% Ni, 2.51% Cu, and 335 g/t Co assayed by inductively coupled plasma mass spectrometry (ICPMS).

Exploration by GSI in collaboration with the Bureau de Recherches Géologiques et Miniéres (BRGM), France, at the Boula–Nausahi area, covered 2000-m long stretches by drilling 17 holes and has established a resource of 14.2 Mt (7.70 Mt Probable and 6.50 Mt Possible category) of ore at a cutoff grade of 0.5 g/t Pt+Pd as of April 2000 (IBM, 2003). The ore contains 155 ppb Au, and Cu values from 0.1 to 0.4%.

6.4.5.3 Reserves and Resourcesof Orissa

The reserves and resources status of platinum, nickel, and chromium in the State of Orissa, India is given in Table 6.4.

6.4.6 Tamil Nadu

The unique layered-intrusive complex of anorthosite–pyroxenites–dunites is observed in country rocks comprising Precambrian gneisses

TABLE 6.4   Status of Reserves and Resources in the State of Orissa, India

Commodity	Category	Quantity	Grade
Platinum + Palladium	Resource (tonnes ofmetal)	14.2	1–2 g/t Pt + Pd
Nickel	Resource (Mt)	175	0.5–+0.9% Ni
Chromite	Reserve (Mt)	53.07	40–52% Cr_2O_3
	Resource	136.95	
	Total	190.02	

at Tamil Nadu. These are spread over from Coimbatore in the west to the adjoining districts of Dindigal, Karur, Namakkal, Salem, Tiruchirappalli, Perambalur, and Tiruvannamalai. The mafic–ultramafic complex hosts small- to medium-size chromite occurrences. The chromate ores were mined in the past. The magnesite and dunites are being mined at present for industrial usage. The three significant platinum prospects of Tamil Nadu were reported more than three decades ago. The belts include (1) the Karungalpatti, Sittampundi of Namakkal district covering the Karungalpatti, Chettiyampalaiyam, and Tasamapalaiym blocks; (2) the Mettupalayam Belt of Coimbatore district covering Solavanur, Mallanayakkanpalaiyam, and Karapaddi blocks; and (3) Salem. Tamil Nadu is likely to be the future host of PGE resources to a large extent.

6.4.6.1 Karungalpatti and Sitampundi Belt

The Karungalpatti (11°15′:78°00′) and Sittampundi (11°14′10″:77°54′30″) chromite belt, Namakkal district, occurs as an intrusive complex during the Archean–Proterozoic transition into an Archean supracrustal group (c. 3000 Ma). It is composed of anorthositic rocks with associated amphibolites and eclogite gabbro, exposed as a sickle-shaped mass with a lateral extent of more than 32 km and maximum width of about 1800 m. The anorthosite, gabbro, norite, and pyroxenite form the most part and have been metamorphosed to amphibolite and granulite facies, with relics of igneous texture and mineralogy.

Chromite lodes occur as discrete steeply dipping layers enclosed conformably within anorthosite and hornblende gneiss. The chromite concentration is in the lower part of the anorthosite zone. The complex trends in the ENE–WSW direction, and can be traced laterally for more than 16 km with maximum width of 400 m. There are four to five subparallel bands occurring as seams, veins, lenses, and pods confined to amphibolized pyroxenite. The chromite bands are 200 m to 2 km in length and thickness from 2 cm to 6 m. These isolated lodes are exposed over a length of about 12.6 km. The length of individual bands varies from 3 to 52 m.

In the Sittampundi area, a chromitite band with width varying from 15 cm to 1 m occurs discontinuously over a strike length of 2 km. The chromite anorthosite-layered complex resembles the layering of the Bushveld Complex. These bands are rich with an average chromite content of 60%. The primary sedimentary (magmatic) features such as cross-lamination, slump, scour, and fill structures are common. The occurrences are small in size and erratic.

The GSI (1988–89) mapped the area around the Sittampundi Complex, Namakkal district (Fig. 6.29). The analyses of chromitite within anorthosite have shown 0.03–0.75 g/t Pt, 0.1–1.5 g/t Pd, whereas amphibolite samples showed 0.03–0.05 g/t Pt and 0.03–0.5 g/t Pd. Norman Page of the US Geological Survey (USGS) reported total PGEs in chip samples from the same area ranging between 0.26 and 0.929 g/t during 1979. Subsequently, GSI carried

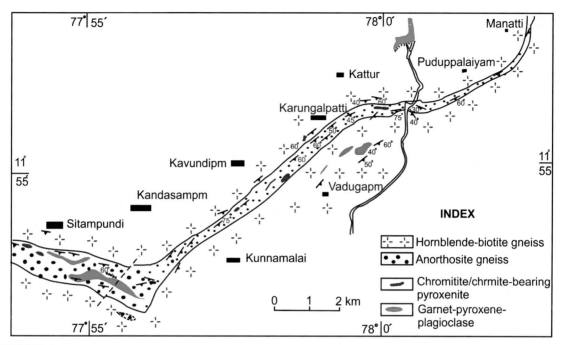

FIGURE 6.29   Simplified geologic map of Cr–PGE occurrences in Karungalpatti–Sittampundi mafic–ultramafic intrusive belt, Tamil Nadu.

out trench sampling during 1984 in Sittampundi and Karungalpatti areas resulting 0.28–1.5 g/t Pt and 0.5–4 g/t Pd over 4 m in the eastern extremity of Karungalpatti village. The samples from 65 m west of the village indicated value of 0.07–0.50 g/t Pt and 0.1–0.5 g/t Pd. The rock types exposed in the area include hornblende anorthosite gneiss, chromitite, and chromite-bearing pyroxenite, metapyroxenite, hornblende biotite gneiss, and pegmatite veins.

Total reserves of the Karungalpatti–Sittampundi block have been estimated at ~0.22 Mt up to 6 m depth of overall low grade (24–28% Cr_2O_3, 22–35% FeO, 20–30% Al_2O_3 and 5–15% SiO_2).

6.4.6.2 Mettupalayam Belt

The Mettupalayam Belt covers Solavanur (11°24′30″N:77°009′00″E), Mallana-yakkanpalaiyam (11°25′55″N:78°013′40″), and Karapaddi (11°26′00″N:19°45′00″E) blocks in the Coimbatore district. The mafic–ultramafic suite of

rocks comprising gabbro, anorthosite, amphibolite, and chromiferous pyroxenite are conformably intruded into Archean migmatite–gneiss. The chromite mineralization occurs as compact, massive, granular, and disseminated texture in layered bands and as float ore. The chromite resource has been estimated at 30,000 tonnes (~33,000 tons) with grade ranging between 22 and 28% Cr_2O_3 (Indian Bureau of Mines, 2013).

6.4.6.3 Salem Magnesite Belt

The Salem magnesite deposits are exposed with crisscrossed, fracture-filled, and minor faulting of magnesite veins at the roadcuts, and exhibit as common phenomena in and around Salem town (Fig. 6.30). Small chromite bodies from dunite host rock have been mined from Chalk Hills and Kanjamlai, situated within and at the periphery of Salem town. The area is under active-mining operation for magnesite and dunite. Large crystals of chromite are

FIGURE 6.30 Crisscross and fracture-filled magnesite veins at roadcuts within Salem town. The magnesite mine with large crystals of chromite is behind the road.

separated and stacked for commercial uses. The layered-intrusive complex of anorthosite–pyroxenites–dunites is reported in various sizes at Coimbatore, Dindigal, Karur, Tiruchirappalli, Perambalur, and Tiruvannamalai districts. These mafic–ultramafic assemblages will be excellent for future targets for hosting PGE mineralization.

6.4.7 Goa

The large ultramafic pluton of Archean greenstone association in the Dudhsagar Belt near Ponda has been reported to be chromiferous. However, no commercial chromite lode could be established. In the Usagaon area, southern Goa, PGM samples up to 0.03 g/t Pt and 0.03–0.15 g/t Pd were reported.

6.4.8 Tertiary Ophiolite Sequence

Small occurrences of chromite and PGEs are reported from the Mesozoic ophiolitic association from Jammu–Kashmir, Manipur, Nagaland, and the Andaman Islands. None of the deposits/occurrences are economically viable at present level of information. But these associations are

technically lucrative to host chromite and PGEs and lucrative for future exploration targets.

6.4.8.1 Jammu and Kashmir

The chromite deposits are found in the Cretaceous volcanics/ophiolites at Dras, Burzil, Bumbat, Tashgam, and Kargil Valley of Ladakh. The chromite is mainly used in metallurgy.

The Brown Hill chromite deposit is located 4 km north of Dras, Kargil district, Ladakh. The chromiferous ophiolite/dunite host package is emplaced into the Lower Cretaceous–Miocene Sangeluma Group volcanic formation. The mineralization extends in the NW–SE direction and occurs in the form of pods, veins, stringers, and disseminations. The metal content of chromite-rich zones varies between 43% and 47% Cr_2O_3. The Kyun Tso–Shurok Sumdo chromite deposits are located in the Nidar valley, Leh district, Ladakh. The ultramafic rocks hosting the chromite mineralization was emplaced within the Sangeluma Group of volcanics and is bound on either side by regional faults. The mineralization occurs in the form of bands, lenses, streaks, and disseminations and is characteristically concentrated in dunite, harzburgite, and orthopyroxenite. The thickness varies between 1 and 1.5 m and is 8–18 m wide. The inferred resources of chromite in the Leh district, Ladakh, have been estimated at 14,000 tonnes (~15,400 tons).

6.4.8.2 Manipur

N–S trending Nagaland–Manipur–Chin-Arakan–Yoma Suture Zone, extension down to Andaman and Nicobar Islands, and the emplacement of ophiolitic serpentinized ultramafics along this line during Cretaceous–Eocene period suggest favorable geological setup to host nickliferous chromite–PGE incidences. Chromite deposits are located in Ukhrul and Chandel districts, Manipur. The host rocks are essentially of ophiolite suite composed of harzburgite, dunite, and serpentinite. The mineralization is Alpine type in the form of small pockets, lenses and pods. The mineralization is typically massive, granular, interbanded and podiform. It also occurs as disseminated in highly fractured dunite.

The major deposits include Sirohi and Gamnom in Ukhrul district, and Moreh area in Chandel district. The chromite loads are small in size, extending few meters, locally prospected, and worked by surface pits. Three large pits are located north of Ranshokhong area with dimension up to 20-m long, 15-m wide, and 6 m in depth. The grade, in general, is high ranging between 44 and 59% Cr_2O_3, and ~15% Fe_2O_3.

6.4.8.3 Nagaland

The ophiolite belt of Phek (Reguri and Washello), and Tuensang (Pang, Pokphu, and Wui) districts host small occurrences of chromite. The presence of PGMs is yet to be established.

6.4.8.4 Andaman Islands

The chromite mineralization in the Andaman Islands is intimately associated with the ophiolite suit of rocks composed of altered dunite–harzburgite–lherzolite, and peridotite assemblage. The chromite occurs as small pods and layers with high grade up to 50% Cr_2O_3, and the very low TiO_2 content in chromite corresponds to the ophiolitic environment. The major bodies are located in places around the Chidiatapu–Bedanabad area in South Andaman, Rutland Island, "Jones" point hillock in Middle Andaman, and Kalighat in North Andaman. The chromite loads are not economically viable at this point due to extremely small size, widely apart, and scattered.

6.5 INDONESIA

Indonesia follows a mixed economy with a significant role between Private Sectors and Government. Indonesia is rich in reserves of copper, nickel, chromite, tin, gold, silver, bauxite, coal, natural gas, and petroleum. The geology of the island chain at the Asian continental margin is primarily composed of mafic–ultramafic volcanic and plutonic rocks. The structure is strongly affected by trenching, faulting, folding, and emplacement of volcanics and with active volcanoes. The lithostructural system and magmatic processes are suitable to host a wide range of mineralization, namely, chromium–nickel–copper–PGEs. Sulawesi Island is the 11th largest island in the world and shares major ophiolite-hosted high-grade chromite deposits, and high-grade laterite nickel–cobalt deposits.

6.5.1 Chromite Deposits

The rich chromite deposits are widely distributed in eastern Indonesia, within mafic–ultramafic rocks, especially, in South Kalimantan, Sulawesi, Maluku, Halmahera, Gebe, Gag, Waigeo, and Papua. These deposits result from weathering of ophiolite rocks as part of the Pacific plate.

The chromite deposits of South Kalimantan Province occurs in the Pelaihari district, Tanah Laut region, and Karang Intan district, Banjar region. Chromite lodes are confined within detached ultramafic rocks trending NNE–SSW. The area has been explored, and reserve has been estimated at 10,000 tonnes (~11,000 tons) grading ~30% Cr_2O_3, and resource of 132,000 tonnes (~145,000 tons) (Ernowo et al., 2010).

Chromium–nickel–titanium deposits are located in the central part of South Sulawesi Province hosted by ophiolite rocks. Chromite lenses occur as in situ primary podiform type and alluvial placer float ore in surrounding areas. Reserve at Barru district has been estimated at 234,000 tonnes (~258,000 tons) grading 43–53% Cr_2O_3, and resource of ~7000 tonnes (~7700 tons) at Laritae. Southeast Sulawesi terrain is thrusted over by an older ophiolite suite that hosts chromite, nickel, cobalt, and lateritic iron. Central Sulawesi Province is dominated by placer-type chromite deposits located at six places. Reserve has been estimated at 88,000 tonnes (~97,000 tons) grading 37–46% Cr_2O_3, and a hypothetical resource of 1.55 Mt (Ernowo et al., 2010).

The North Maluku Province is divided into volcanic and ophiolite belts with mineralization in the ophiolite belt which includes iron sand, cobalt, nickel, chromite, gold, and silver.

6.5.2 Nickel Deposits

Indonesia was among the world's top five producers of nickel ore in the early 1990s. The in situ high-grade nickel–cobalt inventory of Indonesia is globally the largest, and second highest grade after New Caledonia. Laterite ore is predominantly present in Sulawesi and Halmahera islands. Indonesia supplies over 50% of China's demand. The major lateritic nickel-hosting areas include Soroako Island, Gag Island, Weda Bay, and Tanjung–Buli.

The Soroako laterite nickel deposit is the largest open-pit mine in Indonesia and a major producer of nickel ore/matte for refining in Japan. The small mining town is located about 40 km from Malili. The deposit is formed by intense chemical weathering of Cretaceous ultramafic rocks of East Sulawesi ophiolite. This ophiolite is tectonically dismembered and exposed over 10,000 sq. km. The area around Soroako deposit is dominated by harzburgite–peridotite with high Cr, and minor occurrence of lherzolite and dunite (Sufriadin et al., 2011). Soroako represents one of the largest nickel reserves in Indonesia having estimated at 109.4 Mt of ore grading 1.79% Ni and contains 1.95 million tonnes (~2.15 million tons) of nickel metal.

Gag Island deposit is a large mine in east of Indonesia in Sulawesi. Host rock is nickel-bearing laterite and saprolite on ophiolitic serpentinite and melange. It has the largest reserve of laterite nickel–cobalt estimated at 240 Mt averaging 1.35% nickel [3.24 million tonnes (~3.57 million tons) nickel metal], and 0.08% cobalt (including both oxide and silicate laterite zones). Development of the project has been suspended following uncertainty over Indonesian policy relating to Reserve Forest.

Weda Bay Ni–Co deposit is located on Halmahera Island. The deposit forms a part of the largest undeveloped lateritic nickel in the world, and has been under exploration since the late 1990s. Mineralization occurs as nickel hydrosilicates in saprolite over ophiolitic serpentinite. The in situ nickel metal reserve has been estimated at +7 Mt that can sustain mining and extraction for over 50 years at the rate of 65,000 tons of nickel and 4000 tons of cobalt per annum. The trial extraction phase produced 35,000 tonnes (~38,600 tons) of nickel and 1300 tonnes (~1400 tons) of cobalt. Halmahera Island has indicated and inferred resources of 216 Mt grading 1.37% nickel and 0.12% cobalt.

Tanjung–Buli lateritic nickel deposit is situated on Halmahera Island, East Halmahera district covering 390 sq. km surface area. Buli ultramafics is composed of peridotites (dunite, harzburgite/lherzolite) with pyroxenite, serpentinite, saprolite, limonite, basalt, and gabbro. The deposit has estimated resources of 41 Mt (saprolite) containing 2.50% nickel, and 11% iron.

6.6 IRAN

The major minerals/metals 0f Iran include iron ore, bauxite, coal, copper, zinc–lead, gold, uranium, and chromite.

The chromite deposits are the podiform type and grouped into two major discontinuous giant ophiolitic thrust sheets resting on continental substrate. The northeastern series of ophiolite segments is known as the Forumad Meshed belt, and the southwestern series as Esfandagh belt. The ophiolite complexes of Iran are part of the Tethyan ophiolite belts of the Middle East that link to Pakistan in the east and ophiolites in the Mediterranean region such as Turkish, Troodos Greek, and eastern Europe in the west. The ophiolite complexes are uniformly characterized by large Alpine-type ultramafic bodies represented mostly by harzburgite and dunite with secondary lherzolite and pyroxenite. Peridotites and pyroxenites are usually invaded by smaller gabbro masses. The ultramafic rocks were emplaced during the Mid-Jurassic to Upper Cretaceous Era (80–170 Ma), and largely serpentinized. The mode of occurrence of all chromite bodies is podiform in shape of discontinuous pods, lenses, and rods.

Chromite deposits/occurrences/mines are located in east of Sabzevar, Forumad Area, northwest of Sabzevar, Jaz Murian in the state of Sistan and Baluchistan, Minab in the state of Hormozgan, and Robat Sefid in the state of Khorasan. More than 74 chromite potentials have been reported in different ophiolite terrains of Iran; 18 of them are currently under mining and four are in different stages of exploration and development. The three large active chromite operations are at Faryab, located northeast of Bandar Abbas, Esfandegeh, located south of Kerman, and Foroumad–Gaft north of Sabzevar in Khorasan (Fig. 1.20). The total annual production capacity is ~166,000 t of chromite concentrates and mainly for export.

6.7 JAPAN

Japan has scanty mineral/mining resources that include coal, petroleum, iron ore, copper, and chromite. Several ultramafic complexes have been identified from the Sangun zone (Inner zone) from southwest Japan. The ultramafic complexes include Ashidachci, Izushi, Ochiai–Hokubo, Sekinomiya, Tari–Misaka, and Wakasa. The Sangun zone is composed of high-pressure intermediate type of metamorphosed/unmetamorphose Palaeozoic formations on to which many ultramafic complexes have been emplaced by thrust faulting. The ultramafic rocks are composed of harzburgite, dunite, peridotite, pyroxenite, chromitite, and aligned in NNE–SSW trend. The ultramafic units are concentrically arranged in five mineral zones within the complex, and strongly serpentinized up to 90% (Arai, 1994). The euhedral to subhedral podiform-type chromites are densely distributed within dunite.

Several chromite mines (the Wakamatsu and Hirose mines) have been operating in the past and abandoned from time to time.

6.8 KAZAKHSTAN

The mineral resources of Kazakhstan include the second largest chromium reserves in the world after South Africa. The country is second/third highest producer of chromite ore at par with India. There are ~80 small and large chromite deposits in the Kempirsai massif with variable size of reserves totaling ~310 Mt chromite resources. The large chromite and nickel deposits of Kazakhstan are located at (1) Kempirsai massif, lying in Aktyubinsk district; (2) Voshkod chromite, Western Kazakhstan; and (3) Shevchenko Nickel deposit in northwest Kazakhstan. The country is currently producing about 4 Mt of chromium ore with plan to grow between 7 and 9 Mt per year.

6.8.1 Kempirsai Ophiolite Massif/Belt

The Kempirsai chromite belt occurs in the southern Ural Mountains region in Kempirsai massif, lying in the Aktyubinsk district. Kempirsai massif is the western segment of Palaeozoic Ophiolite Belt. The giant ophiolite-hosted podiform-type deposits were discovered in 1936. The mineralized ophiolite belt of Kempirsai massif covers an area of ~2000 sq. km in an elongated shape with the long axis of 90 km trending N–S following the regional shear zone related to the Main Ural fault, and is up to 32 km wide in an E–W direction (Fig. 6.31). The magmatic assemblage comprises a complete ophiolite sequence, ranging from fertile-mantle to depleted-mantle harzburgite tectonites including chromitites, cumulates, a sheeted dike, lava flows, and ocean-floor sediments. The ophiolite sequences are composed of peridotites, banded harzburgite, dunite–wehrlite–pyroxenite, olivine gabbro, overlain by volcanics and sediments of Precambrian to Palaeozoic age.

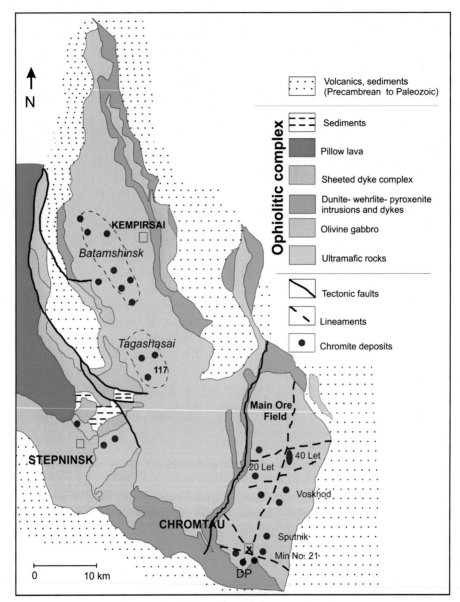

FIGURE 6.31 Geologic map of Kempirsai Massif showing podiform-chromite deposits distributed in total belt. *After Melcher, F., Grum, W., Simon, G., Thalhammer, T.V., Stumpfl E.F., 1997. Petrogenesis of the ophiolitic giant chromite deposits of Kempirsai, Kazakhstan: a study of solid and fluid inclusions in chromite. Journal of Petrology 38 (10), 1419–1458.*

The ultramafic rocks have been serpentinized to variable intensity.

Main Ore Field with a cluster of large chromite deposits is located in southeastern part of Kempirsai massif. The orebodies are primarily hosted by a sequence of dunite and harzburgite within an NNE–SSW-trending anticline. The fault-controlled axial plane can be traced up to 22 km, and reaching a total thickness of 16 km. The large chromite deposits in western part of the structure are Millionnoe, Diamond Pearl, and 20 Years of Kazakh Soviet Socialist Republic (KSSR), all dipping 5–50° to west. The other set of large/small deposits such as Mir, Voskhod–Karagash, 40 Years of Kazakh SSR–Molodezhnoe, and Sputnik are located on the eastern side of the structure and dip 15–75° to east (Fig. 6.31).

Chromite pods forming giant orebodies in Early Palaeozoic ophiolite sequence of the Kempirsai Massif contain a large number of inclusions of silicates, sulfides, alloys, arsenides, and fluids. Chromite orebodies are enveloped by dunite of variable thickness showing transitional boundaries to harzburgite host rocks. The composition of ore-forming chromites in depleted-mantle rocks of southern part of the massif (Main Ore Field) is rather uniform. A diversity of primary and secondary platinum-group minerals is described by Melcher et al. (1997) from chromitites, including alloys, sulfides, sulfarsenides, and arsenides of Ru, Os, Ir, Rh, Ni, Cu, Fe, and Co. The large orebodies and amphibole–chromite veins in southern part formed later from interaction of hydrous, second-stage high-Mg melts and fluids with depleted mantle in a convergent tectonic setting.

Reserve base for the large group of chromite deposits within the Kempirsai massif has been estimated at 310 Mt grading +48% Cr2O3.

6.8.2 Voskhod Chromite Deposit

The Voskhod deposit is one of the 80 deposits in the Kempirsai Massif discovered since 1936, situated 16 km from Khromtau, and ~90 km from Aqtobe Region North West Kazakhstan. The deposit was discovered in 1963. The Voskhod orebody is one large lens of massive to disseminated chromite lying at a depth between 98 and 450 m with northeasterly dip of ~28° in eastern zone of main Kempirsai Massif. The mineralization is classified into several zones from bottom to top. The mineralized zone consists of massive, powdery, orbicular, vein, and disseminated types.

The ore reserve has been estimated at 19.5 Mt at 48.5% Cr_2O_3. The adjacent Karaagash deposit

is expected to increase reserve between 4 and 8 Mt. Mining is done by open pit in combination with underground, accessed through decline and two shafts. Voskhod is a fully mechanized underground operation using modern mining technology in Kazakhstan. The mine has annual capacity of 1.3 Mt of chrome ore and 900,000 tonnes (~992,000 tons) of lump, chip, and concentrate.

6.8.3 Shevchenko Nickel Deposits

Shevchenko lateritic nickel deposits are a greenfield project located close to Voskhod chromite deposit/mine. The deposit developed on ophiolites located in the Trans-Uralian Zone, obducted onto the Kazakh continent during the Uralide Orogeny and deeply weathered up to the Tertiary. The deposit is dominated by nickeliferous smectite with nickel-bearing iron oxides and hydroxides. The weathering process was variably developed on serpentinite. The geological succession and nickel content of lateritic column from bottom upward include fresh rock (0.3% Ni), saprolite–garnierite–serpentinite (1.8–3.0% Ni), transition (1.5–4.0% Ni), yellow limonite (0.8–1.5% Ni), and red limonite (<0.8% Ni). The reserves have been estimated at 21.4 Mt grading 0.85% Ni of proved category, and probable reserves of 83.0 Mt at 0.77% Ni. The total reserve stands at 104 Mt grading 0.79% Ni and up to 0.2% Co.

6.9 MONGOLIA

The principal mineral reserves mined in Mongolia are coal, copper, gold, uranium, and Ni–Cu–PGE deposits. An Ni–Cu sulfide deposit with significant platinum and palladium contents, hosted in a mafic–ultramafic intrusion, has been reported from the Huanghuatan area, Inner Mongolian Autonomous Region.

The Erbutu magmatic Ni–Cu sulfide deposit is located in the southern margin of the Central

Asian Orogenic Belt, Inner Mongolia, and hosted in a small ultramafic intrusion with surface exposure of less than 200 m across. The intrusion occurs as bowl shape with a downward extension up to ~200 m. Regional country rocks are Palaeozoic granite–amphibolite, Paleoproterozoic Baoyintu Group, Mesoproterozoic metamorphosed volcano–sedimentary rocks of Baoyintu Group, and Quaternary sediments. The host ophiolite sequence from bottom to top is olivine orthopyroxenite, orthopyroxenite, and oxidized zone. The magmatic sulfide deposit is related to boninitic magmatism in an arc setting. Sensitive high-resolution ion microprobe (SHRIMP) U–Pb dating of zircons from the intrusion gives a crystallization age of 294.2 ± 2.7 Ma (Peng et al., 2013). The major part of the intrusion contains economic nickel–copper sulfide mineralization as net-texture in olivine orthopyroxenite, disseminated in orthopyroxenite and oxidation overburden zone. The whole-rock analysis shows characteristic enrichments of light Rare Earth Elements (REE).

6.10 PAKISTAN

The mineral resources of Pakistan include coal, copper, gold, chromite, and bauxite. The Southern Baluchistan region is endowed with huge reserves of chromite. The first discovery of chromite was made from Muslim Bagh and Khanozai in Kila Saifullah district in 1901. The deposits of Muslimbagh area of Zhob valley are located ~120 km northeast of Quetta. The other major chromite deposits/mines in Zhob district include Khanozai, Nisai, Gwal, and Jungtorghar. Some other chromite deposits are located near Khuzdar.

The mafic–ultramafic ophiolite complex covers an area of ~2560 sq. km. The igneous complex intrusive was emplaced into Zhob Valley sediments (limestone, shales, and sandstones) of Triassic to Eocene age with ENE–WSW trend. The intrusion is composed of serpentines, dunites,

harzburgite, peridotites, pyroxenites, anorthosite, troctolite, gabbros, lavas, and chromitites. The dunite, harzburgite, and peridotite are extensively serpentinized. Rhythmic banding and layering are common magmatic features between massive and disseminated chromite, dunite, harzburgite, peridotite, and gabbro. Deposits are of podiform type in which chromite occurs as stringers, veins, pods, bands, layers, net-textures, and tubular lenses in serpentinized ultramafic rocks. The average grade of chromite varies between 52% and 59% Cr_2O_3. Cr:Fe ratio of Zhob Valley deposits varies between 3:1 and 2.5:1 with 10–15% Al_2O_3.

Chromite reserves stand at ~4.5 million tonnes (~5.0 million tons). Chromite mining has not been systematic, but random with time and production capacity. Chromite mining adopted both open-pit and underground methods. The daily ore production capacity at Muslim Bagh and Khanozai varies between 300 and 500 tonnes (~330 to ~550 tons). Run-of-Mine ore is transported to Karachi, crushed, and packed in bags for export mainly to China. The tenements operated by Pakistan Chrome Mines have produced an accumulated 1.5 Mt of high-grade chromite and smaller quantities of magnesite ore.

6.11 THE PHILLIPINES

The Philippines is gifted with vast valuable metallic minerals that include the most abundance of nickel–cobalt–PGE deposits, and significant amounts of iron ore and copper. Philippines is the 12th largest in terms of nickel metal reserves at 1.10 Mt and second largest producer of nickel at 230,000 tonnes (~254,000 tons) in 2011, after Russia at 280,000 tonnes (~309,000 tons). The nickel deposits are hosted by laterite. The surface geological conditions of the Philippines are suitable to form rich lateritic Ni–Co–PGE deposits due to abundance of ultramafic rocks with ophiolitic affinity, high annual precipitation with pronounced wet and dry seasons, and leaching of nonnickel elements under plateau and gentle slope of geomorphology.

6.11.1 Acoje Ni–Co–PGE Deposit

The Acoje nickel deposit is at a road distance of 270 km northwest of Manila on the midwest coast of Luzon, Philippines. It forms part of Zambales ophiolite complex, near Santa Cruz in Zambales province. The ophiolite complex forms a suprasubduction zone in two segments separated by a sinistral strike-slip zone. The segments are (1) mid-ocean ridge-related Coto block, and (2) island arc-related Acoje block. This crust–mantle sequence hosts PGEs in the Acoje block. The melts responsible for platinum-bearing nickel sulfide and chromitite deposits are of magmatic origin characterized by high-MgO basalt to boninitic composition which, being second- or third-stage melts, carry higher PGE concentrations (Yumul, 2001). The secondary lateritic nickel formation is a product of chemical weathering.

Phanerozoic primary ophiolitic related mafic–ultramafic rocks in the western margin of the complex are composed of medium- to coarse-grained basal ultramafic cumulate sequence of mantle harzburgite having gradational contacts with a transition-zone dunite to the east. The package extends over 12 km striking north–south and steep dip to east. The olivine dunite is massive and granular, with disseminated chromite and magnetite, and overlain by a mafic sequence of gabbro, olivine gabbro, norite, and pegmatite. The basal ultramafic sequence is largely serpentinized as compared to upper mafic zone. Mineralization can be separated into two groups: (1) Platiniferous chromite and nickel sulfide, and (2) Lateritic nickel horizon.

6.11.1.1 *Platiniferous Chromite and Nickel Sulfide*

The basal dunite horizon hosts two chromite and four platinum-bearing nickel sulfide zones

from base upward. These mineralized layers can be traced over 8 km. The platiniferous sulfide zones are up to 15 m thick, and associated with magnetite and graphite-rich black dunite lenses. The primary sulfide minerals include pyrrhotite, pentlandite, chalcopyrite, platinum, palladium, rhodium, ruthenium, osmium and iridium, gold, and silver. The generation, accumulation, and segregation of platinum-group, oxide, and sulfide minerals from the melts are governed by combined factors of high degrees of partial melting, multiple melt replenishment with concomitant magma mixing, and fractional crystallization (Yumul, 2001).

6.11.1.2 *Lateritic Nickel Horizon*

Lateritic nickel mineralization is developed over the chemically weathered ultramafic bedrock of olivine harzburgite, serpentinized harzburgite, peridotite, and serpentinized peridotite above the 250-m elevation on relatively flat, plateaus/subplateaus with gentle ridge slopes in areas of moderate relief. The laterite zones to the east and south are thin and uneconomic where the topography is steeper. The rich nickel mineralization largely occurs at the base of the Fe laterite, at the top of the saprolite, and in magnesium-rich clays within the ultramafic saprolite. Acoje nickel consists of three nickeliferous horizons:

1. An upper red ferruginous soil or limonite zone up to 4-m thick contains high iron and low nickel grading <0.8% Ni, 0.06–0.12% Co, 50–52% FeO, and 1–3% MgO.
2. An intermediate yellow limonite zone with high iron and medium nickel containing 0.8–1.5% Ni, 0.1–0.2% Co, 40–50% FeO, and 1–5% MgO.
3. An underlying nickel-rich saprolite zone up to ≥4-m thick with low iron and medium–high nickel containing 2–3.5% Ni, 0.01–0.2% Co, 9–15% FeO, and 25–30% MgO.

The ore reserve of has been outlined within the saprolite and upper limonite zone and stands at 33.94 million dry metric tonnes (~37.4 million tons) grading 1.1% Ni, 0.07% Co, and ~2 g/t PGE + Au, as on April, 2013. The inferred resources stands at 2 million dry metric tonnes (~2.2 million tons) grading ~1.04% Ni.

Acoje nickel laterite deposit is a high-iron limonite of a direct-shipping ore project. Conventional load-and-haul surface mining method has been adopted at the open-pit mine. The blended and dried ore is trucked to the port loading facility and directly unloaded onto barges.

6.12 RUSSIA

Russia is endowed with vast mineral resources and ranks amongst the leading producers of aluminum, bauxite, iron ore, copper, gold, magnesium, nickel, cobalt, and PGEs. The major Ni–PGE areas are from magmatic sulfide deposits at Noril'sk, Siberia, and Pechenga, Kola Peninsula. The alluvial type of platinum deposits are located in the Ural Mountains. The other significant polymetallic PGE deposits are from Kondyor mine in the Khabarovsk region and Koryak mine (1994) in Kamchatka. The Kola Peninsula is also known for Ni–Cu–PGE magmatic sulfide mineralization of Monchegorsk Layered Igneous Complex (Bekker et al., 2015; Pripachkin et al., 2015). The discovery of platinum from large alluvial placer deposits in the Ural Mountains in Russia dated back to 1823, and operated from 1824. Since then, exploitation and commercial uses of platinum metals continued and declined around the 1930s.

6.12.1 Noril'sk–Talnakh Group, Siberia Plateau

The Noril'sk–Talnakh group of deposits is the largest Ni–Cu–Palladium resources and ranked the fifth highest reserves and resources of PGEs in the world.

6.12.1.1 Location and Discovery

Exploration for Cu–Ni–PGE deposits in Taymyr Province, northeast Siberia, discovered the Noril'sk deposit in 1935. The Noril'sk deposit is located between Yenisei River and the Taymyr Peninsula, just south of the industrial city of Noril'sk in Krasnoyarsk Krai, northern Russia. Mining was initially done by underground operation with entry through an adit. In the 1940s, two open pits were developed. The Ugol Creek pit was smaller and closed in the late 1960s. Medvezhy pit is still in operation. The Zapolyarny underground mine was in production since the early 1950s. Mining and exploration continued around Noril'sk with identification of few more rich deposits. The high-grade copper–nickel deposits at Talnakh, ~35 km north of Noril'sk, was discovered in 1960. Four mines were developed between 1960 and 1980s, and are still in operation. The Mayak mine started in 1965 and still producing at a depth of 400 m. Komsomolsky and Oktyabrsky mines started in ~1970 and ~1975, respectively. The Oktyabrsky mine is the largest and richest of Talnakh operations, and mining at a depth of ~1100 m. The Tayimyrsky mine started production during the 1980s and is worked to depths of ~1500 m.

6.12.1.2 Geological Setting

The regional geological setting is characterized by vast and thick Permian–Triassic sediments. The ore-bearing sulfide magma was hosted in mafic Kharaelakh intrusions (northwest Talnakh intrusion) that intruded the sedimentary sequence and immediately underlies the Siberian flood basalt volcanic trap sequence (Fig. 6.32). Magma, probably from a large mantle plume, interacted with crustal rocks during its passage to the surface (Arndt, 2011). The Kharaelakh intrusion is one of multiple sill-like, sulfide ore-bearing gabbroic intrusions that are coeval with the overlying flood basalts in the Noril'sk region, Siberia (Li et al., 2009). The tectonics of the region is dominated by the Noril'sk

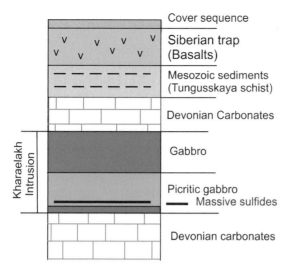

FIGURE 6.32 Geological succession of Noril'sk–Talnakh Ni–Cu–PGE deposits, showing the multiple magmatic events of sulfide-bearing mafic host coeval with the overlying Siberian flood-basalt trap intruded within Permian–Triassic sedimentary rocks.

FIGURE 6.33 Sperrylite embedded in pyrrhotite–chalcopyrite–pentlandite base from Talnakh deposit, Noril'sk Group, Russia. *Avinash Sarin.*

Kharaelakh and Imangda faults. The structure is important from the point of mineralization control in the area.

Deposits are found within and adjacent to gabbro–dolerite sills that represent part of the feeder zone up to 3500-m thick (250 Ma) overlain by Siberian trap basalts. The eruption resulted in development of flat multilayered lava conduits spread over 1 million cubic km. A large portion of it is lying below the Noril'sk and Talnakh Mountains. The ore was formed when the erupting magma became saturated in sulfur by assimilation of evaporites from the country rocks, and led to the segregation of metal-rich sulfides ore that accumulated in the lower parts of shallow-level intrusions. The process resulted into forming globules of pentlandite, chalcopyrite, other sulfides, and the platinum group of minerals (Fig. 6.33). These sulfides were then washed by the continuing torrent of erupting magma, and upgraded their tenor with Ni–Cu–Pt–Pd.

6.12.1.3 Mineralization

Mineralization occurs at several different levels within and below thick mafic sills showing variable differentiation. The ore types are:

1. Disseminated sulfides within differentiated gabbro–dolerite sills as fine drops and sulfide veins forming conformable sheets up to 40-m thick and comprising chalcopyrite, cubanite, pyrrhotite, and pentlandite. The average grades range between 0.5 and 0.6% Ni, 0.6 and 0.7% Cu, and 5 and 6 g/t PGEs.
2. Massive sulfides occur usually on the lower contact of mineralized sills and enclosing rocks. Sulfides are pyrrhotite, cubanite, chalcopyrite, and pentlandite. Individual bodies may be up to 60-m thick as at Oktyabrsky covering an area of 3.5 sq. km. The average grades are 2.8% Ni, 5.6% Cu, and 15 g/t PGE.
3. Dissemination veins 10–20-m thick cupriferous layers between massive sulfide and the lower margins, and within sediments on the upper margin of the sill as skarn-altered brecciated dolerite and marl/argillite. Ore minerals are pyrrhotite, pentlandite, chalcopyrite, cubanite, millerite, pyrite, magnetite, bornite, chalcocite, etc.

6.12.1.4 Reserves and Resources

Reserves and resources have been estimated at 1309 Mt grading 9.50 g/t PGEs (1.84 g/t Pt, 7.31 g/t Pd, 0.19 g/t Rh, 0.09 g/t Ru, 0.03 g/t Ir, and 0.04 g/t Os), 1.77 wt% Ni, 3.57 wt.% Cu, and 0.06 wt% Co (Naldrett, 2004).

PGE reserves are estimated at 81.791 million troy ounces with proven and probable components at 62.183 million troy ounces of Pd and 15.993 million troy ounces of Pt with ore grades ranging between 5.5 and 11.1 g/t. Mineral resources are estimated at 141 million troy ounces of Pd and 40 million troy ounces of Pt.

6.12.2 Pechenga Group—Kola Peninsula

The Pechenga nickel–copper deposits, with associated cobalt, platinum, and palladium, are located on the Kola Peninsula of northwestern Russia. The Pechenga–Varzuga sedimentary–volcanic Greenstone Belt is associated with a major Early Proterozoic Rift System, and extends over 700 km. The deposit is situated between blocks of Archean crust. The sedimentary–volcanic belt is composed of shales, sandstones, and tuffs of the Pilgujarvi Formation. A number of layered mafic–ultramafic intrusions were emplaced between Early Proterozoic rifts and Archean basement rocks. The differentiated sills (intrusions) with composition grading upward from wehrlite through pyroxenite to gabbro–wehrlite association occur within and adjacent to the Pechanga–Varzuga belt. The mafic–ultramafic intrusions host numerous massive and disseminated Ni–Cu-sulfide ore deposits within Ni-bearing gabbro–wehrlite. The Ni-sulfide deposits are concentrated in and around Pechenga and Imandra–Varzuga structures.

Imandra–Varzuga intrusions are characterized by a uniform mineralogical–geochemical signature. It extends over 80 km with width up to 6–7 km. The sheet-like igneous complex trends NNW–SSE and dips toward the SW down to a depth of 4–5 km. The massif body splits into three large blocks by a set of distinct wide transverse faults forming Fedorova, West Pana, and East Pana. The first two blocks are composed of mafic (gabbronorite) and ultramafic rocks. The East Pana Massif is laterally heterogeneous, and characterized by gabbro, gabbronorite, and mafic pegmatite rock. There are six significant Ni–Cu–PGE deposits in Pechenga–Imandra–Varzuga belts that include Pechenga (main), NE Pechenga, Allarechka, Lovnoozero, Monchegorsk, and Pana and Fedorova Tundra.

The ore-forming minerals are pentlandite, pyrrhotite, and chalcopyrite occurring as discordant, disseminated, stratabound massive sulfides, breccia-matrix, net-veined, flow-structured, stringer mineralization in the basal sections of intrusions.

The overall size of the deposits has been estimated at ~150 Mt at 1% Ni, 0.04% Co + PGE + Au and that for the district/belt has been estimated at 339 Mt at 1.18% Ni and 0.63% Cu, 0.045% Co, and 0.32 g/t PGEs (Naldrett, 2004).

6.12.3 Sukhoi Log Gold–Platinum Deposit

Sukhoi Log deposit is a large gold–platinum association located within the Lena gold field, 120 km north of Bodaibo, Eastern Siberia. The prospect with gold quartz–vein lode mineralization was known since 1886 in the Nygri River basin (Distler and Marina, 2005). The rock types comprise sand- and siltstone, calcareous and carbonaceous shales, volcanogenic andesite–basalt and tuff. The host rock is exclusively stratabound (up to 3-m thick) calcareous and organic carbonaceous silt–sandstone, argillite, and shales belonging to the Upper Riphean Formation of Neoproterozoic age. The organic compounds within these sediments have been carbonatized by hydrothermal fluids. The deposit occurs in a complex tectonic fault zone along the axes of overturned anticlines that plunge at 30–40° to the W or NW. The axial zone/core of the fold contains an abundance of quartz–sulfide veinlets with complex shapes. The outer zone on both sides is characterized by disseminated fine-grained pyrite, gold, PGEs, and other sulfides.

Total estimated reserves at Sukhoi Log stand 384 Mt with an average grade of 2.5–2.7 g/t Au. Resources include 165 Mt at 2.0–2.3 g/t Au from a possible pit extension, and 205 Mt at 0.8 g/t Au from mineralization envelope. Total placer type production since 1846 to date has been 1500 tonnes (~1650 tons) of gold.

6.12.4 Volkovysky Cu–PGE Deposit

Volkovysky PGE deposit is located in the northeast part of a large gabbro massif, and was discovered in 1912. Since then, it has been intermittently explored and mined by small surface pits and underground stoping. Palladium-bearing Cu–Fe–V–Ti–P mineralization is hosted by Late Ordovician pyroxenites intruded into a Late Ordovician gabbro massif. Mineralization occurs as many nonpersistent small individual ore lenses concordant with the layering of the host intrusion and can be traced over 3 km and have a thickness of hundreds of meters. The orebodies trend E–W at the northwestern end, become arcuate at Volkovysky, and finally trend N–S at Lavrovo–Nikolaevsky in the southern end with average dip of 60° toward the S in the northwest and 40° toward the W at the southern end.

Copper is the principal economic metal (bornite–chalcopyrite) with vanadium, titanium, and phosphorus as byproducts, and palladium, selenium, tellurium, gold, and silver as value-added products. The average grades are ~1.5% Cu, and 0.1–0.2 g/t Pd + Pt.

6.12.5 Nizhny–Tagil PGE Deposits

The Nizhny–Tagil platiniferous deposits are located within the Ural Platinum Belt of the northern Ural Mountains. These platiniferous mineralizations are the primary source of Ural platinum placers.

The in situ mineralization is hosted by a mafic–ultramafic pluton comprising the largest (14 × 6 sq. km) dunite massif in the area. The mafic–ultramafic pluton exhibits a concentric zonal structure with a central dunite core, gradationally surrounded by successive bands and rims of serpentinite, wehrlite, thin clinopyroxenite, feldspathic pyroxenite, and finally passes into gabbro. Platinum mineralization is closely and exclusively associated with irregular pods, lenses, and branching veins of disseminated and massive chromite within the central core composed of fine-grained protodunite. Platiniferous chromites crystallized at the last phase of the ultramafic magma. Chromite occurs within brecciated dunite, set in a matrix of massive chromite and veins of chromite within dunite. PGEs occur as Pt–Fe alloys, platiridium, and other platinum alloys. The enriched chromite accounts for PGE concentration with minor osmium and ruthenium.

6.12.6 Kondyor Platinum Placer Deposit

The Kondyor mineralization belongs to a classical ultramafic–alkaline massif of Palaeozoic or Mesozoic age. Dunites in the central part of massif are extensively weathered forming a depression creating an alluvial–eluvial placer type of PGE deposits. Alluvial placers from the central depression have been in operation since 1984 with PGE resources of ~60 tonnes (~66 tons).

6.12.7 Guli Placer Deposit

The Guli massif complex is situated at the periphery of the Siberian Craton. The presence of economic PGEs as placers is a specific feature of the Guli massif. The clinopyroxenite–dunite–harzburgite ophiolite-type ultramafic complex forms gold–platinum placer deposits. The unique features of the placers at Guli are the dominance of laurite and osmium-rich alloys over other PGEs and gold. The significant expected resources of noble metals are osmium and laurite. Os-rich alloys are frequently intergrown with Ru–Os sulfides, chromite, olivine, and clinopyroxene at Guli.

6.12.8 Burakovsky Chromite Deposit

The Burakovsky intrusive complex is the largest layered of the Fennoscandian intrusions of Mesoproterozoic age (2430 Ma). The intrusion is located in the Karelia region. The layered sequence consists of a thick ultramafic part dominated by dunite and peridotite overlain by alternating layers of pyroxenite, wehrlite, and gabbro (Transition Zone), followed by a thick sequence of gabbronorite, pigeonite gabbronorite, and magnetite gabbronorite–diorite. The stratified layered-intrusive igneous complex covers an aerial extent of ~700 sq. km and has thickness of 4–6 km. The deposit contains chromite and Ni-silicate, and is a potential source of platinum. The main chromite horizon is located at the interface between the ultramafic part and the layered series. Seam thickness varies between 0.5 and 4 m. The grade of the main chromite seam varies between 49 and 52% Cr_2O_3 and hosts laurite and Os–Ir alloy as tiny inclusions in chromite. The upper layered sequence above the main chromite zone is characterized by disseminated sulfide blebs of chalcopyrite, pyrrhotite, and pentlandite assemblage associated with Pd and Pt arsenides/tellurides.

6.13 TURKEY

Turkey is one of the leading chromite-producing countries after South Africa, India, and Kazakhstan. The deposits are of podiform type and occur as discontinuous masses of lensoid bodies in tectonic belts related to the Alpine orogeny. Most of the chromite deposits are located at high altitude. Turkey ranks fourth after South Africa, Kazakhstan, and Zimbabwe in chromite reserves.

The Guleman area of Elaziğ in Eastern Turkey has been one the important chrome ore-producing districts in Turkey since 1936. The Guleman deposit is located 50 km to the southeast of Elaziğ city in astern Anatolia. The mines include Kefdağ and Soridağ. Host rocks for the chromite deposits are characterized by Guleman allochthonous ophiolitic ultramafic assemblages. They cover a large area and are underlain by sedimentary rocks. The Guleman ophiolite unit is overlain by basal conglomerate of allochthonous Upper Cretaceous and Middle Eocene shallow-sea sediments. The Guleman peridotite unit covers ~200 sq. km area with an east–west elongation. The peridotite unit can be classified in two distinct groups: tectonites and cumulates. The tectonite group comprises harzburgite and dunite. The cumulate group includes alternating layers of dunite, wehrlite, pyroxenite, troctolite, and gabbro. The cumulate rocks cover ~70 sq. km, and tectonites cover the rest in the Guleman peridotite unit.

The Guleman peridotite unit hosts over 500 chromite occurrences with variation in size, continuity in strike, dip, and length of 1 m to a few hundreds of meters. The chromite orebodies of the tectonites are usually larger and rich in chromite content compared to the same in cumulates. The reserves and resources have been estimated at 220 Mt grading up to 48% Cr_2O_3.

6.14 YEMEN

Yemen has few metallic mineral deposits that include nickel (Suwar Ni–Cu), copper, cobalt, gold, silver, and zinc.

The Suwar Ni–Cu deposit is located near Suwar and Nashir villages, 4 km north of Bayt Adhaqar small town, and about 60 km northwest of Sana'a, Yemen. Ni–Cu mineralization is visible at the surface at several gossanous exposures within the intrusive areas and serves as the key geological guide for exploration. The regional geology of the Suwar area covers a part of Proterozoic basement rocks exposed by the headwaters of a local river system. The geological setting of the intrusion is established as a broadly differentiated noritic–gabbroid mafic–ultramafic intrusive host system. The complex intrudes into retrograded amphibolite-facies of the Pan-African Afif lithotectonic terrane, and is usually undeformed, unaltered, and unaffected by the Pan-African orogeny. The emplacement of the complex (639 Ma) occurred during a tensional tectonic regime on the Arabian Peninsula and marked the time of proto-Iapetan rifting (Greenoush et al., 2011). The mafic–ultramafic component of the intrusive at Suwar is exposed over a length of about 3 km, along the Suwar Wadi (gorge), a maximum width of some 2 km across the wadi, and a depth (thickness) of 200 m.

The area has been surveyed by aerial photography, magnetics and electromagnetics, surface mapping, and drilling. The principal sulfide mineralization at Suwar consists, essentially, of chalcopyrite and a pyrrhotite–nickel sulfide intergrowth in which the nickel sulfide component is wholly or mostly pentlandite.

The estimated size of the mafic–ultramafic intrusion, its noritic composition, Archean subcontinental lithospheric-mantle signature, and its position in thin Proterozoic lithosphere abutting Archean cratonic rocks, give it the key characteristics of known, mostly Proterozoic, intrusions that host world-class Ni–Cu–Co ore deposits (Greenoush et al., 2011).

Deposits of Australia

Platinum-Nickel-Chromium Deposits
http://dx.doi.org/10.1016/B978-0-12-802041-8.00007-9

191

7.1 BACKGROUND

Australia, New Zealand, New Caledonia, and Papua New Guinea belong to the Oceania group of continent/islands, and have been placed in one chapter.

7.2 AUSTRALIA

The Commonwealth of Australia is a continent and a country too. The neighboring countries include Papua New Guinea, Indonesia, and East Timor to the north, the islands of Solomon, Vanuatu, and New Caledonia to the northeast, and New Zealand to the southeast. The mineral resource-based industries and mining-related exports are the key pillars of the Australian economy. Australia is the leading producer of bauxite and iron ore; the second-largest producer of alumina, lead, and manganese; the third of brown coal, gold, nickel, zinc, and uranium; the fourth of black coal and silver; and, finally, the fifth-largest producer of tin and platinum in the world. It has significant mineral endowment supported by strong advanced mineral exploration technology along with scientific and mechanized-mining tradition. The country is equipped with technical institutions and vocational training centers, well-trained engineers, and workforces. The exploration and mining companies are spread over the globe at different capacities.

Western Australia and the Northern Territory hold major resources of the platinum group of metals (PGMs). The total identified reserves and resources of the platinum group of elements (PGEs) stand at 318 tonnes (~350 tons) (2009). The production is exclusively from nickel sulfide deposits hosted by Archean komatiitic rocks in the Yilgarn Craton of Western Australia. A large, low-grade lateritic chromium–nickel–platinum deposit has been discovered at shallow depths in the Flemington Intrusion, part of the Tout Intrusive Complex near Fifield. The intrusion contains a core of ultramafic rocks, primarily dunite, with an outer shell of gabbro and hornblende-rich rocks. The majority of the chromite deposits are located in New South Wales (NSW). The deposits are relatively small and of the podiform type associated with serpentinite and other ultramafic intrusive rocks. The largest deposit of this type in NSW is near Bingara, associated with a large serpentinite body. Several chromite pods and float ore occur over a considerably large area. A significant deposit has been discovered at Coobina, in the north of Western Australia.

7.2.1 Munni Munni Igneous Complex

7.2.1.1 Location

The Munni Munni PGE-bearing mafic–ultramafic magmatic complex is located within the world-class iron ore/zinc/copper Pilbara mining region of north Western Australia. The deep seaport, rail hub, regional administrative and central service center of Karratha town is at a distance of about 55 km to north of the deposit. The deposit is extremely well located to another port at Dampier at a distance of 8 km from rail access. The layered igneous complex was described in the 1960s, but the platinum potential was first discovered by Dr. John Ferguson in the 1980s, and accordingly, the identified mineralized horizon is referred to as "Ferguson Reef."

7.2.1.2 Geology

The regional geology is represented by Pilbara Block of the Archean Pilbara Craton covering an area of ~61,000 sq. km. Pilbara Craton consists of a series of east–west trending granite–greenstone terranes of dome-shaped granitic batholiths separated by synclinal belts of metasediments and metavolcanic rocks. The regional basement stratigraphy is collectively called the Pilbara Supergroup (2800–3000 Ma). It is unconformably overlain by late Archean to early Proterozoic sediments and volcanics of the Mount Bruce Supergroup (Fig. 7.1).

The differentiated Archean layered mafic–ultramafic complexes occur throughout the Pilbara Craton. Munni Munni Igneous Complex is relatively the largest (25 × 9 km) and the best preserved among these complexes. Munni Munni Complex (2925 ± 13 Ma) is a boat-shaped (similar to a tilted canoe) intrusion emplaced

into the Archean Whundo Group and Cheratta Granitoid Complex of Western Pilbara Craton. It is composed of an alternating sequence of ultramafic rocks overlain by a thick mafic package of predominantly gabbroic rock. The layering dips toward the center of the intrusion and trends laterally into a narrow and variably contaminated chilled margin. The samples from the chilled margin show evidence of in situ contamination, and indicate that the parent magma to the ultramafic portion of the intrusion was a high-Mg, low-Ti basalt with similarities to typical Archean siliceous high-Mg basalts (Barnes et al., 1994).

The intrusion is +5000-m thick with the bottom of ultramafic unit of 1800-m thick, and the upper gabbroic package of 3600-m thick. The ultramafic unit is composed of websterite, olivine websterite, lherzolite, wehrlite, clinopyroxenite, and minor dunite. The uppermost 100-m layer of the ultramafic unit is characterized by medium to coarse porphyritic websterite.

The ultramafic layers are thin to the west of the intrusion and bulge down in one locality to join with the Cadgerina dike, which is interpreted as a feeder to the complex that was active at the time of the formation of the porphyry websterite zone (Naldrett, 2004).

7.2.1.3 Mineralization

Ferguson Reef has the largest estimated platinum resources in Australia. The thickness of the PGE mineralization in Ferguson Reef ranges between 1 and 9 m with an average of 2.9 m. The mineralization extends over 2 km in strike and continues from surface over 1 km depth at an average dip of 45° toward the S. The mineralization is characterized by two types of ore dominance, such as distinctly high and low sulfides. The dominant sulfides are chalcopyrite and pyrrhotite with traces of pentlandite, and no chromite. The mineralized layer is a stratiform accumulation of PGE-enriched disseminated sulfides over an interval of 2–10 m within the top 20 m of the porphyritic websterite zone along its entire exposed strike length,

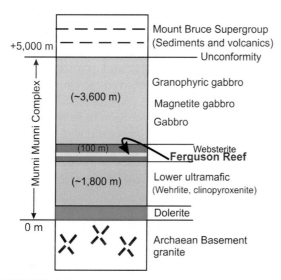

FIGURE 7.1 Schematic stratigraphic column of Munni Munni PGE–Ni–Cu-bearing layered mafic–ultramafic complex showing the Ferguson Reef within porphyritic websterite at the contact between ultramafic and mafic layers.

with maximum total PGE grades up to 8 g/t over 50-cm widths. The main Ferguson Reef has been subdivided in two reef types referred to as Coincident and Offset Reef recognized by core drilling (Cawthorn, 2005b). The offset layer comprises upper disseminated PGE-poor and lower massive PGE-rich units. The Coincident Reef is of higher metal grade than that of Offset Reef. The Coincident Reef occurs at the base of the mineralization.

7.2.1.4 Reserve Base

Reserve and resource for Ferguson Reef have been estimated at 23.6 Mt grading 1.1 g/t Pt, 1.5 g/t Pd, 0.2 g/t Au, 0.1 g/t Rh, 0.1 g/t Cu, and 0.09 g/t Ni at 1.9 g/t PGE + Au cutoff. The reserve and resources have been reported under the Measured, Indicated, and Inferred category following Joint Ore Reserve Committee (JORC) Code 2004. The estimation includes all drilling completed at Cherratta, Pinderi, Central, Maitland, and Yannery Zones of the Northern Domain (Platina Resources Limited, 2010).

The Munni Munni megaproject is being focused on a smaller venture with scope for selective mining of higher-grade mineralization in combination with metallurgical advances suited for small-scale processing.

7.2.2 Panton Mafic–Ultramafic Sill

Exploration for PGEs in Australia in the mid to late 1980s resulted in discovery of the potentially mineable Panton deposit in the Kimberley. PGE exploration in Australia is currently undergoing resurgence with active projects in the Kimberley region.

Panton PGE deposit is located in the North Kimberley of Western Australia. It contains one of the largest resources and the known highest grade of PGEs–Cr in Australia. The deposit occurs in a differentiated and tectonized, layered mafic–ultramafic intrusion with stratabound PGE–chromitite layers immediate below the mafic–ultramafic contact. The drilling in the late

1980s had established high-grade resource of 2 Mt grading 6.02 g/t combined Pt + Pd + minor Au in a chromitite layer about 1 m of thickness. The chromitites have been mapped at surface over a strike length exceeding 12 km. Subsequent shallow drilling program 5500 m in 60 holes (Diamond Drilling + Reverse Circulation) during 2000–2001 confirmed substantial mineralization in the shallow zone above the known resource. Drilling has covered a strike length of ~3 km with generally two holes at every 100-m interval along strike. In addition, surface trenches have been excavated along drill section lines at every 200 m along strike. The drilling through a series of drill sections has confirmed that the chromitites extend to the surface with high grades similar to those drilled at depth and have demonstrated wide zones of lower-grade Pt–Pd mineralization disseminated in rock adjacent to the chromitites. These drilling results are likely to substantially increase the resource and open the way for low-cost open-pit mining from surface to ~60-m depth. The drill cores indicate that the mineralization is essentially unoxidized. Metallurgical testing will determine the recovery from this lower-grade shallow mineralization.

The reserves and resources have been estimated at 14.3 Mt grading 2.19 g/t Pt, 2.39 g/t Pd, 0.31 g/t Au, 0.27% Ni, and 0.07% Cu of Measured, Indicated, and Inferred categories (March 2012). The encouraging results obtained from the shallow drilling over 3 km of strike, suggest that similar results are expected from drilling of further strike extensions of the known mineralization. The Panton project is on hold pending improvement in PGE prices. On May 21, 2012, M/s Panoramic Resources Ltd announced that the Company had purchased the Panton PGE deposit from Platinum Australia Ltd.

7.2.3 Fifield Platinum Province

The Fifield Platinum Province, situated about 380 km WNW of Sydney, contains a number of Alaskan-type mafic–ultramafic complexes intruding the metasedimentary rocks of Cambrian–Ordovician Girilambone Group in central NSW. The Fifield Platinum Field (deposit), located 100 km WSW of Dubbo, NSW, was the largest producer of platinum and gold in Australia from Tertiary gravel placer remnants. The Fifield is the only locality in Australia with records of alluvial platinum mining dating back to the 1880s. The sources for the alluvial platinum are several ultramafic intrusive complexes in the region including the Murga Complex. The mafic–ultramafic intrusions are composed of monzodiorite, gabbronorite, hornblendite, hornblende/olivine/biotite/magnetite-clinopyroxenite, peridotite, and dunite emplaced during the Silurian–Devonian periods as conduits to volcanoes. The exploration interest in the area follows from Black Range Minerals defining 107 Mt at 0.66% Ni, 0.11% Co ± Pt in a lateritic deposit over the Tout ultramafic intrusion at Syerston, NSW.

The Murga intrusive is central to the largest mined alluvial platinum leads in the Fifield area. The drilling conducted over the Murga intrusive in the late 1980s intersected mineralization of 1 m at 0.93 g/t Pt and 20 m at 0.37 g/t Pt. The Kars intrusive complex defined a 500-m long by 50-m wide zone of platinum mineralization averaging 0.42 g/t platinum through trenching and drilling in the early 1990s. The mineralization is hosted by olivine-rich wehrlite, dunite, and olivine pyroxenite. A dunite core, the source rock type for laterite Ni–Co, has been mapped at Avondale, south of Fifield. The higher platinum results are also reported over a 2500 × 300 m zone. These results include 8 m at 0.32 g/t Pt. A 50-m wide zone of platinum mineralization grading ~0.45 g/t Pt over a strike length of 500 m has been defined at the Kars Prospect. The diamond-drill intercepts indicate mineralization continuing to depth.

The resource has been estimated at 10.2 Mt at 0.61 g/t Pt, 2 Mt at 0.18% Co, and 0.35% Ni. The Platina Resources Ltd announced Indicated and Inferred reserves and resources totaling 12.7 Mt

grading 0.7 g/t Pt for its Owendale North, Cincinnati, and Milverton deposits at Fifield region (2012). Platina Resources Limited is drilling the Owendale deposit and reassaying some of the core from historic drill holes.

7.2.4 Weld Range Complex–Parks Reef

The Weld Range PGE–Au open-pit property is located within the Archean Yilgarn Craton, Western Australia. The orebody occurs in a truncated lateritic profile overlying low-grade primary platinum group mineralization in ultramafic rocks. The Weld Range PGE deposit is adjacent to the very large Weld Range lateritic nickel–cobalt deposit. The Weld Range lateritic nickel–cobalt deposit is also associated with a low-grade chromium resource.

Inferred resources for Weld Range ultramafic complex have been estimated in three blocks: Weld Range Chromite block (63.5 Mt grading 5.2% Cr and 38% Fe), Weld Range Ni–Co block (330 Mt grading 0.75% Ni and 0.06% Co), and Weld Range PGE + Au block (14.76 Mt grading 1.1 g/t Pt + Pd + Au).

Weld Range Metals Ltd conducted a scoping study in August 2010 and concluded that Stage 1 of the project is technically and economically feasible using processing equipment and technology currently being adopted by the steel industries. The PGE resources are present in prospects associated with komatiitic nickel–cobalt sulfide and lateritic nickel deposits.

7.2.5 Kambalda Nickel Field

The Kambalda Nickel District in Western Australia is one of the great nickel provinces in the world. The deposit has produced over 47 Mt of ore, containing more than 1.4 million tonnes (~1.54 million tons) of nickel metal since its discovery in 1968 by the Western Mining Companies Resources Ltd. The District has consistently generated more than 35,000 tonnes (~38,600 tons) of nickel metal per annum, apart from a brief period in the late 1990s when it was partly shut down.

7.2.5.1 Location

The Kambalda Nickel Field is located about 60 km from the mining city of Kalgoorlie, ~75 km south east of Coolgardie, and 616 km east of Perth in Western Australia. The area is situated within the Kalgoorlie gold field and was discovered as a part of the flourishing gold mining boom that triggered prospectors all over Western Australia to move into Kambalda and settle at the base of nearby Red Hill in 1897.

Systematic exploration for nickel deposits started during the early 1970's. The Western Mining Companies (WMC) was the first company to develop a major mining and processing unit at Poseidon with a total resource of 8.5 Mt of ore at 2.02% Ni. The company continued mining until 1989. The first shipment of nickel concentrate was made in 1974. The fate of nickel mining passed through bad weather due to market fluctuation, and WMC was taken over by Broken Hill Proprietary (BHP) Billiton.

7.2.5.2 Geology

The Kambalda nickel sulfide deposits are located discontinuously along the 600-km long NNW–SSE strike trend of the Kalgoorlie Eastern Goldfields superterrane of the Yilgarn Craton. The northern and central portions of the terrane are characterized by the occurrence of both large low-grade and smaller high-grade deposits associated with komatiite units within felsic volcanic rocks. The Kambalda dome lies within the southern Kalgoorlie superterrane, where nickel sulfide deposits are associated with thick komatiite flow units in stratigraphic contact with a thick pile of pillowed and massive tholeiitic basalts (Barnes et al., 2013a).

The stratigraphy of the Kambalda–Tramways area comprises the Kalgoorlie Group, Black Flag Group, and Merougil Beds Group of rocks. The Kalgoorlie Group consists of komatiites of the Kambalda Komatiite Formation and basalts of the Lunnon Basalt, Devon Consols, and Paringa Basalt Formations. The komatiite–basalt sequence is overlain by the felsic volcanic–volcaniclastic rocks, and clastic sedimentary rocks of the Black Flag Group. The Black Flag Group is unconformably overlain by the Merougil Group (Stone and Archibald, 2004) (Fig. 7.2).

The Kambalda nickel field comprises the Archean volcanic peridotite association covering an area of about 400 sq. km. It is consists of two sequences of ultramafic, mafic, and felsic volcanics, and sedimentary rocks. Most of the komatiitic iron–nickel–copper–PGE-sulfide ores occur in the high-magnesium lower (Kambalda) sequence at the base of the Kambalda ultramafic rocks. The ultramafic rocks of the upper (Bluebush) sequence contain very little mineralization (Gresham and Loftus-Hills, 2008).

The trough structural deformation at the Kambalda Dome is marked by repetition of stratigraphy, older sequences emplaced on the younger (Foster Thrust), upright anticlinal structures (Kambalda Anticline), and the Lefroy fault zone (Fig. 7.2). The Kambalda Dome (D3) is a doubly-plunging anticline on the crest of the major regional Kambalda Anticline (D2). The core of the dome is composed of Lunnon Basalt, intruded by granitoid (2662 ± 6 Ma), and flanked by the Ni-sulfide-bearing Kambalda Komatiite (2709 ± 4 Ma) (Fig. 7.2). The dips on the east flank are steeper (up to subvertical) than that of the west flank (45–60°). The plunges are steeper in the north (40°) than in the south (10–20°), and steeper than the Kambalda Regional Anticline (5–10°). The Lunnon Basalt is thickest (~2 km) in the core and the Kambalda Komatiite is thickest (~1 km) on the north (above Juan shoot) and south (above Beta shoot) flanks. The Kambalda Komatiite on Lunnon Basalt sequence is repeated on the west flank by the Loreto Thrust, and truncated on the east flank by the Lefroy fault zone (Fig. 7.2) (Stone and Archibald, 2004).

The ultramafic rocks have been intensely serpentinized due to seafloor alteration, and progressive and retrogressive regional metamorphism. The komatiitic peridotite has been hydrated to serpentine-dominated mineral

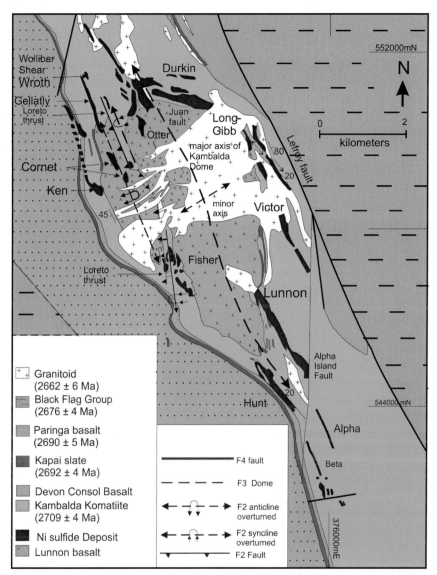

FIGURE 7.2 Geological map of Kambalda Dome showing the stratigraphic distribution of rock units, major structures, and Ni–Cu–PGE deposits. *After Stone, W.F., Archibald, N.J., 2004. Structural controls on nickel sulphide ore shoots in Archaean komatiite, Kambalda, WA: the volcanic trough controversy revisited. Journal of Structural Geology, 26, 1173–1194.*

assemblages preserving relict volcanic textures and minerals. The serpentinite subsequently altered to talc–carbonate mineral assemblages synchronous with D1, and most intense near major faults. The talc–carbonate alteration and progressive metamorphism involved porphyroblast overgrowth and development of cleavage overprinting relict primary textures and depositional relationship under the conditions of 480–530°C at 2–3 kb (Stone and Archibald, 2004).

7.2.5.3 Mineralization

The NiS ore and wall rocks are characterized by four geologic structures: (1) Trough structure in the upper surface of the Lunnon Basalt; (2) NiS ore partly confined by the trough structure (contact ore), and rarely at the base of overlying flows (hanging wall ore); (3) lack of sedimentary rocks in the trough; and (4) thickened, MgO-enriched ultramafic rocks or channel-facies komatiite in the immediate hanging wall.

The orebodies occur in more than 24 clusters each containing multiple ore surfaces caused by structural dislocation of a few, grouped, ribbon-like original orebodies. The majority of the contact ores occur at the base of the lowermost ultramafic flow and generally occupying as elongate troughs in the footwall basalt–ultramafic contact. The remaining mineralization belongs to the hanging-wall ore typically directly above contact ore at the base of the second or third ultramafic flow. The ores contain varying proportions of massive, matrix, and disseminated sulfides.

The contact-type ore constitutes more than 80% of the nickel resource and occurs as tabular to linear ore shoots up to 2.5-km long, 300-m wide, generally, 5–10-m thick. The reserve of individual orebody ranges between 0.6 and 12 Mt in size. The dominant sulfide minerals are pyrrhotite + pentlandite ± pyrite ± chalcopyrite bodies displaying vertically zoned structure. The massive ore contains more than 80% sulfide at the base and overlain by/or blocks of matrix-type ore or disseminated–blebby ore <40% sulfides.

The sulfide mineralization is characterized by three types: Massive, Matrix, and Disseminated.

The massive-type zone is 0.2–1.5-m thick, and consists of essentially 100% sulfides. The most abundant minerals are pyrrhotite, pentlandite, and minor amounts of pyrite, and chalcopyrite. The massive sulfide zone forms the lowermost unit, and lies directly on the underlying basalt unit. The grade of this unit varies between 10 and 14% nickel.

The matrix-type zone overlies the massive sulfide unit, and consists of a net-textured rock composed of intermixed sulfides and other non-sulfide minerals. The texture reflects the original interstitial distribution of sulfides and olivine grains in the original rock. The matrix sulfides range in grade between 3 and 8% nickel, and may be up to 1.5 m thick.

The disseminated zone consists of a fine-grained (0.5–2 mm) of sulfides scattered throughout the ultramafic host rock. It usually grades from 0.5% to 2.0% nickel, and forms the uppermost unit of the ore zone. The upper boundary is often gradational against the unmineralized overlying host rock.

The reserves and resources of Kambalda Ni–Cu–Co–PGE deposits stand at 67 Mt grading 2.90% Ni, 0.21% Cu, 0.21% Co, and 1.13 g/t PGEs.

7.2.6 Mount Keith Ultramafic Complex

The Mount Keith Ultramafic Complex is one of the largest komatiite belts in the world to host a group of nickel–copper– PGE- sulfide deposits.

7.2.6.1 Location

The Mount Keith group of Ni–Cu–PGE-sulfide deposits are located in the northern end of the 800-km (500-mi) long Agnew–Wiluna Greenstone Belt, which forms part of the northern section of the Eastern Goldfields Province of the Yilgarn Craton of Western Australia (Fiorentini et al., 2007). The deposit is situated 720 km northeast of Perth, and 450 km north of Kalgoorlie in the Northern Gold Field (Fig. 7.3). The nearest town of Wiluna is 85 km (53 mi) to the north, and well connected by Highway. Since 1892, the area around the Agnew–Wiluna Belt was explored for gold. The Wiluna gold mine is at a distance of 3 km from the town. The Mount Keith nickel deposit was discovered by Mr. J. T. Jones in 1968 while drilling a gossanous outcrop 1200 m south of Mt Keith Deposit-5, and the open-pit mining operation established by 1971 to provide 10% of the world's demand for nickel. The property was acquired by Broken Hill Proprietary (BHP) Billiton in June 2005 from its erstwhile owner WMC Resources Limited.

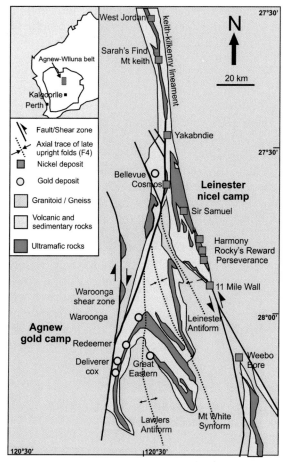

FIGURE 7.3 Geological map of Agnew–Wiluna Greenstone Belt in the Eastern Goldfields province of Western Australia, showing the location of Mount Keith Leinster camp, and other nickel sulfide and Agnew gold deposits within the belt and proximal to the NNW–SSE-trending Mt. Keith–Kilkenny lineament. *Modified after Hill, R.E.T., Barnes, S.J., Gole, M.J., Dowling, S.E., 1995. The volcanology of komatiites as deduced from field relationships in the Norseman–Wiluna greenstone belt, Western Australia. Lithos 34, 159–188.*

7.2.6.2 Geological Setting

Agnew–Wiluna Greenstone Belt is an N- to NW-trending stratigraphy and associated NNW shears localized on lithological contacts. The greenstone belt (2700 Ma) comprises a sequence of volcanic and volcaniclastic (felsic

and ultramafic) rocks, sulfidic cherts, carbonaceous pelites and shales, and laterally variable komatiites (olivine adcumulates, olivine mesocumulates, and olivine orthocumulates) including cumulates, thin spinifex-textured units, and komatiitic basalts (pillowed, tholeiitic, and high Mg). The olivine mesocumulates that host the Ni-sulfide mineralization are the thicker and lenticular part of the komatiitic sequence. The internal architecture of the Mount Keith Ultramafic Complex comprises seven distinct internal units that range from extreme adcumulate dunite to relatively fractionated pyroxenitic and gabbroic horizons. The ultramafic complex is divided into three komatiite horizons from stratigraphic bottom to top: the Mount Keith Ultramafic unit, the Cliffs Ultramafic unit, and the Monument Ultramafic unit (Fig. 7.4). The main Mount Keith Ultramafic adcumulate unit is situated at the base of the complex, contains the bulk of the disseminated nickel sulfides, and is dominated by coarse adcumulate olivine textures (Rosengren et al., 2007). The overlying Cliffs unit is an upper fractionated zone dominated by mesocumulate-to-orthocumulate peridotite containing pyroxenite and gabbroic lenses. The major sequences vary in thickness and strike continuity, and dip at 80–85° toward the W. The disseminated mineralization occupies a curvilinear thickening in the olivine adcumulate core of the host Mount Keith Ultramafic unit. The orebody plunges at low angle toward the S for a distance of 3.5 km. The bending sharply plunges steeply to the NNW at depth. The thickest part and the best nickel grade are associated with the bend in the overall ore trend (Perring, 2015a). The Monument Ultramafic unit at the top of the sequence does not contain any nickel sulfide mineralization.

The Agnew–Wiluna Domain is one of the most highly Ni-endowed belts in the world (Barnes et al., 2011c), and contains several world-class Ni–Cu–PGE deposits, including two of the largest-known komatiite-hosted ore deposits at

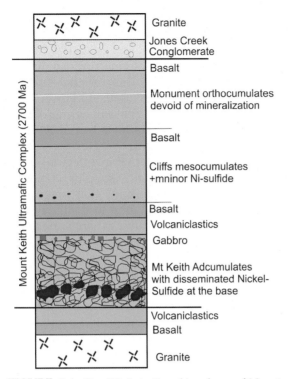

FIGURE 7.4 Simplified stratigraphic column of Mount Keith Ultramafic Complex. Mount Keith komatiite unit located at the base is composed of coarse-grained olivine adcumulate and contains the bulk of the disseminated nickel sulfides. The overlying Cliff unit of medium-grained olivine mesocumulate contains less nickel sulfide mineralization. The Monument ultramafic unit at the top of the sequence does not contain any nickel sulfide mineralization.

Mount Keith and Perseverance. The other nickel deposits are Sarah's Find, Honeymoon wells, Cosmos, Rocky's Reward, and Harmony.

A Ni–Co–As–Pd geochemical halo identified around the Sarah's Find orebody, 4.5km north of Mount Keith (Fig. 7.3), is interpreted to form by syndeformation circulation of As-rich hydrothermal fluids dissolving base metals, Pd and Pt from the orebody and redepositing them along the sheared footwall contact. Similar geochemical haloes could potentially exist around any magmatic nickel sulfide mineralization that has undergone a phase of arsenic metasomatism. The geochemical signature will be an

exploration target as a proximity indicator for undiscovered nickel sulfides in hydrothermally altered terranes (Vaillant et al., 2015a).

7.2.6.3 Mineralization

The disseminated Ni–Cu sulfide mineralization extends over 2km at Mount Keith with 300-m wide and continue down depth for 600 m and beyond. The dominant sulfide minerals are pentlandite, pyrrhotite, millerite, violarite ($Fe^{2+}Ni_2^{3+}S_4$), and trace amount of pyrite, chalcopyrite, heazlewoodite (Ni_3S_2), gersdorffite (NiAsS), polydymite (Ni_3S_4), sphalerite, Co sulfide, and tochilinite [$6Fe_{0.9}S·5(Mg,Fe^{2+})(OH)_2$].

7.2.6.4 Reserve Base

The Mount Keith deposit is the world's largest accumulation of magmatic sulfide-hosted nickel within rocks of komatiitic affinity, with a premining resource of 503 million tonnes (~554 million tons) of ore at a grade of 0.55% Ni for 2,770,000 tonnes (~3,050,000 tons) of contained Ni (Barnes et al., 2011b).

7.2.7 West Jordan Nickel Deposit— a New Discovery

West Jordan nickel deposit is a new discovery in the known Agnew–Wiluna greenstone belt, Yilgarn Craton, Western Australia. The deposit is located about 52km south southeast of Wiluna, between the Honeymoon Well nickel complex and the Mt Keith mine (Figs. 7.3 and 7.5). The large low-grade nickel sulfide deposit is defined by exploratory drilling during between 2009 and 2012.

The Archean greenstone sequences are dominated by basaltic and felsic volcanics and volcaniclastics, komatiites, large cumulate ultramafic komatiitic bodies, and minor sedimentary rocks The belt forms part of the northern section of the Kalgoorlie Terrane, and is dominated by rocks of 2700 Ma age. The West Jordan area forms part of the Albion Downs domain. The nickel-bearing belt trends in a north-northwest to

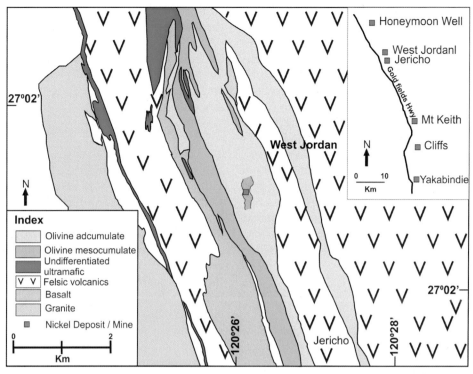

FIGURE 7.5 Surface geological map of West Jordan nickel sulfide deposit in Agnew–Wiluna Greenstone Belt—a new disseminated large volume and low grade deposit. *After Grguric, B.A., Seat, Z., Karpuzov, A.A., Simonov, O.N., 2013. The West Jordan deposit, a newly-discovered type 2 dunite-hosted nickel sulphide system in the northern Agnew–Wiluna belt, Western Australia. Elsevier. Ore Geology Review 52, 79–92.*

south-southeast direction, steeply dipping east, and is bound by large terrane-scale faults and Archean granitoid bodies. The deposit is hosted by serpentinized Type-2 dunite (olivine adcumulate), located in the core of a large dunite body olivine mesocumulate, and undifferentiated ultramafics (Fig. 7.5).

The mineralization at West Jordan is dominated by intercumulus disseminated sulfide blebs (20 μm–6 mm across) in assemblages containing pentlandite, pyrrhotite, heazlewoodite (Ni_3S_2) and locally native nickel, sphalerite and chalcocite. The mineralization grades between 0.2% and 2% Ni, with the majority of samples ranging between 0.35% and 0.7% Ni with an average of ~0.58% Ni. The mineralization is consistent with most komatiitic Type-2 systems. It is

deeply weathered, and shows hypergene alteration of the host ultramafic rocks caused by retrograde metamorphic fluids, resulting in extensive serpentinization as well as structurally controlled late magnesite alteration (Grguric et al., 2013).

The whole-rock chemistry of samples obtained from a series of 12 diamond-drill holes suggests the igneous stratigraphy to be west-younging, and consistent with deposits to the south along strike, such as the Mount Keith, and Yakabindie Type 2 nickel deposits (Grguric et al., 2013).

7.2.8 Leinster Ultramafic Complex

7.2.8.1 *Location*

The Leinster Ni–Cu–PGE Camp includes three significant deposits of the Harmony, the Rocky's

Reward, and the Perseverance with the Sir-Samuel to the north and the 11 Mile-Well to the south (Fig. 7.3). The Leinster Ni operations are located ~330 km north of Kalgoorlie gold mining camp, and about 90 km south of Mt Keith Ni mining camp. The group of deposits are located at a few km north of the Leinster town, established in 1976 by Agnew Gold Mining for its nickel operations. The Perseverance and Rocky's Reward nickel deposits are about 2 km apart, and the smaller Harmony deposit, a further 2 km to the north of Rocky's Reward. The Perseverance is a large komatiite hosted world-class Ni–Cu– PGE deposit. The Leinster area is situated toward the southern end and along the eastern margin of the Agnew–Wiluna Greenstone Belt in the Eastern Gold Field Province, Yilgarn Craton of the Western Australia (Fig. 7.3).

7.2.8.2 Geological Setting

The regional greenstone belt comprises felsic volcaniclastic sediments and lavas, with intermittent komatiite and black sulfidic, graphitic shales. The host komatiite of the Leinster group of Ni deposits are the largest type of basal sulfide-rich accumulation in the world that extend over a strike length of 20 km adjacent to the Perseverance fault. The mineralization occupies the overturned eastern limb of a regional anticline. The layered and banded komatiites display a wide range of compositions and textures, ranging from highly magnesian, olivine ortho-meso- to adcumulate rocks composed of olivine with forsterite contents of up to 94.5%, and capped by sequences of thin, and spinifex-textured komatiite flow units (Perring, 2015b). The core of the komatiite is called the Perseverance ultramafic complex, which is a thick intact accumulation of olivine-rich ultramafic rocks in which the primary igneous cumulate textures are preserved.

The ultramafic sequences in the Leinster are characterized by structurally repeated eastern and western successions. The mineralization (deposit) is exclusively confined at the stratigraphic base within the komatiites of the eastern felsic–sedimentary package dated at ~2700 Ma. The western section comprises a thick tholeiitic pillowed basalt with abundant sulfidic interflow sediments, spinifex-textured komatiite and intercalated felsic sediments, are devoid of any economic targets.

7.2.8.3 Mineralization

The orebodies at Perseverance deposit occur as a series of north-striking, steep dipping, individual fault-bounded sheets because of physical remobilization of the fault-related lodes and fold hinges. The main mineralization is a series of vertically stacked lenses. The surrounding low-sulfide envelope shows disseminated sulfide forming a distinct shoot, plunging 70° toward the S at more than 1100 m below the surface, and striking north, dipping steeply to the west. The extremely disseminated ores in the central zone contain >1.5% Ni, and 20% of the volume of the ore is predominantly associated with pyrrhotite and pentlandite. The weakly disseminated sulfide sheets contain 0.5–0.7% Ni. The principal sulfide minerals are aggregates of pyrrhotite and pentlandite with trace chalcopyrite.

The lateritic weathering of massive and disseminated sulfides, consisting largely of pyrrhotite and pentlandite, formed a supergene pyrite–violarite assemblage that has been replaced by goethite, carbonates, and quartz near the surface.

The mineralization at Rocky's Reward is localized in three ultramafic layers. The two eastern layers contain high disseminations and lesser massive sulfides and are folded about a steep, north-trending fold axis to form two vertically stacked mineralized layers that plunge at 10–15° toward the N. These are truncated to the west by a north-trending, steeply dipping high-strain zone that contains abundant remobilized massive sulfide. The disseminated and massive sulfides are hosted by two thin komatiite horizons within a package of thick felsic volcanics and sediments. The ultramafic rocks at Rocky's Reward have been hydrothermally altered, metamorphically recrystallized, and serpentinized, often with primary igneous textures.

The Harmony deposit/mine, small in size (3.2 Mt at 2.31% Ni), trends NNW dipping at 30–80° toward the W, overturned, and metamorphosed. The stratigraphic footwall to the west is composed of volcano-sedimentary and volcaniclastic rocks (2720–2725 Ma), and overlain by the host nickel-bearing komatiitic unit. The immediate footwall is characterized by pyrite-rich pelitic units. The stratigraphic hanging wall comprises reworked sedimentary or volcaniclastic rocks interlayered with barren spinifex-texture komatiite units. The host komatiite is thickest in the center of the deposit, thinning to both the north and south. The massive (~85%) and lesser disseminated (~15%) sulfides are largely concentrated near the basal sheared footwall with ubiquitous low concentrations in host komatiite. The principal sulfide minerals include pyrrhotite and pentlandite with lesser pyrite, violarite, chalcopyrite, and magnetite.

7.2.8.4 Reserve Base

The orebody at Perseverance is one of the largest single accumulations of komatiite-hosted massive to highly disseminated nickel sulfides in the world.

The mine production at Perseverance was 10.6 Mt at 2.1% Ni up to 1997, with 31 Mt at 1.65% Ni remaining in the deposit. The mine production at Rocky's Reward was of the order of 9.6 Mt at 2.4% Ni up to 1997 with remaining proved reserve base of 3.2 Mt at 2.85% Ni, and probable reserve of 6.4 Mt at 2.22% Ni. In addition, the low-grade indicated open cut resources of 79 Mt at 0.8% Ni and inferred resources of 144 Mt at 0.7% Ni are estimated.

Leinster Nickel operations had reserves + resources of underground ore totaling 53.2 Mt at 2.15% Ni and open-pit ore amounting to 157 Mt at 0.6% Ni at the end of 2004. The mine processed 2.73 Mt of ore at a head grade of 1.88% Ni to produce 44,577 tonnes (~49,140 tons) of Ni metal during 2004.

TABLE 7.1 Published JORC-Compliant Ore Reserves and Mineral Resources at Leinster Nickel Complex Estimated as of June 30, 2012

Category	Reserve + Resource (Mt)	Grade (% Ni)
MASSIVE SULFIDES		
Open pit	7.0	1.40
Underground	21.0	2.40
Total	28.0	2.15
DISSEMINATED SULFIDES		
Open pit	173.0[a]	0.52
Grand total (Massive and disseminated)	201.0	0.74

[a] Includes Open-pit reserve—3.1 Mt at 1.3% Ni; Underground reserve—10 Mt at 1.8% Ni; making a total reserve—13.1 Mt at 1.7% Ni. BHP Billiton Annual Report 2012.

The status of reserves base and grade as of June 2012 is given in Table 7.1.

The Perseverance mine (Leinster mine) is named after a nearby well. It is an underground mine to reach more than 1400 m below the surface and is one of the deepest in Australia.

7.2.9 Widgiemooltha Dome Komatiite Complex

The Archean komatiite-hosted nickel–copper–PGE sulfide deposits are located within the Eastern Goldfields Superterrane of the Yilgarn Craton in Western Australia. The major deposits are situated on the eastern flank of the Widgiemooltha dome within the fault-bounded Coolgardie domain of the Kalgoorlie terrane (Vaillant et al., 2015b). The komatiite-hosted Kambalda nickel sulfide deposit is ~50 km to the north.

The Widgiemooltha dome structure is a large regional anticline dome with surface coverage of 450 sq. km (174 sq. mi). The dome sequence comprises Archean ultramafic, mafic, and felsic metavolcanic rocks, and younger metasediments. The stratigraphic sequence around the

Widgiemooltha dome can be explained in two parts. The lower part contains two cycles of ultramafic lavas: the lower low-magnesium Mount Morgan komatiite, and the upper high-magnesium Widgiemooltha komatiite. The komatiite cycles are confined within extrusive and intrusive Mount Edwards tholeiitic basalts. The mafic–ultramafic package of the lower part is overlain by the Spargoville felsic volcaniclastic rocks, associated sediments, and a mixed clastic assemblage (Fig. 7.6). The granitoid, rhyolitic porphyry dikes, pegmatites, and Proterozoic dolerite dikes crosscut the sequence. Both komatiite sequences are exposed all around the margins of the Widgiemooltha dome as concentric rims centering on the large massive granitoid core.

The massive nickel sulfide bodies are located exclusively at the contact between the Mount Edwards tholeiitic basalt and the overlying Widgiemooltha komatiites in the upper cycle. The nickel mine includes: Mount Edward, Miitel, Marines, and Redross mines in the eastern flank, and the Wannaway mine in the western flank. In addition, a number of anomalies/prospects have been identified in both the flanks for exploration.

The Miitel orebody is elongated, near vertical, with gently plunging configuration, making it ideal for underground mining. The channel structure that hosts the ore lenses has been proved well continuous. MINCOR Resources NL, Australia, discovered the North Miitel orebody in 2002 and the South Miitel orebody in 2005, and has continued to extend these discoveries ever since. The overall mineralized system remains open to the north and south, and is subject to ongoing exploration. The orebodies extend in arcuate shape for 3.5 km and occur as discontinuous lenses with steep dips. The width varies between 2 and 10 m. The total postproduction reserves and resources at Miitel Mine stand at 629,000 tonnes (~693,000 tons) garaging 3.1% Ni [19,500 tonnes (~21,500 tons) of nickel metal] as of June 2015. The nickel Reserve base for the Belt has been estimated at 3.276 Mt at 3.6% Ni, equivalent to 117,100 tonnes (~129,000 tons) of nickel metal (Source: MINCOR Resources NL, Australia).

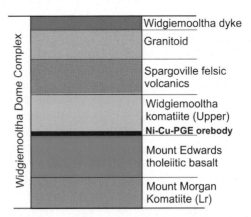

FIGURE 7.6 Stratigraphic column of Widgiemooltha structure showing the ultramafic–mafic and felsic sequence from base to top and characteristic localization of nickel–copper sulfide orebodies at the contact between the Mount Edwards tholeiitic basalt and the magnesium-rich upper komatiite cycle.

7.2.10 Forrestania Greenstone Belt

The Forrestania Greenstone Belt, the southern extension/end of the Southern Cross Greenstone Belt, is located about 360 km east of Perth. The Forrestania Greenstone Belt had been regionally explored for nickel since the late 1960s involving geological mapping, ground magnetic and electromagnetic (EM) surveys, and extensive percussion Rotary Air Blast (RAB) pattern drilling (Box 7.1). The exploration efforts led to the discovery of rich nickel sulfide mineralization at Flying Fox (1976), New Morning, Spotted Quoll (2007), Cosmic Boy, Digger Rocks, Mount Hope, Liquid Acrobat, and Seagull during the 1970s over a strike length of over 50 km following a nearly N–S direction.

The Forrestania Greenstone Belt comprises two main volcano-sedimentary associations of 2900 Ma age. The lowermost sequence is composed of basalt (tholeiitic)–ultramafic (komatiite)–Banded Iron Formation (BIF)±felsic sediments. The overlain upper sequence comprises felsic metasediments and black shale.

There are about six parallel ultramafic layers within the Forrestania Greenstone Belt (Fig. 7.7). The ultramafic layers extend over 90 km in N–S strike and dips between 40° and 80° toward

---

BOX 7.1

RAB / DTH

RAB (Rotary Air Blast) **Drill** or Down-The-Hole **(DTH) Drill** use a pneumatic reciprocating piston-driven "hammer" to energetically drive a heavy drill bit into the rock. RAB and Reverse Circulation (RC) noncore drills are used most frequently in the mineral exploration industry for faster and low-cost sampling with respect to diamond drilling. The samples can quickly delineate approximate shape, size, and grade for larger orebodies. The exact mineralization contacts and grades have to be confirmed by diamond drilling.

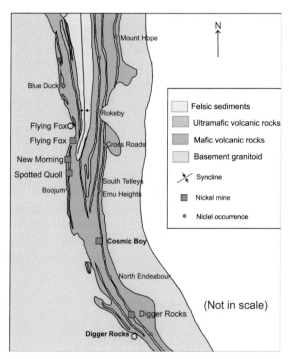

FIGURE 7.7 Surface geological map of Forrestania Greenstone Belt showing the six major ultramafic (komatiite) layers with the tholeiitic basalt, and approximate locations of Ni–Cu–PGE mines and occurrences in the belt. *After Perring, C.S., Barnes, S.J., Hill, R.E.T., 1995. The physical volcanology of Archaean komatiite sequences from Forrestania, Southern Cross Province, Western Australia, Lithos, 34, 189–207.*

the E. The layers include Eastern, Mid-Eastern, Central, Takashi, Mid-Western, and Western Ultramafics. The nickel sulfide mines/deposits/occurrences occur predominantly in the two komatiite layers (Eastern and Western). The granitic rocks intrude and exploit preexisting faults, and E–W-trending dolerite dikes crosscut the stratigraphy. The regional stratigraphy experienced mid-amphibolite facies metamorphism and polyphase deformation.

The Flying Fox and Spotted Quoll PGE-bearing nickel sulfide deposits are hosted within the Western Ultramafic belt. The Western Ultramafic belt comprises four distinct units. The footwall sequence is dominated by pelitic schists, quartz–feldspathic metasediments, and basaltic rocks. The footwall layer is overlain by cumulate-rich differentiated complex komatiite-flow sequence comprising a channel facies grading upward from ortho- and mesocumulates to noncumulates, with massive sulfides located at or near the base of the package. The overlying unit is a komatiite-basalt thin-flow facies sequence, in which noncumulate komatiites and high-MgO basalts dominate. This differentiated suite is sulfidic and associated with a continuous conductive horizon well known from downhole and surface transient Electro-Magnetic (EM) surveys. The hanging wall is represented by metasedimentary assemblages.

The Flying Fox and the Spotted Quoll nickel sulfide deposits are the two highest-grade nickel mineralizations in the world.

7.2.10.1 Flying Fox

The Flying Fox nickel sulfide deposit is situated at the extreme northern part of the Western

Ultramafic layer of the Forrestania Greenstone Belt. The mineralization is locally hosted in a layered quartz-rich footwall felsic metasedimentary sequence at the contact with the lower komatiite unit. The sheet-like massive orebodies continue in the footwall felsic sediments at depth. The discrete and independent Flying Fox orebody extends from 250 m to 1.2 km below the surface. The Flying Fox comprises several mineral zones, extending downward from T Zero to the T7 zone extending more than 700 m below T1, in a zone up to 600-m wide.

The Flying Fox reserve base has been estimated at 1.665 Mt grading 5.4% Ni as of June 2014 following JORC Code. The ore grade increases at depth from 3.9 to 6.2% nickel. The Flying Fox is the first producing mine in the area and is one of the highest-grade nickel nines in the world.

7.2.10.2 Spotted Quoll

The Spotted Quoll Ni sulfide deposit is located 6 km south of the Flying Fox, and 2.8 km south of New Morning mine. The mineralization is locally hosted in layered quartz-rich footwall metasedimentary sequence (>200 m thick) and 30–50 m below hanging wall komatiite. The PGE-bearing Ni sulfide orebody forms a sheet structure striking north–south, dipping at 50° to the east, and extends to a depth of more than 1000 m. The orebody is open at depth. The orebody is often disrupted by faulting. The deformed ore has been subjected to a hydrothermal event resulting in concentration of the PGE, Au, and As at the edge of the Ni ore. The PGM minerals include sudburyite (PdSb), sperrylite (PtAs$_2$), and irarsite (IrAsS). The PGEs and minor Au are hosted in gersdorffite (NiAsS). The arsenic in the magma system acted as a stimulator for the distribution of PGMs as solid solution in Ni sulfarsenides, and Pd in pentlandite (Prichard et al., 2013). The mineralization occurs as structurally hosted massive, matrix, and stringer sulfide in lode parallel to layering in local sequence.

The ore reserves stand at 2.90 Mt at 4.2% Ni, containing 121,400 tonnes (~133,800 tons) of nickel metal. The mine was initially designed as Tim King low-cost open pit, which concluded as planned in February 2012 after producing around 27,000 tonnes (~29,800 tons) of contained nickel. The property is now operating as a top–down underground mine using paste fill.

7.2.11 Nebo–Babel Large Igneous Province

7.2.11.1 Location and Discovery

Nebo–Babel (Nebo and Babel) mafic complex forms a part of the Large Igneous Province (LIP) located in the Mesoproterozoic West Musgrave Block, Western Australia. The area was subjected to modern exploration by BHP Billiton in the mid 1990's that culminated in the discoveries of the Nebo and Babel deposits in 2000, the Succoth deposit in 2013, and the large Cu–Ni–PGE magmatic sulfide deposit at Manchego Prospect in 2013. The Nebo–Babel is the first major discovery in the West Musgrave region, Western Australia. The initial drilling and exploration programs focused on the development of a large scale, low-grade open-pit mining operation.

7.2.11.2 Geological Setting

The Musgrave province is a Late Mesoproterozoic orogenic belt effected by several major tectonic episodes. It is an LIP close to the state borders of Northern Territory, South Australia, and Western Australia. The tectonic block covers an area of ~800-km long and 350-km wide with coherent geophysical and geologic characteristics (Karykowski et al., 2015).

The Nebo and Babel LIP forms two parts of an originally continuous gabbronorite intrusion. The mafic intrusion was emplaced into sulfide-free felsic orthogneiss (1502 ± 14 Ma) and subsequently offset by a km with a north–south shift by the Jameson fault. The fault-offset Nebo–Babel chonolith extends for ~5km in east–west direction and 1.5-km wide. The intrusion is formed by multiple, distinct, and chemically cogeneric magma pulses, emplaced along a linear weakness of the country rock, associated with the Giles Complex layered intrusions in the Warakurna LIP. The emplacement of Nebo–Babel at 1068±4.3Ma was part of a major magmatic event, which included the Giles Complex and the Warakurna LIP (Seat et al., 2011). The parental tholeiitic magma contains 8–9% MgO. The Nebo–Babel complex is characterized by a concentrically zoned, olivine-free, tube-like gabbronorite intrusion.

The Nebo–Babel intrusion is divided into three main lithostratigraphic units (Seat et al., 2007) and depicted in Fig. 7.8. The divisions are:

1. The chilled margins and marginal microgabbros forming the upper and lower contacts of the mafic intrusion, and gradually grading into the variably textured leucogabbronorite representing the first magma pulse.
2. The mineralized gabbronorite that intruded into the core of the variably textured leucogabbronorite as a second magma pulse with a composition broadly similar to unit (1).
3. The large volume of barren gabbronorite and the oxide–apatite gabbronorite (Nebo block) representing the final magma pulse(s) emplaced into the core of the intrusion.

The Nebo–Babel mafic intrusion and associated Ni–Cu–PGE sulfide mineralization formed between the intrusions of the low- and high-Ti basalts and may have originated from either mixing between these two magma types, or because of continuous change in the melting conditions between the magma cycles (Godel et al., 2011).

The Manchego Ni–Cu–PGE sulfide intrusion is situated in a magma conduit-type system and assimilated crustal sulfur (Karykowski et al., 2015).

7.2.11.3 Mineralization

The sulfide minerals in decreasing abundance include pyrrhotite, pentlandite, chalcopyrite, and traces of bornite and pyrite. The pentlandite occurs as coarse segregation along grain boundaries of pyrrhotite, and as exsolution flames within. The chalcopyrite occurs as irregular tiny grains and occasionally as coarse lamellae in pyrrhotite. The platinum group minerals are predominantly associated with the disseminated type of mineralization as tiny irregular grains of moncheite ($PtTe_2$), merenskyite ($PdTe_2$), michenerite (PdBiTe), and melonite ($NiTe_2$).

The sulfide mineralization at Nebo–Babel occurs mostly in two major types: massive and disseminated type.

7.2.11.3.1 MASSIVE SULFIDES

The massive and brecciated sulfides occur as isolated and discontinuous lenses close to or along the hanging-wall contact between the marginal breccia zone-chilled margin of gabbronorite and the felsic orthogneiss wall rock,

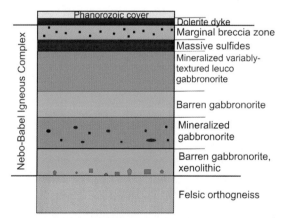

FIGURE 7.8 Stratigraphic succession at Nebo–Babel mafic complex showing the concentration of disseminated and massive Ni–Cu sulfide ore.

mainly consist of subrounded orthogneiss xenoliths in the massive sulfide matrix. The thickness of mineralized lenses vary between <3 m (Babel) and <27 m (Nebo).

7.2.11.3.2 DISSEMINATED SULFIDES

The disseminated sulfides are hosted by fine-grained variably textured leucogabbronorite, predominantly along the upper hanging-wall contact with the felsic orthogneiss country rocks and adjacent to the mineralized gabbronorite, melagabbronorite, and minor troctolite units. The sulfides occur as fine-grained bleb-like aggregates, angular sulfide blocks, and laths at interstices of silicate minerals (Seat et al., 2009).

All lithologic units at Nebo–Babel are variably depleted in PGE, indicating crystallization from magmas that have experienced various degrees of sulfide segregation prior to emplacement. The $\delta^{34}S$ values (0.0–0.8 per mil) are consistent with mantle-derived sulfur (Seat et al., 2009).

7.2.11.4 *Exploration and Reserve Base*

A systematic drill program comprising 147 RC holes for 23,135 m was completed in 2014. M/s Cassini Resources Limited estimated the Indicated and Inferred resources of Nebo and Babel deposits at 0.3% Ni cutoff in February 2014. The reserves and resources of Nebo stand at 28.9 Mt at 0.53% Ni, 0.46% Cu and 217 g/t Co. The reserve base for Babel stands at 174.2 Mt at 0.39% Ni, 0.41% Cu and 137 g/t Co. The total reserve and resource of Nebo–Babel have been estimated at 203.1 Mt grading 0.41% Ni, 0.42% Cu, 148 g/t Co, 2.1 g/t As, and 15% Fe_2O_3.

Phases 1 and 2 RC drilling program of 2340 m in 15 holes at the Manchego Ni–Cu–PGE Prospect has been completed by Phosphate Australia Limited till the end of 2013. A significant occurrence of massive to disseminated sulfide mineralization reaching up to 0.62 wt% Cu, 0.47 wt% Ni, and 1 ppm Pt + Pd was identified at the Manchego prospect (Karykowski et al., 2015).

7.2.12 Radio Hill Ni–Cu Deposit

The Radio Hill Nickel–Copper Deposit/ Mine is located 35 km south of Karratha. The nickel–copper sulfide mineralization at Radio Hill was discovered in 1972, and the Ni–Cu deposit status was recognized in 1983 by Abu-Ghazaleh Intellectual Property (AGIP), Australia. The mine made historic production of about 25,000 tonnes (~27,600 tons) of nickel metal and 22,000 tonnes (~24,200 tons) of contained copper for four years, and kept in abeyance in 2008 due to adverse fall in commodity prices.

The geology of the area comprises mafic–ultramafic assemblages that present from bottom to top as hornfels–intermediate and mafic metavolcanics, gabbro–gabbronorite–plagioclase websterite (containing massive Ni–Cu sulfide zones), interlayered dunite–lherzolite–olivine websterite–clinopyroxenite–plagioclase websterite, melanocratic olivine gabbronorite–olivine gabbro–gabbro, and websterite (Fig. 7.9). The Radio Hill orebodies are hosted by a thin gabbro unit (subzone "A") at the base of the Radio Hill layered-intrusive complex and the underlying basalt. The various massive sulfide orebodies have strike lengths up to 600 m, widths up to 60 m, and thickness of 15 m, with significant disseminated-sulfide mineralization overlying the massive orebodies, the source of present production. The orebodies occur as discontinuous subhorizontal lenses dipping at a shallow angle to the east.

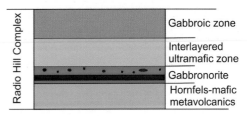

FIGURE 7.9 Stratigraphic succession of mafic–ultramafic complex at Radio Hill Ni–Cu–PGE deposit showing the massive and disseminated sulfide mineralization (red) at the basal contact between thin gabbro subzone-A and underlying mafic metavolcanics.

The subsequent exploration of the nearby Sholl Archean mafic–ultramafic intrusive complex discovered the Sholl B2 deposit with an indicated and inferred resource of 5.96 Mt at 0.53% Ni and 0.62% Cu, and a few smaller satellite deposits including Sholl B1 and Razerline. M/s Fox Resources Ltd (2010) reported the remaining reserve and resource for Radio Hill mine as 4.22 Mt at 0.65% Ni and 0.76% Cu. The company also reported that the remaining Indicated and Inferred Resources of palladium amounted to 1.275 Mt at 0.493 g/t Pd.

7.2.13 Waterloo Nickel Deposit

The Waterloo Nickel Deposit is located about 5 km north of the Thunderbox Gold Mine and in close proximity to the Goldfields Highway, 30 km south east of Leinster.

The Waterloo Nickel Deposit is set within a narrow, NNW-trending greenstone belt in the northeastern Goldfields region of the Archean Yilgarn Craton in Western Australia and bounded by granitoid and granitic gneisses on all sides. The deposit is hosted within a folded remnant of ultramafic stratigraphy on the eastern limb of a tightly folded southerly plunging synclinal structure. The nickel sulfide mineralization occurs at the basal contact of a 20–100-m thick serpentinized olivine mesocumulate unit. The overlying mafic–ultramafic sequence comprises dacite, interbedded thin differentiated komatiite flows, pillow basalts, mafic tuffs, and metasedimentary rocks. The host sequence dips steeply toward the W and is disrupted by low-angle shearing and associated folding. The host ultramafics have been truncated at depth by a large flat-dipping shear zone.

The komatiite-hosted magmatic nickel sulfide deposit occurs as elongated ribbon and comprises matrix- and disseminated-style primary sulfides with local structurally remobilized massive sulfides. The primary sulfide assemblage consists of pyrrhotite–pentlandite and minor violarite ($Fe^{2+}Ni_2^{3+}S_4$). The ore lenses are discontinuous and blind, and occur at depths between 100 and 400 m below the surface.

Noril'sk Nickel Company previously mined the Waterloo Nickel project by open-pit and underground methods between 2005 and 2008.

The ore reserves and mineral resources of the Waterloo deposit have been estimated as of December 31, 2009 (Noril'sk Nickel webpage, 2013) as:

1. Proven and probable ore reserves—0.011 Mt at 2.37% Ni.
2. Probable mineral resources—0.386 Mt at 2.09% Ni,
3. Indicated resources—0.286 Mt at 1.76% Ni.

7.2.14 Horn Nickel Deposit

The Horn nickel sulfide deposit is located at Wildara, in the Northeastern Gold Field, Western Australia. The geological mapping confirmed the presence of nickeliferous gossans within a structurally bound ultramafic unit with high-magnesium oxide (MgO) located at immediate south of the known mineralization, and serves as an excellent guide for exploration. The deposit is located toward the upper contact of a subhorizontal ultramafic unit and overlain by a flat covering of basalt. The host ultramafic occurs with a stacked series of ultramafics and is underlain by an ultramafic unit containing disseminated nickel sulfide. The deposit extends over 500 m in a NNW–SSE trend. The deposit is extensively drilled and open along the strike to both north and south, and along its western margin.

M/s Breakaway Resources Limited reported a small Inferred Resource for the Horn Nickel Deposit at 600,000 tonnes (~661,000 tons) grading 1.39% Ni, and 0.30% Cu, equivalent to 8300 tonnes (~9150 tons) of contained nickel, 1800 tonnes (~1980 tons) of contained copper, and 0.5 g/t Pd + Pt [Australian Securities Exchange (ASX) announcement on April 14, 2008]. The Revolution prospect, located immediately north of the Horn Nickel Deposit includes PGE-bearing disseminated-nickel mineralization.

7.2.15 Nyngan Nickel–Cobalt– Scandium–Platinum Deposit

The Nyngan (Gilgai) Nickel–Scandium Deposit is located 5 km south of Miandetta, off the Barrier Highway connecting the town of Nyngan to the west at a road distance of 20 km, and about 450 km northwest of Sydney, NSW, Australia. The deposit was discovered in the late 1980s on a routine exploration program searching for platinum.

The area is dominated by Cenozoic alluvial plains derived from the Darling River Basin with minor colluvium and outcrop. The region is situated on the shallow southern margin of the Surat Basin. The varying degree of lateritization in the area is frequently visible. The Nyngan complex is covered by 8–50 m of alluvial material. The intrusive complex, is the source of the scandium, nickel, cobalt, and PGMs in a regolith type of surface cover. It is an Alaskan-type ultramafic complex composed of a range of rock types including hornblende monzonite, hornblendite, pyroxenite, olivine pyroxenite to dunite–peridotites of Ordovician age. The intrusives are included within the "Fifield Platinum Province."

Jervois Mining Limited reported (2005) a resource of 16 Mt at 0.87% Ni, 0.06% Co, including 3 Mt at 290 g/t Sc, and 0.22 g/t Pt. The scandium-rich area of this deposit was updated (2009) as the Measured Reserves of 2.718 Mt at 274 g/t Sc, and Indicated Resources of 9.294 Mt at 258 g/t Sc.

7.2.16 Syerston Ni–Co–PGE Deposit

The Syerston complex is located 2 km from the regional town of Fifield, and 350 km northwest of Sydney, NSW. The Fifield district remains the location of Australia's only historic source of platinum production of about 20,000 ounces of the metal extracted between 1887 and the mid-1960s.

The Syerston orebody represents a typical surficial deposit hosted within a Tertiary age lateritic weathered profile. The metal enhancement of the minerals of economic interest occurred during a secondary process attributed principally to chemical weathering of the underlying metal-rich ultramafic rocks. The selective leaching of the relatively soluble elements, such as magnesium and silica, occurred during the continued and prolonged weathering process, leaving a highly iron-enriched residue or laterite rich in base and precious metals. The enrichment can enhance further during mechanical weathering or erosion.

The Tout Ultramafic Complex is one such intrusive body that underlies the laterite at the Syerston complex. The complex is concentrically zoned, ultramafic in the core grading to mafic material on the outer edges. The mafic layer is composed primarily of dark magmatic minerals in the core that diminish outward. The preferential weathering accelerates over the ultramafic core resulting in the formation of the laterite profile. The thickness of laterite reaches its maximum of 35–40 m over the core and thinning out laterally over the surrounding less-mafic rocks. The precious metallic mineralization occurs from the alluvial layers in the uppermost part of the deposit to the goethite zones further below. The open-pit mine strip ratios are very low due to the shallow depth of mineralization.

M/s Black Range NL announced a total platinum reserves and resource of 108.3 Mt grading at 0.21 g/t Pt which occurs partly within the Syerston nickel–cobalt deposit.

7.2.17 Coobina Chromite Deposit

The Coobina chromite deposit is located ~50 km southeast of Newman, and 585 km south of Port Hedland within the Sylvania inlier, Pilbara Craton, northern Western Australia. The chromite deposit was first described in 1924 at

Jimblebah, and named Coobina thereafter with the identification of some gold workings in the Peak Hill gold field area. It has also been referred to as Murramunda after a homestead about 10 km to the east. The Coobina is the only economic chromite deposit/mine in Australia and is a rare example of a chromite-bearing intrusion intruded into an Archean greenstone belt.

The Coobina chromite deposit occurs at the western end of a 10-km long ultramafic–mafic layered magmatic complex that intruded into the Archean granites of the Sylvania inlier. The Archean ultramafic–mafic complex is extensively deformed and consists of predominantly serpentinized olivine cumulates, overlain by gabbroic rock. The top 40 m show evidence of alteration by surface fluids to varying degrees. This ultramafic body forms part of the Jimblebar Greenstone Belt, and is a relatively late-stage ultramafic sill intruded into an Archean greenstone sequence. The intrusive ultramafic unit was large and hot enough to allow efficient cumulus processes, involving olivine and chromite, to occur. The chrome contents and Cr/Al ratios are indicative of a parent magma of komatiitic affinity. The absence of Pt and Pd enrichment is attributed to formation from uniformly S-under-saturated magmas at high temperatures of formation (Barnes and Jones, 2013).

The mineralization occurs as massive chromite accumulations in the form of lenses or pods at the extreme western end of the ultramafics. There are about 150 tectonically dislocated and folded chromite lenses or a series of chromite seams, typically less than 2-m wide, in the Coobina belt. The largest orebody extends for ~340 m in length and up to 6-m wide zones of multiple lenses.

The Coobina chromite deposit contains resource of 1.96 Mt to a depth of 30 m. Extensive alluvial/float resources of chromite have also been identified in the nearby area. The annual production capacity at Coobina operation is planned for 250,000 tonnes (~276,000 tons) of chromite ore. The overall annual contribution is

little in international standard. The deposit was mined sporadically from a series of small open pits over the following decades, and is currently being mined for metallurgical chromite grade. The majority of the pits are less than 70 m deep.

M/s Consolidated Minerals acquired the Coobina project in 2001 and commissioned a new mining and processing plant in February 2002. Since the start of production, the company has become well established as an independent and reliable supplier of chromite ore to the world market.

7.2.18 New South Wales Chromite Belt

The chromite resources and production has been comparatively little in Australia due to the paucity of economic deposits. The sparse and intermittent production was mainly from ~169 small recorded occurrences in the New England region and NSW up to the mid-1900s. All of the chromite occurrences are associated with serpentinites and/or other ultramafic rocks in the form of pods, narrow veins, and lenses up to 30-m long by 10-m across. The deposits occur in three significant parallel belts trending NNW–SSE that include from west to east: the Coolac Serpentinite Belt, the Great Serpentinite Belt, and the Gordonbrook Serpentinite Belt.

7.2.18.1 Coolac Serpentinite Belt

The Coolac Serpentinite Belt extends over 130 km from Young in the north to Tumut in the south and hosts a number of chromite deposits/occurrences. The prominent deposits are Wallendbeen, Gundagai, and Tumut. The host rock represents a deformed and disrupted oceanic crust of the Late Silurian to Early Middle Devonian. The chromite occurs as pods throughout the Coolac Serpentinite Belt. The closely packed pods form massive chromite ore that changes into open-textured disseminated mineralization. The chromite pods are of two types: olivine-bearing chromitites containing Cr-rich chromite ores, and clinopyroxene-bearing chromitites rich in alumina.

7.2.18.2 Great Serpentinite Belt

The Great Serpentinite Belt extends ~120 km hosting four major deposits from north to south: Bingara, Barraba, Attunga, and Nundle. The chromite pods developed along the Peel Thrust and extend southeasterly from Bingara, through Barraba, to Bowling Alley Point near Nundle. All the deposits occur in schistose serpentinite as subvertical single or multiple lenses in linear stacks. Some of the stacks are connected by thin vein-like aggregates of chromite. The chromite occurrences are small and widespread. The three deposits produced ~2000 tonnes (~2200 tons) each per annum.

7.2.18.3 Gordonbrook Serpentinite Belt

The Gordonbrook Serpentinite Belt extends for about 100 km from Fine Flower to Ewingar near Baryulgil in northeastern NSW. The main chromite deposits are concentrated in the southern parts of the Gordonbrook Serpentinite Belt, although minor occurrences are noted elsewhere.

7.3 NEW ZEALAND

New Zealand, an island country in the southwestern Pacific Ocean, is located 1500 km (900 mi) east of Australia. It comprises two main islands—North and South. The surface area covers 268,201 sq. km (103,483 sq. mi) with 4.60 million population. New Zealand enjoys a modern, prosperous, and developed market economy mainly based on trade and infrastructure.

The minerals include coal, limestone, iron ore, gold, silver, and small chromite deposits. The podiform chromite occurs mainly in the South Island near Nelson and in West Otago. The known chromite deposits are of small size. The chromite ore was mined in the past. About 6000 tonnes (~6600 tons) of chromite ore at a grade of 20–54% Cr_2O_3 were mined in East

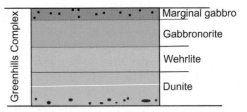

FIGURE 7.10   Generalized sequence of layered ultramafic-to-mafic complex at Greenhills, South Island, New Zealand. The basal dunite contains disseminated chromite (red) with minor platinum group of minerals.

Nelson at Dun Mountain, Croisilles Harbor, and D'Urville Island between 1859 and 1902. The mining of chromite took place intermittently in the latter half of the 19th century in the South Island. The current needs for chromite ore are attained by import.

7.3.1 Greenhills Igneous Complex

The Greenhills mafic–ultramafic Complex is a composite ultramafic-to-mafic layered igneous intrusion located on the Bluff Peninsula, 30 km south of Invercargill, Southland, New Zealand. The Greenhills Complex occurs as two main lobes that range from dunite at the base passing through wehrlite, hornblende gabbronorite, and marginal gabbro at the top (Fig. 7.10) of Permian age in the volcanic arc-derived Brook Street Terrane (Mossman, 2000). The layered complex formed from primitive hydrous basaltic magmas in feeder chambers overlain by an island-arc volcanic complex as part of the Permian Brook Street Terrane. The dunite is partially serpentinized with magnesium content varying between 10 and 28 wt% MgO. A series of dikes of varying composition crosscut the dunite.

The basal dunite of the south lobe is well exposed in the Greenhills quarry. The main primary minerals are cumulus olivine with minor cumulus chromite and rare PGEs, intercumulus clinopyroxene, and rare orthopyroxene and plagioclase.

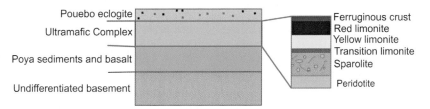

FIGURE 7.11   Regional stratigraphic sequence of New Caledonia (left) with elaboration (right) of lateritic-nickel formation by tropical weathering that resulted in extensive lateritization in the upper part of the ultramafic sequence. New Caledonia contains 30% of global nickel resources from lateritic and fresh ore, and fifth-largest producer in the world.

7.4 NEW CALEDONIA

New Caledonia, comprising dozens of islands, is located in the southwest Pacific Ocean, 1210 km (750 mi) east of Australia. The land area covers 18,576 sq. km (7172 sq. mi) with population of 268,767 (August 2014 census). New Caledonia is ensured one of the largest economies in the South Pacific region primarily based on nickel industries. The weathered lateritic soils of New Caledonia contain about 25% of the nickel ore reserves in the world.

The lateritic nickel oxide occurrences, resulting from the extensive weathering of peridotites in the source ultramafic massifs that cover about one-third of the land surface of the main island of New Caledonia, were discovered in 1864 by Jules Garnier. The parent peridotite contains 0.2–0.3% Ni.

The New Caledonian islands are composed of continental crust fragments thrusted away from Australia caused by the breakup of the eastern margin of Gondwanaland in the Late Cretaceous period. The basement rocks are undifferentiated felsic components. A number of thin sheets of ultramafic complex represent obduction of lithospheric mantle materials onto the ocean basin sediments and basalts of the Poya Terrane during continent–arc collision in the Miocene Epoch. The ultramafic massifs occur all over the islands trending in the NW–SE direction. The ultramafic units consist of predominantly harzburgite and minor amounts of dunite, pyroxenite, and gabbro (Fig. 7.11).

The tropical weathering resulted in extensive lateritization in the upper part of the ultramafic sequence, exposed on the ridges of the coastal range (Yang et al., 2013). The ultramafic complex is overlain by the high-pressure medium-temperature Pouébo Terrane of the Pam Peninsula, northern New Caledonia, which includes eclogite and variably rehydrated equivalents.

The richest Ni silicate concentration is localized in the fractures and joints within the thin layers of bedrock and overlying saprolite horizon in New Caledonia. The Ni-rich veins are activated by dissolution and precipitation processes at the level of the water table. The reworked mineralization is characterized by chemical and mineralogical supergene concentric zoning/riming. The concentric zonation results from evaporation–precipitation processesrelated to alternating periods of hydration and drying induced by fluctuations of the water table and made the lateritic nickel deposits of New Caledonia richer than the average magmatic sulfide deposits (Cathelineau et al., 2016).

The nickel-bearing silicates are liberated by hydrolysis and become concentrated toward the base of the weathering profile attaining exploitable concentrations of 2–5% Ni. The nickel content locally attains 10–15% Ni in small quantities at the weathering front, adjacent to the bedrock and in rock fissures. At present, the mean concentration of mined nickel ore is about

2.6% Ni, and reserves with at least this concentration are estimated at 200 million tonnes (~220 million tons). The superficial laterites also contain nickel ore of lower concentration (1.3–1.6% Ni). New Caledonia contains 30% of the known nickel mineral resources in the world. New Caledonia is the fifth largest producer of nickel (107,000 t) after Russia (266,000 t), Indonesia (189,000 t), Canada (181,000 t), and Australia (167,000 t) as on 2009.

7.4.1 Goro Nickel Mine

The Goro nickel project is located on the Goro Plateau on the South Pacific island of New Caledonia and considered one of the best-known laterite orebodies in the world. The nickel production commenced in August 2010 with annual capacity of 60,000 tonnes (~66,000 tons) of nickel and 4300 to 5000 tonnes (~4700 to 5500 tons) of cobalt.

The Goro deposit comprises lateritic iron oxides near the surface and is underlain by magnesium silicate-bearing saprolite. The nickel and cobalt contained result from chemical concentrations within this residual formation.

The Goro prospect contains one of the richest nickel grades among the lateritic nickel deposits in the world. The Proven reserves in the initial mining area of the Goro Plateau has been estimated at 124 million tonnes (~137 million tons), with an average nickel content of 1.48% and a cobalt content of 0.11%. In addition the measured reserves and indicated resources around the initial mining area have been estimated at 55 million tonnes (~60 million tons), giving a nickel content of 1.49% and a cobalt content of 0.12% as of 2009.

7.4.2 Koniambo Nickel Mine

The Koniambo nickel deposit is located on the 20-km long and 5-km wide Koniambo massif in the North Province of New Caledonia. The deposit is close to the small town of Voh on the west coast of the island at an elevation of 930 m above sea level and 15 km away from the shore. The area has been exploited prior to the Second World War by the *Société Le* Nickel Smelting Company (SLN), the French-owned company. The Koniambo deposit is one of the biggest and highest-grade undeveloped nickel laterite deposits in the world.

The economic potential laterite is exposed along the ridge of the massif as scattered pockets containing nickel, cobalt, and iron ore varying with depth. The high-grade nickel laterite mineralization of the Koniambo orebody covers an area of 21 sq. km. The laterite at Komiambo is divided into five distinct horizons, and three zones based on lithology and chemistry (Yang et al., 2013). The horizons from bottom upward include Bed rock, Lower saprolite, Upper saprolite, Transition, and Limonite. The Bed rock represents the unweathered ultramafics dominated by harzburgite or dunite. The Lower rocky saprolite comprises harzburgite protolith, commonly less altered or fresh harzburgite. The Upper saprolite is primarily earthy saprolite, intensely weathered from harzburgite and dunite, and often includes limonite as a thin horizon or breccia and includes fresh saprolite. The transition horizon is transitional between saprolite and limonite. The limonite horizon comprises three units in sequence of yellow limonite, red limonite, and ferricrete (Fig. 7.11). The rocky and earthy saprolite units with 15–30% MgO contain the maximum concentration of nickel and cobalt minerals. The earthy and rocky saprolite are the dominant types of laterite in zone 1, which are overlain by reddish limonitic laterites exposed on the two upper levels of bench wall.

The Koniambo has one of the largest nickel reserves in New Caledonia. The proved and probable ore reserves of the open-pit nickel mine have been estimated at 61.5 million tonnes (~67.8 million tons) grading 2.36% Ni, as of December 2013. The measured and indicated resources stood at 70 million tonnes

(~77 million tons) grading 2.47% Ni. The inferred resources has been estimated at 83 million tonnes (~91 million tons) grading 2.5% Ni.

The open-pit nickel mining method involving truck and shovel operations from multiple pits in the massif, and the pyrometallurgical plant started production in November 2014 at a time of low nickel prices, and production targets have not yet been met. The overburden soil is removed before the underlying saprolite host rock is weakened by explosives and mined using hydraulic shovels. The annual production at Koniambo is planned for 60,000 tonnes (~66,000 tons) of nickel in ferronickel over its mine life of 25 years.

7.5 PAPUA NEW GUINEA

Papua New Guinea in the southwestern Pacific Ocean occupies the eastern half of the island of New Guinea. The total surface area covers 462,840 sq. km (178,703 sq. mi) with a population of 7.06 million (2011 census). Papua New Guinea is richly endowed with natural resources, including mineral and renewable. The main mineral deposits include oil, gold, copper, and nickel–cobalt, and account for 72% of the export.

7.5.1 Kurumbukari (Ramu) Nickel–Cobalt Mine

The Kurumbukari is a world-class Nickel–Cobalt project located in the Madng Province, on the north coast of the Papua New Guinea Island. The project is planned as a low-cost operation treating laterite ores to produce nickel and cobalt as a high-grade intermediate product. This is one of the largest and most ambitious mining and processing projects to come into production during the previous decade. The Ramu project is premised on mining and beneficiating the nickel–cobalt ore at Kurumbukari, located in the foothills of the Bismark Ranges, approximately 75 km southwest of the provincial capital of Madang.

The parent rocks comprise dunite, gabbro, and pyroxenite of the Miocene Epoch at the Marum Basic Belt. These mafic–ultramafic rocks intruded into the Asai Shale and Kumbruf Volcanics in the southeast. The mafic–ultramafic complex has extensively been chemically weathered with an average laterite profile of ~15 m. The laterite thickness changes locally due to active erosion by streams, attains higher thickness along major fracture zones in the dunite, and promotes downward leaching by surface water. The laterite profile from bottom to top includes: bedrock of parent dunite, rocky saprolite, earthy saprolite, limonite, and overburden. The earthy saprolite is enriched in nickel and cobalt with significantly higher MgO. The top of the saprolite marks the boundary between acidic weathering and alkaline weathering conditions in the profile. The yellow limonite ore hosts the bulk of the nickel–cobalt resource. The overburden is composed of a transported humic layer, and red limonite contains low nickel (<0.5% Ni) and cobalt grades.

The reserves base of Kurumbukari Resource Block, Ramu West, and Greater Ramu Block covering 18 sq. km surface area has been estimated at 143 million tonnes (~158 million tons) grading 1.01% Ni and 0.10% Co. This resource was reported in accordance with the Australasian Code for the Reporting of Identified Mineral Resources and Ore Reserves (JORC compliance). The project has been planned for production of 31,150 tonnes (~34,300 tons) of nickel and 3300 tonnes (~3600 tons) of cobalt annually over a 20-year mine life. The mine came into production in 2012 and has produced 6 million tonnes (~6.6 million tons) of ore and processed 2.273 million tonnes (~2.505 million tons) during 2014.

Deposits of Europe

8.1 BACKGROUND

The continent of Europe comprises the westernmost part of Eurasia, the combined natural continental landmass of Asia and Europe. Europe is bordered by the Arctic Ocean to the north, the Atlantic Ocean to the west, and the Mediterranean Sea to the south. The east and southeastern borders of Europe are arbitrarily set incorporating the watershed divides, cultural, and political essence of the Asian continent, and not observing the primary physiographic definition of "Continent." The separation between Europe in the west and Asia in the east follows watershed divides of the Ural and Caucasus Mountains, the Ural River, the Caspian and Black Seas, and the waterways of the Turkish Straits.

Europe is the home to some of the largest mining companies in the world, without having any major metal producer in the region. The relatively large metal-producing countries in Europe include: Albania (chromium and iron–nickel ore), Norway (titanium), Portugal (zinc–lead–copper–tin), and Poland (silver). The top five metal production of the continent in terms of worldwide percentage are: chromium (10.7%), silver (8.5%), zinc (7.7%), lead (7.5%), titanium (7.1%), and nickel (2%).

8.2 ALBANIA

The Republic of Albania, in a small mountainous country on the Balkan Peninsula, Southeastern Europe. The economy is primarily based on agriculture, tourism, and large deposits of petroleum and natural gas. The rich mineral resources, exploration, mining, and processing are the key components of the country's economy. The metallic minerals in order of abundance include chromium, nickeliferous iron ore, nickel silicate, copper, and bauxite.

The podiform type of chromite deposit occurs along the Balkan part of the Alpine mountain chain formed because of the Alpine orogeny. Albania is a leading producer of chromite ore with steady production between 1980 and 1990 attaining its peak of ~1.20 million tonnes (~1.3 million tons) in 1987, declined between 1991 and 2005 at ~0.10 million tonnes (~0.11 million tons), and steady increasing thereafter at ~0.35 million tonnes (~0.39 million tons) of chromite ore. The Bulqiza Massif has been the dominant producer since production began in the 1940s with supporting ore from the Kalimpash area, and other small podiform orebodies. The small producers have been an increasingly important contributor since 2000. The nickel laterite deposit is located at Bitincke in the southeast of Albania.

8.2.1 Bulqiza Ultramafic Massif

8.2.1.1 Location

The Bulqiza chromite mining district is about 40 km NE of Tirana, the capital city of Albania, and about 70 km east of the principal port city Durres on the Adriatic Sea. Bulqiza is the local administrative center and provides all the necessary services as well as skilled labor. Electrical power and water are available in the area. The topography in the Bulqiza area is mountainous with elevations range between 600 and 2000 m above mean sea level. The conventional road access from Bulqiza chromite collection point to the port at Durres is 150 km with transport time of 12 h through steep mountain roads. The roads have been redesigned to reduce journey time by 2–3 h.

8.2.1.2 Geology

The regional geology of central and northern Albania is dominated by the Mirdita ophiolite sequences formed during the development of oceanic lithosphere in Triassic–Jurassic time. The ophiolite sequence was subsequently thrust westward over the younger Mesozoic platform carbonates. The Albania ophiolite massifs form two distinct arcuate north–south-trending belts and designated as: (1) the Eastern Ophiolite Belt of 150-km long and 20–50-km wide, and (2) the Western Ophiolite Belt of 200-km long and 10–20-km wide. The Eastern Ophiolite Belt is 12–14-km thick and composed of lower harzburgites, interbedded harzburgites, and dunites. The sequence is intruded by a lower plutonic unit with ultramafics and layered gabbro complex, followed in turn by an upper plutonic unit of thick intrusive bodies including gabbros, quartz diorites, and plagioclase granites. The Western Ophiolite Belt is up to 4-km thick and composed of clinopyroxene-bearing harzburgites, intruded by plutonic mafic gabbros. The upper half of the ophiolite sequence comprises volcanic rock, predominantly basaltic pillow lavas with minor interlayered cherty sediments. The Albanian ophiolite belts can be traced south into Greece, and further east into Turkey and Cyprus.

Three chromite-mining districts are recognized, all in the eastern ophiolite belt—the northeastern district including the Tropoja and Kukes ultramafic massifs, the central district with the Bulqiza ultramafic massif, and the southeastern district with the Shebenik–Pogradec ultramafic massif.

The Bulqiza ultramafic massif consists primarily of tectonized harzburgite and overlying dunite cumulates. It extends over an area of 370 sq. km, and includes about 400 known occurrences of chromite mineralization. The Bulqiza ultramafic massif is an allochthonous structure on the upper plate of a major thrust fault. The Bulqiza massif is divided into northern and southern sectors by a valley filled with Quaternary sediments. The total thickness of the massif is more than 4000 m and composed of a layered sequence of Lower harzburgite, Upper harzburgite, and an overlying Cumulate Series mainly of dunite, pyroxenite, and gabbro (Fig. 8.1). The harzburgite is serpentinized throughout the

Bulqiza ultramafic massif, except toward the eastern margin.

8.2.1.3 Mineralization

The lower and upper layers of harzburgite, differentiated by increasing dunite facies and chromite lodes mainly dominating in the upper sequence. The chromite mineralization in the Bulqiza–Batra system extends continuously along NNW–SSE strike for 4.5 km with variable southwesterly dip down up to 1.6 km.

The Bulqiza Massif contains about 65 chromite deposits under mining and advance development stage, and over 370 occurrences. The principal mineralization occurs in the Bulqiza–Batra ore structure. The chromite ores consist of tabular–concordant layers of banded, semimassive to massive mineralization within serpentinized-dunite alteration envelopes containing disseminated mineralization. Ore thickness averages about 3 m and is usually thickened by folding. Subconcordant, pipe-like, and podiform styles of mineralization are common features. The size of the deposits range between several hundred thousand tonnes and several million tonnes and occur throughout the Massif. The grade varies between 33 and 48% Cr_2O_3. The majority of the chromite lodes are for metallurgical uses.

8.2.1.4 Reserve Base

The preproduction reserves and resources as estimated during 1986 and 1987 by different agencies are referred for all purposes (Table 8.1).

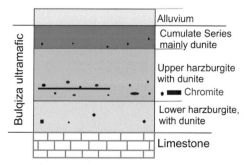

FIGURE 8.1 Generalized overview of total stratigraphic sequence at Bulqiza Massif displaying preferential distribution of banded, massive, and podiform chromite lodes in the Upper Harzburgite layer, and some scattered podiform chromite in the Lower Harzburgite layer.

TABLE 8.1 Preproduction Reserves of All Categories for the Bulqiza Group of Chromite Deposits (1986–1987)

Deposit	Reserve (Mt)	% Cr_2O_3	Resource (Mt)
Bulqiza	5.22	40.16	3.01
Batra	2.82	33.80	0.86
Thekna	0.41	28.86	0.24
Total	8.45	37.49	4.11

8.2.1.5 *Mining*

M/s Empire Mining Corporation has acquired exclusive exploration and development rights to four exploration licenses, totaling 64.5 sq. km. The licenses are the Bulqiza–Batra (38.8 sq. km), Bulqiza Veriore (6.9 sq. km), Qafe Burreli (6.1 sq. km), and the Liqeni I Dhive (12.7 sq. km).

The Bulqiza Massif is the most productive of 17 chromite-bearing ultramafic bodies in eastern Albania. The orebodies have been developed and mined semicontinuously for a strike length of 5 km and through a vertical depth of 900 m in the Bulqiza mine and 450 m in the Batra sector. The highly mechanized mines adopted open-pit and underground methods and gain access through adits and shafts, with a rated annual capacity of 0.36 million tonnes (~0.4 million tons). The mine has produced about 20 million tonnes (~22 million tons) at Run-of-Mine grade of 40–42% Cr_2O_3, and concentrate at 50% Cr_2O_3 since the 1940s.

8.2.2 Kalimash Ultramafic Massif

The Kalimash Ultramafic Massif hosts many small chromite deposits, partially mined through small-scale traditional local crafts. The area is rugged and mountainous and located in the Kukes district, Kukes Region, northeast Albania, and close to the border of the Kosovo Republic. Kukes is 200 km by road from Tirana to the south. The new road from Morine to port of Durres passes through Kukes which reduces the travel time to the port.

The ultramafic rocks of the Kalimash area include harzburgite, dunite, and undifferentiated peridotite around the margins of the massif. The podiform chromite bodies occur within the ultramafic rocks, particularly along the contacts between dunite and underlying harzburgite. The main Kalimash chromite horizon is 1000–12,000-m thick and characteristically located within the upper harzburgite–dunite zone with minor chromite occurrence in the overlying main dunite and underlying lower harzburgite layers (Fig. 8.1).

There are at least 10 separate chromite zones identified in the Kalimash ultramafics with thickness varying between 5 and 10 m that can reach 40 m. The chromite lodes are concordant with the host rocks and dip at 15° to 30° toward the S.

The chromite lodes occur in combination with higher-grade podiform lenses of massive, banded, or lenticular pods, and broader disseminated zones of lower-grade nodular spotty chromite. The grade of the pod-type ranges between 15 and 25% Cr_2O_3 with a typical average of 20%. The grade of the disseminated mineralization ranges between 4 and 8% Cr_2O_3.

The area has been explored extensively during the 1970s amounting to 50,000 m in 210 holes. The present-day exploration methods include trenching, Diamond and Reverse Circulation drilling, borehole geophysics, sampling, and data validation by Quality Control/Quality Assurance studies.

The reserves and resources at Kalimash chromite project have been estimated at 2% Cr_2O_3 cutoff as 6,721,425 tonnes (~7,409,000 tons) grading 4.36% Cr_2O_3, in accordance with Joint Ore Reserves Committee Code of reporting 2004.

The larger deposits had been partly mined by the State between 1980 and the 1990s. About 1 million tonnes (~1.1 million tons) of chromite ore were extracted by underground methods and concentrated in an onsite processing plant. About 300,000 tonnes (~331,000 tons) of concentrate were produced. Local mining is currently active in the area.

8.2.3 Bitincke Nickel Laterite Deposit

The Bitincke nickel laterite deposit is located in the southeast of Albania. The nickel silicate and iron–nickel are mainly located in Devoll (Bilisht and Bitincke), Pogradec (Guri Kuq and

Cervenaka), Librazhd (Prrenjas, Skroskě, Xixillas, and Bushtrice), Kukes (Mamez, Trull Surroi, and Nome), and many more (Fig. 8.2).

The nickel-bearing laterite in Albania formed on the ultramafic sequence of the ophiolites representing the large NNW–SSE-trending suture zone extending about 1000 km from Croatia to Greece. The ophiolites occur within a nappe structure and are divided into East and West Ophiolite Belts. The deposit developed on peridotites in Late Jurassic ophiolites that were obducted in the Early Cretaceous and now form part of the Albanian Mirdita ophiolite zone. The limestones and conglomerates overlying the deposit restricted the minimum age of lateritization.

The lateritic deposits contain Ni±Co formed by prolonged intensive weathering of peridotites in warm and humid climates. The weathering process removes the principal components of the peridotite, namely MgO and SiO_2, leaving 5–7% residue of the original rock enriched in the other major components, such as Fe, Al, and Cr, and minor constituents such as Ni, Mn, Cu, and Co. The laterite is composed of two separate distinct mineralized zones characterized by a nickel silicate zone and a residual iron oxide–nickel phase (Fig. 8.3). The silicate zone contains Ni concentrations reaching up to 1.5 wt.% with lateral and vertical variation. The silicate zone is composed of olivine and serpentine at the base of the horizon and is gradually replaced by secondary silicates and iron oxides toward the upper horizon. The boundary between the weathered nickel-bearing silicate and iron oxide horizons is sharp and characterized by an increase in Fe_2O_3 between 10 and 80 wt.% with a reduction in MgO. The oxide horizon is dominated by goethite with little variation in texture or geochemistry. The nickel enrichment is highest at the base of this zone.

The thickness of the weathering profile at the Bitincke deposit varies between 2 and 20 m, and dips ~20°→E. The variation in thickness is caused by topography, displacement, and protolith fracture density. The oxide zones, formed on topographic highs, are subject to increased

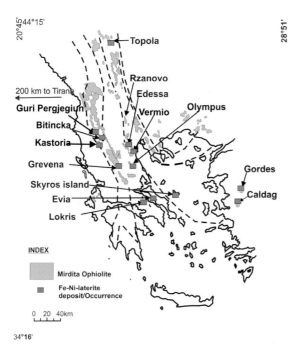

FIGURE 8.2 Sketch map showing the distribution of ophiolites and the location of the Fe–Ni-laterite deposits in the Balkan Peninsula and W. Turkey. *After Eliopoulos, D.G., Economou-Eliopoulos, M., Apostolikas, A., Golightly, J.P., 2012. Geochemical features of nickel-laterite deposits from the Balkan Peninsula and Gordes, Turkey: the genetic and environmental significance of arsenic. Ore Geology Review 48, 413–427. Elsevier.*

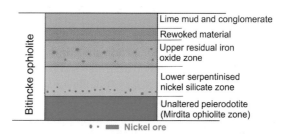

FIGURE 8.3 Typical nickel laterite profile displaying the weathering and compositional variation upward from unaltered peridotite at depth to nickel-rich silicate zone, residual iron oxide phase covered by carbonate muds and conglomerate at the Bitincke lateritic nickel deposit, Albania.

rates of erosion. The laterite profile within topographic lows and in areas of relatively high fracture density, tends to be thicker due to increased permeability. The most functional sections of the Ni-laterite weathering profile developed in small fault-controlled basins and preserved by the deposition of a sequence of limestones and mudstones.

The nickel ore occurs in various forms, such as nickel silicate of the remaining/primary crust, and iron–nickel ore, redeposited or secondary-crust nickel-silicate crust ore.

The estimated ore resource at Bitincke stands at 35.6 Mt of nickel ore with an average grade of 1.2% Ni (Thorne et al., 2012). The total iron–nickel resources at Albania stand at 230 Mt and resources for nickel-silicate ore are estimated at 81 Mt.

8.3 FINLAND

The Parliamentary Constitutional Republic of Finland is a Northern European nation. The major sectors of the economy are services, manufacturing (electronics, machinery, vehicles, and metals), and refining.

Finland is a European hub for mineral exploration and mine development with its geoscientific base data being the best in the world. The exploitation for copper, nickel, cobalt, chromium, zinc, lead, vanadium, and iron ore deposits has provided the raw-material base for the country's metal industry. The country has significant technology for processing and refining of copper and nickel concentrates at Harjavalta, chromium at Kemi, zinc at Kokkola, and iron ore at Raahe. The major industrial minerals mined in Finland are carbonates, apatite, and talc.

There are ~270 metal mines in operation with total annual production of ~250 million tonnes (~276 million tons) of ore per annum. The main commodities include copper, nickel, cobalt, chromium, zinc, iron ore, and vanadium.

A large area of Finland was affected by a rift system during the Proterozoic era (2500–2450 Ma) resulting in a series of layered mafic–ultramafic intrusions hosting nickel–copper, chromium, and the platinum group of metals. The localization of the platinum group of elements (PGEs)–Ni–Cr-hosted layered intrusions can be broadly grouped into northern (Kemi chromite deposit, Lomalampi, and Kevitsa Ni–Cu–PGE deposits), central (Rytky nickel deposit), and eastern (Vaara PGE deposit) Finland hosting PGE, nickel, and chromite mineralization. The northern intrusives include the Central Lapland Greenstone Belt (CLGB), the Peraepohja schist belt, and the Kuusamo schist belt in Finland, and extend to NW Russia as the rich nickel deposits of Pechenga–lmandra–Varzuga rift in the Kola Peninsula.

The promising komatiite-hosted PGE–Cu–Ni sulfide–Cr deposits in northern and eastern Finland are associated with both Archean (Tainiovaara, Hietaharju, Peura–aho, Vaara, Ruossakero, and Sarvisoaivi), and Paleoproterozoic (Lomalampi and Hotinvaara) komatiitic to komatiitic–basaltic magmatism. The deposits are divided into three main groups based on their metal content: (1) deposits enriched in chromium±PGE (Kemi), (2) deposits enriched in PGE (Pd+Pt>500 g/t) and Cu (Ni/Cu<13) (Lomalampi, Kevitsa, Vaara, Hietaharju, and Peura–aho), and (3) deposits enriched in Ni (Ni/Cu>15) with low-PGE contents (Sarvisoaivi, Ruossakero, and Hotinvaara).

8.3.1 Kemi Chromite Deposit, Northern Finland

The Kemi chrome deposit (Elizarvi) is situated about 7 km northeast of the town of Kemi on the coast of the Gulf of Bothnia in northern Finland. The deposit was discovered in 1959 followed by detailed geological exploration and metallurgical investigations, and the mine was established in 1968.

The stratiform and layered type of chromite deposits at Kemi is associated with mafic–ultrabasic sill-like early Proterozoic intrusion (2400 Ma) into the Archean basement complex comprising granitoids. The mafic–ultramafic intrusion extends over 15 km and width varies between 0.2 and 2 km with average of 1.5 km. The layered Intrusion strikes northeastward along the Svecokarelidic Peräpohja schist belt. The chromite-rich horizon dips 70° northwest and is 40-m thick. The orebodies occur continually along the strike length.

The lower part of the intrusion is composed of peridotite and pyroxenite cumulates and layers of chromitite at the base. The upper part of the intrusive complex is devoid of olivine and chromite, and comprises predominant plagioclase mafic cumulus phase (gabbro–anorthosite) occurring either alone or together with pyroxenes (Alapieti et al., 1989). The layered-intrusive complex is successively overlain by mafic volcanics and subvolcanic sills (2100–2200 Ma), dolomite, quartzite, and phyllite (Fig. 8.4). The lower and upper parts of the complex were metamorphosed during the Svecokarelidic orogeny, and completely altered to chlorite, serpentine, talc, amphiboles, and carbonates. The ore zone is intensely fractured, but the middle part is well preserved. The chromite grains, too, suffered from alteration, but the cores of the grains still exhibit primary compositions.

The chromitite layer can be continuously traced for the entire length of 15 km of the intrusion. The thickness of the chromite lode varies between a few mm and 90 m. The main chromitite layer is thickest in the middle of the intrusion and extends for 4.5 km. The mineralized envelope was originally funnel shaped and tilted later to the current lenticular shape of 15-km diameter. The most massive and thickest mineralization was possibly located over the magmatic conduit as feeder to the intrusion. The main chromitite layer is overlain by a sequence of thin chromite-rich layers which continues

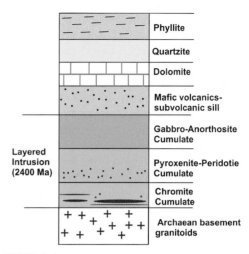

FIGURE 8.4 Generalized stratigraphic overview of the Kemi mafic–ultramafic layered complex showing massive chromite lodes (red) at the basal contacts that have been gradually overlain by a sequence of thin chromite-rich pyroxene-peridotite layers that continue upward to the stratigraphic level of 500 m above the basal contact of the intrusion.

upward to the stratigraphic level of 500 m above the basal contact of the intrusion. The upper part of the main chromitite is layered, but the lower part is more massive and brecciated.

Alapieti et al. (1989) suggested that the chromitite was deposited during the influx of magma into the Kemi intrusion. This early-phase magma was likely contaminated by sialic material from the underlying basement complex mixed with the fresh input of primitive magma resulting in chromite saturation. The chromite crystals formed during mixing in a plume and accumulated preferentially around the magmatic conduit.

The ore reserves for open-pit mining stand at 40 million tonnes (~44 million tons) with an average grade of 26.6% Cr_2O_3 and a Cr/Fe ratio of 1.53. The whole area has additional reserves of about 110 million tonnes (~121 million tons) of ore—estimated to a depth of 1 km. The chromium ore-bearing intrusion has been traced to a depth of 2–2.5 km by geophysical survey.

The conventional shovel-and-truck bench mining started in 1968 at the present Main Pit. Satellite pits have been mined out in the intervening period. The mine is owned by M/s Outokumpu with an annual production capacity of over 1.3 million tonnes (~1.4 million tons) of ore. The underground development to replace the open pit and increase capacity started in 1999; production started in 2003 at 150,000 t/y, and in 2006 the underground mine became the sole source of ore production. The underground entry system is by an asphalt-surfaced 1:7 access ramp descending from a portal in the footwall side of the pit to a skip-loading station on the 600 level, via workshops, pumping stations, and a gyratory crusher. The ore is extracted from the 275 down to the 500 level by sublevel stoping with drifting and production rigs. The production of upgraded lumpy ore totals around 200,000 tonnes (~220,000 tons) and fine concentrates about 400,000 tonnes (~440,000 tons). The chromium oxide content of the concentrates varies between 35.0 and 45.0% Cr_2O_3.

8.3.2 Lomalampi PGE–Ni–Cu Deposit, Northern Finland

The Lomalampi PGE–Ni–Cu–Au deposit was discovered in 2004 by the Geological Survey of Finland. The deposit is located in the municipality of Sodankylä, about 75 km northeast from the town of Sodankylä. The nearest all-service town of Rovaniemi is 135 km further to the south from Sodankylä. The town of Rovaniemi has a railroad connection and several daily flights to Helsinki. The deposit occurs on the NW-sloping side of a hill above a swamp 250–280 m above mean sea level. The discovery of the Lomalampi deposit is the first significant find, which is associated with the Sattasvaara–Peurasuvanto–Mertavaara formations, indicates further potential for komatiitic units in northern Finland.

Geochemically, the Lomalampi PGE–Ni–Cu–Au deposits are an unusual PGE-enriched, low base-metal mineralization hosted by a komatiitic olivine cumulate (peridotitic cumulate) body of the Sattasvaara Formation in the Paleoproterozoic (2060 Ma) CLGB (Fig. 8.5). The host rock is a chromite undersaturated Al-undepleted high-Mg basalt or low-Mg komatiite (Törmänen et al., 2015). The bedrock is mostly composed of metasedimentary rocks belonging to the Matarakoski formation (Savukoski group) and various types of ultramafic (komatiitic to komatiitic basalts) volcanic and cumulate rocks of the Peurasuvanto formation (Savukoski group). The ultramafic volcanic rocks comprise fine-grained lavas, breccias, and tuffs. Associated cumulate rocks are mostly fine to medium-grained olivine orthocumulates with subordinate amounts of pyroxenitic cumulates, olivine- and pyroxene-lavas, and gabbros. The most abundant metasedimentary rocks are phyllites, black schists, graywackes, and quartzite belonging to the Matarakoski Formation.

The Lomalampi area has been mapped by the Geological Survey of Finland, followed by systematic ground geophysical surveys using magnetic and electromagnetic (EM) VLF-R measurements, and six Induced Polarization (IP) Survey lines. The Mobile Metal Ions (MMI) soil sampling has been conducted along two lines, one across to known mineralization and the other further toward NE. The Geological Survey of Finland has drilled a total of 6157 m in 48 holes in the prospective area.

The PGE–Ni–Cu–Au mineralization is hosted by 30–60-m wide, steeply SE dipping olivine orthocumulate body. The mineralization is generally 10–20-m thick and has been traced along strike for 390 m, and to a maximum depth of about 140 m below the surface. The mineralization consists of disseminated base-metal sulfides, mainly of magmatic pentlandite, pyrrhotite, and chalcopyrite (1–6 vol.%), without substantially modified by metamorphic processes. The main platinum phase is sperrylite that occurs mostly with silicates (80%), and the main Pd carrier is an unusual Pd–Ni–Te–Sb–Bi phase (s). About

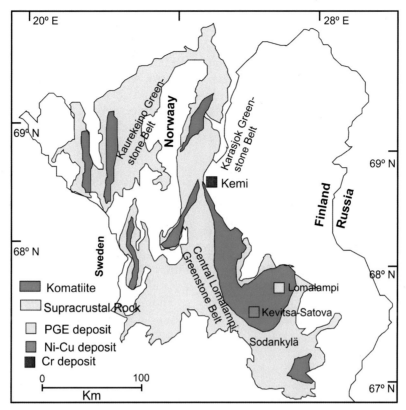

FIGURE 8.5 Schematic geologic map of Northern Finland showing the komatiite intrusion in supracrustal rocks with chromite, platinum, and nickel deposits.

40% of the Pd phases occur with sulfides and sulpharsenides.

The reserve base of the Lomalampi deposit at 0.1 g/t Pt cutoff has been estimated totaling 3.06 Mt with average grade of 0.269 g/t Pt, 0.122 g/t Pd, 0.074 g/t Au, 0.17% Ni, 0.06% Cu, and 117 g/t Co.

8.3.3 Kevitsa Ni–Cu–PGE Deposit, Northern Finland

The Kevitsa–Satovaara Ni–Cu–PGE deposit was discovered by the Geological Survey of Finland in 1987. The deposit is located about 40 km southwest of Lomalampi PGE–Ni–Cu deposit, and 142 km north of Rovaniemi, the capital of Lapland, Northcentral Finland. It is a large volume mafic–ultramafic magmatic and sulfide-bearing sedimentary rocks that provide the region a great potential for Ni–Cu–PGE mineralization. The large undeveloped nickel deposit is suitable for low-cost open-pit mining.

The bedrocks are mostly composed of metavolcanics and metasediments belonging to the Matarakoski formation (Savukoski group) and various types of ultramafic (komatiitic to komatiitic basalts) volcanic and cumulate rocks of the Peurasuvanto formation (Savukoski group). The total geological system belongs to the CLGB of Finland. Minor olivine cumulates associated with thin volcanic flows mostly occur as sheet-like cumulate bodies up to several tens of meters in thickness that can be traced for 500–1500 m along strike. The host mafic–ultramafic package (2058 Ma) of a plutonic–volcanic sequence is represented from the bottom upward by footwall dunite, marginal olivine websterite, gabbro, olivine pyroxenite with inclusions of dunite, plagioclase-bearing websterite, gabbronorite, microgabbro, pyroxenite, central dunite, and gabbro. The mineralized sequence occurs exclusively within the olivine pyroxenite and the inclusions (Fig. 8.6), forming massive lodes and discrete lenses in an overall synclinal structure. The ore zones occur as layers.

The Kevitsa orebody is a large-volume low-grade deposit, and the total reserves and resources haves been estimated at 240 Mt grading 0.30% Ni, 0.42% Cu, 0.21 g/t Pt, 0.15 g/t Pd, and 0.11 g/t Au. The mineable reserve estimated in September, 2009, stands at 165 Mt

grading 0.28% Ni, 0.42% Cu, 0.20 g/t Pt, 0.15 g/t Pd, 0.10 g/t Au, and 148 g/t Co. In addition, the deposit has 43 Mt of inferred resources grading 0.28% Ni, 0.43% Cu, 0.16 g/t Pt, 0.13 g/t Pd, 0.10 g/t Au, and 148 g/t Co. The resources are estimated down to a depth of 730 m below average surface elevation.

The production at Kevitsa open-pit mine started in 2012.

8.3.4 Rytky Nickel Deposit, Central Finland

The Rytky nickel deposit is located 40 km south of the city of Kuopio in the municipality of Leppävirta in Eastern Finland. The Rytky exploration project is located about 1 km SE of the Kotalahti nickel deposit. The Geological Survey of Finland initiated exploration in the NW–SE-trending Svecofennian–Kotalahti Nickel Belt as a part of an extensive research in 1992, and the Rytky occurrence was identified in 2000.

The Archean basement in the central part of the Fennoscandian Shield extends from Russia into central Finland. The regional geological setting comprises Archean basement rocks of granitoids and gneisses, quartzite, schist, and migmatite.

The Rytky mafic–ultramafic intrusion occurs in the SW part of the Archean tonalite–quartz diorite gneisses of the Kotalahti Dome and is cut by gabbro–amphibolitic dikes. The Archean rocks are overlain by Paleoproterozoic quartzites, limestones, calc-silicates, black schists, and banded diopside amphibolite (craton margin sequence). The Rytky intrusion (1900 Ma) is composed of medium- and coarse-grained lherzolite, websterite, melagabbro, and gabbronorite containing the nickel sulfide mineralization. The intrusion is located in the contact zone between the Archean craton and Proterozoic supracrustal rocks (Mäkinen and Makkonen, 2004). The mafic–ultramafic rocks are affected by the amphibolite facies of metamorphism and four events of deformation. The emplacement of the intrusion took place at the culmination of D2

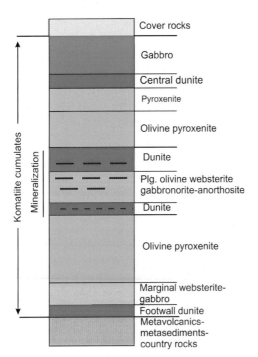

FIGURE 8.6 Generalized stratigraphic sequence of country rock and Paleoproterozoic komatiite-hosted Ni–Cu–PGE mineralization of massive and layers (pink) at the Kevitsa deposit/mine.

deformation as shown by the gneiss inclusions with S2 schistosity within the intrusions. The originally subhorizontal intrusions were rotated into a subvertical position during the third deformation event (D3) forming the present stratigraphic top to the west. The marginal part of the intrusion is contaminated by the country rock, as indicated by high TiO_2, P_2O_5, Rb, Zr, and light rare earth elements.

The main nickel and copper-bearing minerals are subhedral to euhedral pentlandite and chalcopyrite. The pentlandite is associated with the lherzolite and websterite.

The exploration included geological mapping, petrological and lithogeochemical studies, geophysical surveys, and diamond drilling. The measured and indicated reserves at Rytky deposit have been estimated at 1.633 million tonnes (~1.800 million tons) grading 0.58% Ni, 0.19% Cu, and 0.023% Co.

8.3.5 Vaara Ni–Cu–PGE Deposit, Eastern Finland

The Vaara Ni–Cu–PGE-deposit is located in the central part of the Archean Suomussalmi Greenstone Belt, representing the northernmost segment of a ~220-km long and 10-km wide, N–S-trending, semicontinuous zone of supracrustal ultramafic rocks in Eastern Finland. The central and southern segments of this zone are assigned to the Kuhmo and Tipasjärvi Greenstone Belts, respectively. The Geological Survey of Finland discovered the deposit in 1998.

The Vaara subeconomic deposit is hosted exclusively by komatiites in the Archean Suomussalmi greenstone belt (2800 Ma). It occurs in the central part of the Cr-poor serpentinized olivine cumulate zone of a komatiitic extrusive body and is composed of disseminated interstitial sulfides consisting of pyrite, pentlandite, millerite, violarite, and a very abundant chalcopyrite accompanied by plentiful magnetite (Konnunaho et al., 2013). The presence of millerite and violarite-bearing pyrrhotite-free sulfide

assemblages are due to postmagmatic, low-temperature hydrothermal oxidation of the primary magmatic pyrrhotite–pentlandite–chalcopyrite assemblages at Vaara. The loss of sulfur in the process led to a significant upgrading of the original metal enrichment. The platinum-group and tellurium minerals are mainly associated with Ni-bearing sulfides, chalcopyrite, and pyrrhotite. The platinum-group and tellurium minerals occur as small grains inside or at the edges of sulfide grains.

The deposit is nearly north–south-trending and has been delineated by drilling for a strike length of ~450 m with steep dip to the east, and to a depth of 50–180 m below the surface. The deposit consists of three separate parallel and overlapping mineralized-sulfide horizons. The thickness varies between ~2–3 m at the northernmost end and ~50 m at the southernmost end. The nickel–copper sulfide and platinum-group mineralization at Vaara occurs mainly disseminated (Type-II), and occasionally massive form (Type-I). The mineralization is mostly associated with relatively Cr-poor serpentinite cumulates at the lower horizon, and occasionally in olivine orthocumulate (metapyroxenite) in the upper horizon (Fig. 8.7). The metal contents are characterized by strong-positive correlation among each other.

The indicated reserve of the Vaara deposit is estimated at 2.62 Mt of ore at 0.49% Ni, 0.04% Cu, 0.01% Co, 0.11 g/t Pt, and 0.28 g/t Pd. The inferred resource stands at 0.14 Mt grading 0.45% Ni, 0.04% Cu, 0.01% Co, 0.10 g/t Pt, and 0.24 g/t Pd (Altona Mainlining, 2013).

8.4 GREECE

The Hellenic Republic of Greece is a democratic developed country in southeastern Europe consisting of two mainland peninsulas and thousands of islands throughout the Aegean and Ionian Seas. The key mineral resources include iron ore, lignite, zinc, lead, bauxite,

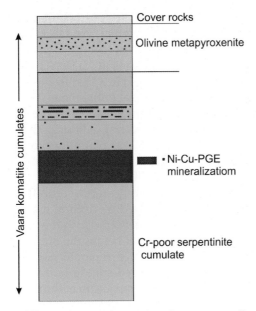

FIGURE 8.7 Generalized stratigraphic sequence of komatiite cumulate showing the distribution of massive and disseminated PGE-bearing nickel–copper sulfide mineralization in successive layers. The mineralization is mainly associated with relatively Cr-poor serpentinite cumulates in the lower horizon, and occasionally in olivine orthocumulate in the upper horizon.

petroleum, and magnesite. The resources of chromium, nickel and platinum group are not large enough for economic mining, and found as small occurrences.

8.4.1 Nickel

Greece shares about 2% of the world production and export in the form of ferronickel alloy in stainless steel to the European industries. The Fe–Ni-bearing lateritic occurrences are more than 110 in number. The deposits are mainly located in the SW Balkans. The economically viable deposits are few such as Kozani, Pella, Mt Parnitha, Skyros, and Mytilini. The deposits are formed by deep crustal weathering of ophiolites containing extra Cr and Co.

Detailed geochemical, petrological, and mineralogical studies indicated a typical natural successive growth of the weathered zones from bottom upward including unaltered primary bedrock, saprolite, a clay horizon in the form of massive goethite, and ferruginous pisolite. The associated Fe–Ni-bearing mineralization is composed of ferruginous spheroidal particles, silcrete, saprolite, and mineral fragments, and is considered clastic sediment.

The economically significant Fe–Ni-bearing lateritic orebodies are extensively exposed at Artaki Evia, Agios Ioannis Viotia, Lokrida Fthiotida, and Kastoria located in the municipality of Mesopotamia, and Ieropigi. The large surface pit mines in Greece are solely operated by the Larco General Mining and Metallurgical Company (LARCO). There are five operational surface mines at Evia having been equipped with modern earthmoving machinery. A total mine production of 2.2 Mt of ore was achieved from the group of mines of Evia, Viotia, and Kastoria in 2011 extracting 18,500 t of ferronickel alloy. The metallurgical plant is located at Larymna Fthiotida.

The total reserves (Proved and Probable) of nickel at Greece are ~1.392 million tonnes (~1.534 million tons).

8.4.2 Chromium

The ultramafic (dunite)- hosted small chromite occurrences are over 200 in number. The chromite orebodies are found in tectonically disrupted ophiolite sequences and belong to the suprasubduction-zone type of ophiolites. The mineralization occurs as podiform, disseminated, nodular, and massive textures. The existence of multiple magma chambers has been established by petrological studies. The host harzburgite represents mantle residue remaining after a high degree of partial melting. The largest ore deposits include: Mt Vourinos, Mt Pindos, and Mt Othrys. The most important chromite ores of metallurgical type are from Kozani (Vourinos, Xerolivado, and Rodiani), Veria, Edessa, and Chalkidiki (Gerakini, Ormylia). The refractory type of ores is from Eretria Larissa, and Domokos Fthiotida. The both types are economically significant. The chromitite ore of the Veria ophiolite complex, and many others, contains PGEs.

Mt Vourinos Kozani is the largest chromite deposit. The mine was under regular production until 1991 and produced ferrochromium during the period between 1983 and 1991. The chromite mines and ferrochrome metallurgical plant were closed in 1991 due to low market price, and inability to produce fine-grained chromite concentrate through flotation.

The indicated reserves of chromium from Mt Vourinos Kozani are estimated at about 1.2 Mt.

8.5 NORWAY

The Kingdom of Norway is a sovereign and unitary monarchy. The main economy is based on export revenue from oil and gas, agriculture, transport, and tourism. The mineral resources include energy, and metallic minerals (iron ore, nickel, ilmenite, titanium, and molybdenum).

8.5.1 Bruvann Ni–Cu–Co Deposit

The Bruvann Ni–Cu–Co deposit/mine is hosted by the Råna Intrusion, located about 20 km southwest of Narvik municipal town, northern Norway. The mafic–ultramafic Råna Intrusion covers about 70 sq. km of surface area. The eastern part of the Bruvann orebody was discovered in 1912, followed by the discovery of Eiterdal deposit in 1913.

The mafic–ultramafic Råna Intrusion consists of a peripheral zone of norite containing bands and lenses of peridotite and pyroxenite, and a core mainly of quartz norite having a total stratigraphy conforming to many layered intrusions in the world. The nickel sulfide mineralization in the Bruvann deposit is hosted by olivine-rich magmatic peridotite cumulates and related ultramafic rocks close to the contact with the surrounding regional granitic gneisses. The Råna Intrusion was emplaced in a synorogenic magmatic environment associated with ridge subduction and crustal-scale thrusting.

The deposit consists of several lenses separated by faulting. The main fault zone, orientated nearly N–S, divides the nickel deposit into Ostmalmen (eastern orebody) and Vestmalmen (western orebody). The vertical distances between the two orebodies varies between 75 and 200 m. The Ostmalmen orebody is subdivided into Sydost dagbrudd, Kronpillar, and Dypmalmen. The Vestmalmen is subdivided into Dinosaurmalmen, Sydmalmen, and Nordmalmen. The mineralized zone extends over 900 m in an east–west direction, and 700 m north–south. The Ostmalmen orebody dips toward the south, and the Vestmalmen orebody dips toward the west.

The mineralization at Råna Ni–Cu–Co intrusion characteristically includes high-grade massive sulfides associated with magmatic breccia and low-grade disseminated sulfides hosted in ultramafic rocks. The mineralization is mainly disseminated (Type-2). The massive and semi-massive ore (Type-1) exists near the southern contact with the gneisses. The Type-1 mineralization exists on both sides of the main fault zone. The interstitial sulfide dissemination passes into massive mobilized sulfides locally, and associated with certain deformation zones. The nickel carrier minerals in order of abundance include pentlandite, pyrrhotite, chalcopyrite, and pyrite occurring interstitially to olivine and orthopyroxene in peridotite and grading up to 0.8% sulfide nickel (Boyd and Mathiesen, 1979). Olivine contains ~0.09% Ni in its lattice, with the Ni/Cu ratio of 4:1.

Surface mining at the Bruvann deposit started in 1989, expanded with the underground method, and finally closed in 2002. The 100,000 m of drill core, obtained during mining between 1989 and 2001, has established the remaining measured reserves of 9.15 Mt grading 0.36% Ni, 0.08% Cu, and 0.015% Co at 0.30% Ni cutoff or alternatively 5.5 Mt @ 0.39% Ni at 0.35% Ni cutoff.

8.6 SPAIN

The kingdom of Spain is a separate state largely located on the Iberian Peninsula in southwestern Europe. The main economy is based on agriculture, energy, tourism, transport, and water supply. Spain has abundant reserves of lead, uranium, tungsten, mercury, magnesite, fluorspar, gypsum, sepiolite clay, iron ore, nickel, coal, crude oil, and natural gas. The country's significant mineral products in terms of value include copper, zinc, gold, steel, coal, cement, and alumina.

8.6.1 Aguablanca Ni–Cu Deposit

8.6.1.1 Location

The Aguablanca Ni–Cu–PGE magmatic sulfide deposit is located about 100 km northwest of Seville in the Extremadura region of southern Spain. The deposit was discovered in 1993 by Presur-Atlanta Copper Holding SA during a regional survey in the Ossa–Morena Zone. The mafic–ultramafic pluton is situated in the northern limb of the Olivenza–Monesterio antiform, a major Variscan structure of the Ossa–Morena Zone (Suárez et al., 2010). The Aguablanca mine lies ~30 km south of the town of Monesterio. The deposit/mine can be accessed by an 11.5 km paved road that is connected to the four lane N630 national highway from Seville.

8.6.1.2 Geological Setting

The host unit is a small stock (Aguablanca) formed along the northeast part the Santa Olalla plutonic complex. The calc-alkaline Variscan pluton intrudes Early Cambrian volcanic, volcanoclastic, and carbonate rocks (Bodonal–Cala Complex) overlying Late Precambrian metasediments (Serie Negra Formation). The intrusion (338–334 Ma) is dominantly composed of medium-grained diorite, massive gabbro, gabbronorite, pyroxene gabbro, and norite (Gabbronorite Unit) and a fine-grained gabbro (Contact Gabbro Unit). The plutonic complex is surrounded by country rocks containing a hypersthene hornfels aureole.

The basaltic magma, formed by the partial melting of the mantle, emplaced within the crust, cools slowly at depth to form gabbro and gabbronorite unit, the main host rock for the Aguablanca mineralization.

8.6.1.3 Mineralization

The deposit consists of two subparallel ellipsoidal orebodies elongated in an E–W direction with subvertical dip. The Main orebody in the south is larger and can be traced for 400 m and dips between 75° and 80°→North. The thickness is over 100 m, extends to a depth of over 600 m, and plunge between N65° and N80°E. The North orebody is about 125-m long along strike direction, 50-m thick, and extends to a depth of over 300 m. The orebodies are open at depth.

The ore-hosting breccias signify fragments of an ultramafic cumulate that was transported to the near surface along with a molten sulfide melt (Tornos et al., 2006). The sulfide-bearing host rocks have been weathered with supergene oxidation to a large extent forming a 10–20-m thick gossan capping. The in situ Ni–Cu-bearing gossan caps are more predominant over the main South orebody. The upper part of the gossan is composed of massive goethite extending to a depth of a few meters. The lower part of the gossan is less affected by leaching keeping the original textures and the sulfide mineralogy better preserved.

The sulfide mineralization is predominantly composed of pyrrhotite, pentlandite, chalcopyrite, and pyrite with minor marcasite, covellite, violarite, magnetite, and ilmenite. The platinum group of metals, gold, and silver are mostly associated at the edges of these base-metal sulfides, notably with pyrrhotite and pentlandite, less with chalcopyrite, and occasionally pyrite.

The mineralization is characteristically hosted within a chaotic magmatic breccia rich in fragments of unmineralized clinopyroxenite and gabbro (Breccia ore). There are three distinct types of mineralization at Aguablanca orebodies:

1. Massive to semimassive ore dominated by pyrrhotite and pentlandite, and is limited to the central core of the southern orebody with Ni/Cu ratio between 3 and 5.
2. Disseminated ore with Ni/Cu ratio between 1 and 1.5 within or surrounding the massive grade brecciated zones. The disseminated ore is reasonably continuous mineralization characterized by sulfides interstitial to silicates and composed mainly of pyrrhotite, chalcopyrite, and pentlandite.
3. Irregular ore with variable sulfides in patches, lenses, aggregates, or bands within the breccia and in distal parts of the deposit.

The contact between the massive/semimassive mineralized breccia in the central part of the Aguablanca orebodies and the disseminated mineralization along both sides of the central breccia is transitional (Fig. 8.8).

8.6.1.4 Reserve Base and Mining

A total of 96,515 m core drilling in 362 holes was conducted during 1984–1986. The Measured and Indicated reserves stand at 24.73 Mt grading 0.41% Ni, 0.35% Cu, 0.22 g/t Pt, 0.20 g/t Pd, 120 g/t Co, and 0.11 g/t Au at 0.01% Ni cutoff. The Inferred resources has been estimated at 1.83 Mt grading 0.18% Ni, 0.14% Cu, 0.11 g/t Pt, 0.10 g/t Pd, 70 g/t Co, and 0.06 g/t Au (Golder Associates, 2009).

The Aguablanca Ni–Cu deposit is a conventional open-pit operation using drill and blast, and a truck-and-shovel fleet with 8 m benches, and final slopes designed with double bench configuration. The open-pit mining will continue till reaching the optimum pit bottom,

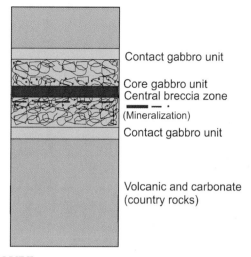

FIGURE 8.8 Overview of the stratigraphic at Aguablanca Ni–Cu–PGE massive, semimassive, and disseminated mineralization in the central brecciated-core gabbronorite unit. The mafic unit is intrusive into the volcanic and carbonate country rocks.

with simultaneous conversion to underground mine accessed through the existing exploration incline. The production from the initial sublevel cave is due to commence following closure of the open pit. The production from deeper sublevel open stoping will commence in 2017. The production capacity is ~1.9 million tonnes (~2.1 million tons) of nickel and copper ore per annum.

The run-of-mine ore is stockpiled, blended, and processed through crushing/grinding/flotation to produce a bulk nickel–copper concentrate. The dry concentrate is transported to Huelva port at a distance of 125 km for final shipment to customer smelter facilities. The tailings are pumped to a fully lined tailings pond to recover the water and reuse in the process plant as "0" discharge.

Genetic Model: PGE–Nickel–Chromium

9.1 CONCEPT

The formation of mineral/ore deposits within the Earth's crust in general, and the platinum group of metals (PGMs), nickel–copper–cobalt and chromium in particular, has engaged the interest of geoscientists, mineral explorers, planners, and naturalists over many decades in developing complex genetic models to simulate this process. The thought process starts, concepts arise, theories form, model tests begin, and new genetic models are then placed in the public domain. The concepts are challenged with new data, debated for hours, days, months, and years in seminars, journals, books, and other publications. Thereby, one or more new genetic model(s) become available for a single mineral or an ore body.

Ore genetic theories largely focus around three basic components including: the unique source of the metals/elements, the conduit/path and transport mechanism, and the move to a final destination/trap. The metal or group of metals/element or group of elements must pre-exist in some source lying deep within or on the crust and liberate by some tectonic process. The liberated mass in fluid or solid state performs the act of physically moving the metal, as well as chemical and/or physical phenomena which encourage movement through accessible conduits or channels. The trap acts to concentrate the metal via some physical, chemical, or geological mechanism into an enriched mineralization which forms the mineable ore.

The biggest mineral or orebodies are formed when the original source is large enough, the

Platinum-Nickel-Chromium Deposits
http://dx.doi.org/10.1016/B978-0-12-802041-8.00009-2

233

transport mechanism is efficient, and the trap is active and ready at the right time. The type of metallic and/or nonmetallic deposits will depend on the type of parental source material (magma). The three-dimensional configuration of the deposits will be based on the shape and size of the trap or depositional environment, and poststructural, physical, and chemical activities. The enrichment and concentration of metals is controlled by the content of the primary source and physical–chemical formation mechanisms.

Ore deposits seldom occur as a single metal or element, rather more likely found as a multi-metallic deposit. The chromium–nickel–copper–cobalt–PGMs originate from the same primary mafic–ultramafic magma and resulting rocks. The chances of coexistence in wide-ranging proportion are extremely high—making one metal predominant over the others. The explorer must expect this phenomenon. The best example may be that of the Bushveld Igneous Complex which was known as the largest chromite resource in the world and where chromite mining has continued for a century. Moreover, the same Bushveld Igneous Complex after many decades has become the largest resource of PGMs to satisfy the global demand of precious metals for succeeding centuries.

9.2 GENETIC MODELS

Deposits of the platinum-group elements (PGEs), nickel, cobalt, and chromium usually coexist in the Earth's crust due to a common genetic cause. The PGMs generally occur in mafic–ultramafic plutonic/volcanic rocks and are often associated with nickel, copper, chromium, and cobalt deposits. The source of platinum and palladium deposits is mafic–ultramafic rocks having enough sulfur to form sulfide minerals while the magma is still in the liquid phase. The sulfide mineral (usually pentlandite, pyrrhotite, chalcopyrite, and

pyrite) attaches chalcophile PGEs by mixing with the bulk of the magma. The PGE alternatively occurs in association with chromite either within the chromite mineral itself or within sulfides associated with it. The sulfide phases form in mafic–ultramafic magmas when the magma reaches sulfur saturation. This sulfur saturation is ensued by contamination of the parental magma with crustal material, especially sulfur-rich wall rocks or sediments, magma mixing, and volatile gain or loss.

Nickel deposits are closely associated with copper–cobalt–chromium and PGMs. The deposits are generally found in two forms, either as magmatic sulfides or the lateritic type. The sulfide-type nickel deposits are formed in essentially the same process as platinum-group deposits. The nickel is a chalcophile element and favors forming as sulfide minerals in the sulfide phase of a mafic–ultramafic magma. The best nickel deposits are formed where sulfide accumulates in the base of lava or volcanic flows, especially the komatiite lavas. The komatiitic nickel–copper sulfide deposits are formed by a mixture of sulfide segregation, liquid immiscibility, and thermal erosion of sulfidic sediments. The sediments are necessary to promote sulfur saturation.

The process of forming a lateritic nickel deposit is fundamentally the product of a deep weathering profile and the oxidation of nickel-rich mafic–ultramafic rocks. The lateritic nickel deposits require very large olivine-bearing parental ultramafic intrusions. One mineral formed in lateritic nickel deposits is gibbsite.

Chromitite deposits are closely associated with nickel and PGMs hosted preferably by layered and stratiform mafic–ultramafic intrusions. The chromite is formed by early differential crystallization from mantle-derived magma, gravity concentration of chromite layers at the base, and settling of the mafic–ultramafic intrusion in a large magma chamber with repetitive addition of new magma cycles. Chromitite also occurs as podiform deposits within complex oceanic crust and upper mantle, emplaced at convergent plate boundaries. The chromitites form where melts react with sections of subducted old lithospheric mantle at the plate. The podiform deposits are not as large as those found within mafic–ultramafic layered intrusions. The major ophiolite-related chromitite deposits are Phanerozoic in age ranging between 500 Ma (Thetford, Canada) and 10 Ma (New Caledonia).

The mafic–ultramafic-hosted mineral formation systems can also involve hydrothermal activity, lateritization, as well as placer-type deposition.

9.2.1 Magmatic Process

Magmatic PGE–Nickel–Chromium deposits form because of the preferential crystallization, segregation, and concentration of crystals or immiscible droplets of liquid sulfides from a mafic–ultramafic parental magma, and the partitioning of chalcophile elements from the silicate magma. The grain size and metal content in the deposits and the ratios of the economic metals vary widely.

The key events necessary for the development of a magmatic sulfide deposit in sequence include: partial melting of the mantle, ascent into the crust, development of sulfide immiscibility because of crustal contamination and interaction, ascent of sulfide-enriched magma to higher crustal levels, concentration of the sulfides, their enrichment through interaction with fresh magma, cooling, and crystallization with the end result of forming a deposit. The processes depend on the solubility of sulfur in silicate melts, the partitioning of chalcophile metals between sulfide and silicate liquids, effective sulfides concentration, and the processes that occur during the cooling of the sulfide liquid.

The magmatic process involves various components such as: (1) Fractional crystallization, (2) Liquid immiscibility, (3) Assimilation, and (4) Partitioning of chalcophile metals.

1. Fractional crystallization
   Fractional crystallization separates ore and non-ore minerals according to their crystallization temperature. The early-crystallizing minerals incorporate certain metallic elements. These crystals will settle at the bottom of the intrusion with concentration of ore-forming minerals. The chromite and magnetite are early crystallizing metallic minerals that form layers or seams in this way.
   The fractional crystallization and accumulation of crystals formed during the differentiation process of a magmatic event are known as cumulate rocks, and those parts are the first which crystallize out of the magma. Therefore, a cumulate is an igneous rock formed by accumulation of cumulate crystals either by settling or by floating within a body of magma. The cumulate occurs in plutonic bodies and thick volcanic lava flows. The chemistry of the cumulate indicates the temperature and pressure of the melt from which it originated.
   The cumulate are subdivided on the basis of the proportion of cumulate crystals relative to crystals formed from the trapped intercumulus liquid. Adcumulates comprise >95% cumulate crystals. Mesocumulates have 95–85% cumulate crystals. Orthocumulates consist of <85% cumulus crystals.

2. Liquid immiscibility
   Liquid immiscibility is a state in which two liquids with different compositions coexist in equilibrium with each other. The liquids do not mix and form an emulsion of droplets or networks of one liquid within the other. In mafic–ultramafic magmas, the Ni–Cu sulfides tend to separate on cooling and sink below the silicate-rich part of the intrusion or be injected into the surrounding rocks. These deposits are found in mafic and ultramafic rocks. The immiscibility between sulfide and silicate liquids is significant in the genesis of

magmatic sulfide deposits. The immiscibility between alkaline silicate and carbonate melts is thought to result in the generation of carbonatite magmas. The immiscibility between Fe-rich basaltic liquids and silica-rich liquids may occur during tholeiitic crystal fractionation and be important in the genesis of peralkaline granitic magmas. Magma mixing does not necessitate liquid immiscibility because the magmas are not in equilibrium and are in the process of mixing.

3. Assimilation

The process of assimilation assumes that a hot primitive melt intruding into a cooler felsic crust will melt the crust and mix with the resulting melt, thereby altering the composition of the primitive magma. Mafic and ultramafic magmas on rising through the felsic crust will assimilate it resulting in felsification of the primitive magma.

4. Partitioning of chalcophile metals

The participation of nickel between sulfide and silicate melts: The Fe, Ni, Cu, and Co atoms are bonded to oxygen in the silicate magma (FeO) and to sulfur in the sulfide melts (NiS).

The participation of nickel between olivine and sulfide melts: The Fe and Ni are bonded to Si in the silicate magma (FeSi) and to sulfur in the sulfide melt (NiS).

The participation of PGEs between sulfide and silicate melts: If the concentrations are due to segregation of the immiscible sulfide melts from the magma, and if these sulfides have not been significantly upgraded in their PGE contents subsequent to their segregation, the coefficients controlling the partition of the PGEs and Au between liquid sulfide and liquid silicate must be very high (Naldrett, 2004).

9.2.1.1 *Magmatic Plutonic Model*

Magmatic PGE–nickel–chromite deposits are formed by differential magmatic crystallization, segregation, gravity settling, or by late magmatic gravitative liquid accumulation from mafic–ultramafic magma. These layered deposits are mostly hosted by Archean ultramafic intrusions with a differential gravity mechanism. The metal collection trap is formed by SiO_2 concentration and mixing of mantle-derived hot primitive magma to saturate in chromite only by density segregation and accumulation. The multiple layering and stratification of the mineralized sequence is due to influx of repetitive new magmatic cycles (Fig. 9.1). The podiform-type deposits are formed by liquid injection. The structural relations of chromite with the associated minerals indicates that the chromite was the earliest to crystallize or it crystallized only in part simultaneously with olivine. All the early magmatic chromite deposits are of sufficiently large size to be of economic interest. The primary chromite separated from a magmatic body are the first minerals to crystallize. The basic principles of crystallization ensure that the first formed spinellids would be relatively rich in MgO and Cr_2O_3, and would be most stable.

The multimetals comprising Ni–Cu–Co–PGE sulfide deposits with/without chromium are formed based on availability of metals in the parent magma and/or in the country rocks. The mechanism of formation is extension of the same concept with portioning of chalcophile metals in sulfides and silicate melts.

The distinct characteristic features of many layered mafic–ultramafic intrusions are the formation of Contact-type and Reef-type Ni–Cu–PGE mineralization.

The "Contact"-type Ni–Cu–PGE mineralization occurs at the margin between the country rocks and the layered mafic–ultramafic intrusion. The grade is not uniformly concentrated in the country or intrusive rocks along the margin. The grade is usually low. The deposits include Duluth, River Valley, and Coldwater complex (United States/Canada).

Thin horizons of sulfide minerals enriched in platinum-group elements are known as PGE

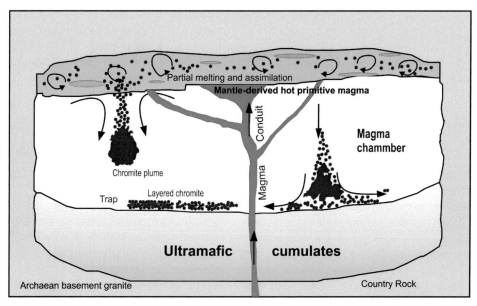

FIGURE 9.1   An overview of the magmatic plutonic model of PGE–Nickel–Chromium deposits hosted by mafic–ultramafic intrusive complexes. The multiple layers of mineralization are formed by repetitive influx of new magmatic cycles. The trap is formed by SiO_2 concentration and mixing of mantle-derived hot primitive magma to saturate solely in chromite by density segregation and accumulation. The layered and stratiform deposits mostly belong to Archean age.

"Reefs. " The thickness of individual reef varies between <1 m and 25 m. The average grade is rich and varies between 3 g/t and 20 g/t total PGEs. The Reef-type deposits include: the Stella intrusion (South Africa), the Great Dyke (Zimbabwe), the Munni Munni and Kambalda Intrusions (Western Australia), Pedra Blanca Complex (Brazil), and Sonju Lake intrusion (Midcontinental rift, Minnesota).

The Bushveld Igneous Complex (South Africa), the Stillwater Complex (North America), the Finish Intrusion (Finland), and Fedorova Pana (Russia) represent both "Contact" and "Reef"-type mineralization.

The magmatic Ni–Cu–PGE sulfide deposits are associated to the subcontinental lithospheric structural design and its proximity to craton and paleocraton margins. These deposits formed by segregation of a magmatic sulfide melt from a large volume of parental mafic–ultramafic melt and are usually coeval to a Large Igneous Province (LIP). Many of the large magmatic sulfide deposits are associated with intracontinental settings or with passive margins at the edge of small marginal basins. A craton-margin model for the genesis of magmatic Ni–Cu–PGE deposits provides a framework for further examination of ore deposit formation (Begg et al., 2010).

9.2.1.2 Magmatic Volcanic Model

Flood basalt is formed by an eruption or series of eruptions of large volcanic episodes that cover vast stretches of land or ocean floor with flows of mafic igneous rocks (basalt lava flows). A flood-basalt province is popularly known as Trap (Deccan Traps in India). Eleven distinct flood-basalt episodes have occurred in the past 250 million years, resulting in large volcanic provinces. In addition, many more volcanic mafic lava flows of medium to small dimensions have occurred that include: the Keweenawan Province of Lake Superior (Duluth complex), the Karoo Province of South Africa, the Parana of South America, and the Coppermine River area of northern Canada.

Similarly, an ophiolite is a section of the earth's oceanic crust and the underlying upper mantle that has been uplifted and exposed above sea level, and often emplaced onto continental crustal rocks. Ophiolites are usually green-colored rocks of spilites and serpentinites. Spilite is a fine-grained igneous rock resulting particularly from the alteration of oceanic basalt. Ophiolites occur in the orogenic belts of the world.

Both these groups of ophiolite rocks originate from relatively primitive mantle-derived mafic–ultramafic magma emplaced on the ocean floor or as lava flows on the Earth's crust by the mechanism of volcanic eruption. These large volumes of relatively primitive magma, often with evidence of chalcophile depletion, are regarded as favorable and prospective environments for hosting nickel–copper sulfide ± PGE deposits within flow-dominated volcanic assemblages. The source of sulfur is assumed to be the surrounding country rocks.

Mineralization occurs when komatiitic magmas, having low silicon, potassium, and aluminum, and high- to extremely high-magnesium content, reach the surface and attain sulfur saturation through melting or assimilation of sulfur-rich footwall rocks. The mineralization can occur if the lava is extruded onto a sulfur-rich deep marine facies. The massive sulfide deposits form in restricted channels at the base of the flows, or disseminated nickel mineralization can occur within sill or dike intrusions. The mineralization is restricted to two facies: compound sheet flows with internal pathways or dunitic compound sheet flows. The key to finding rich concentrations in this environment would seem to be the identification of conduits through which such flow has moved (Naldrett, 2004).

9.2.1.3 Orthomagmatic Model

The Orthomagmatic model is applied to the stage during which the main mass of the silicates crystalize from the mafic–ultramafic mantle-derived magmas that have undergone a high degree of partial melting. The crystallization can occur during the early (anhydrous silicate minerals) or the late (anhydrous and hydrous minerals) Orthomagmatic stage. The deposits usually occur in various tectonic settings including continental rifts, continental rifted margins, continental LIPs, and mid-ocean ridges. The mineral system forms Ni–Cu, Ni–Cu–Co, Ni–Cu–PGE sulfide, chromite, and Fe–Ti–V oxide deposits.

The ideal geological setting comprises layered mafic–ultramafic tholeiitic, komatiitic, or ophiolitic intrusions emplaced easily through a tectonic path in Archean cratons or Proterozoic orogens. Some of the largest Ni–Cu–PGE sulfide deposits in Australia have been identified. Most intrusions are differentiated into a lower ultramafic zone having Ni–Cu ± PGE sulfides and chromitite deposits with an overlying mafic zone in which Fe–Ti oxides crystallize, with PGE deposits located close to the mafic–ultramafic boundary. The PGE reefs are formed during upward accumulation of mineral deposits by downward settling of immiscible sulfides through a silicate melt.

Mantle-derived magmas are the source of fluid, metals, and energy in Orthomagmatic mafic–ultramafic volcanic deposits.

9.2.2 Hydrothermal Model

The hydrothermal model activates on the concentration of selective mineralization during advancing flow and deposition in hot fluid media. The chromite mineralization forms from hydrothermal solutions, for the most part immediately preceding or contemporaneous with serpentinization.

Hydrothermal processes are physicochemical phenomena and reactions caused by movement of hydrothermal fluid (hot water) within the Earth's crust. The processes usually occur because of magmatic intrusion, tectonic disturbances (faults, shear zone, orogenic rifting), or both. The basics of hydrothermal processes are the source–transport–trap mechanism in a continuous sequence. The sources of hydrothermal solutions include seawater and meteoric water circulating through fractured rock, formational water trapped within sediments at deposition, and metamorphic fluids created by dehydration of hydrous minerals during metamorphism.

The magma chamber is the heat source for the hydrothermal solutions which contain both magmatic and meteoric matters. The primary metal sources of economic importance include precious trace elements from the country rocks through which the magma flows, liberated by hydrothermal processes, and/or carried from the source magmas or meteorites. The transport of trace elements by hydrothermal solutions requires soluble media to form a metal-bearing composite. These metal-bearing composites facilitate transport of metals within aqueous solutions, generally as hydroxides.

The chromite forms from hydrothermal solutions, for the most part, immediately preceding or contemporaneous with serpentinization. The chromite of such origin form nests and vein structures.

The hydrothermal process is specifically well applicable in gold metallogeny in which various thiosulfate, chloride, and other gold-carrying chemical composites (tellurium chloride/sulfate or antimony chloride/sulfate) are involved. The majority of metal deposits formed by hydrothermal processes include nickel, copper, zinc, lead sulfides, indicating sulfur is an important metal-carrying component.

Sulfide deposition within the hydrothermal trap zone occurs when metal-carrying sulfate, sulfides, or other composites become chemically unstable. The instability condition arises due to fall of temperature, loss of pressure, and reaction with chemically reactive wall rocks, usually of reduced oxidation, such as iron-bearing rocks, mafic or ultramafic rocks, or carbonate rocks.

The hydrothermal process improves the stability of nickel–copper sulfides and the association of PGEs.

9.2.3 Lateritic Model

The lateritic model is a surficial/near-surficial weathering/oxidation system resulting from physicochemical phenomena. It causes concentration of ore materials (Ni–Cu–Co–PGEs, Al, Fe, etc.) by the action of paleoclimatic environment and occurs within the regolith. Regolith is a layer of soft/fragile/loose, heterogeneous superficial material covering solid rock. It includes dust, soil, broken rock, and other related materials, and is present on/near the Earth's surface. Laterite forms because of supergene enrichment, i.e., weathering via oxidation, chemical, and/or biological changes of mafic–ultramafic rocks. The changes are preferably intensified in fractured and liberated rock fragments creating chemically deposited minerals and clays.

The geological setting of lateritic Ni–Cu–Co–PGE deposits is associated with ophiolites and mafic–ultramafic complexes in mobile belts and Archean/Proterozoic greenstone terranes. The rocks are required to be uplifted and exposed to weathering processes. Secondary mineralization forms when the primary sulfide minerals are oxidized at the surface, and acidified meteoric water carries the Cu and Zn down to the water table where Cu is usually deposited. The formation of lateritic bauxite deposits require intense weathering conditions and is best developed in tropical climates. Low-grade lateritic iron ore forms with the enrichment of Fe by chemical weathering/leaching.

The supergene routes of laterite formation predominantly involve periodic meteoric water circulation derived from precipitation of snow and rain. The meteoric waters flow in the weathering profile under moderate topography with concomitant oxidation and chemical weathering of preexisting ore minerals. The descending meteoric waters oxidize the primary sulfide (pyrite) and other minerals in an acidic condition, percolate downward, and redistribute ore elements below the water table. The lateritic process is most common in arid or semiarid regions. The sources of the metals are the protolith (parent rocks) that is being weathered. Mafic and ultramafic rocks are the most suitable sources for Ni, Cu, Co, Cr, Sc, and PGEs in ore-bearing lateritic deposits. Carbonatites are appropriate sources for light REE, U, Th, Sr, Ba, Nb, Ta, Pb, Zn, V, and P.

Lateritic Ni–Cu–Co and PGE deposits are the result of supergene processes. Supergene enrichment occurs at or near the base of the weathering profile developed over deep-seated sulfide deposits or rocks enriched in Ni, Cu, and Co. The Ni–Cu–Co and other metals are leached from the oxidized part of the laterite profile above the groundwater table, carried downward by percolating groundwater, and deposited either by a change in oxidation–reduction (redox) at the water table or through reactions with hypogene sulfide minerals. These reactions produce secondary oxides and/or sulfides with metal contents higher than those of the primary ore or protolith.

The paleogeological history establishes the process of nickel laterite formation. The laterite formed by intense chemical weathering of a peridotite/dunite/serpentinite protolith in a region with a high water table and relatively low topography, results in the formation of an oxide deposit. In addition, silica precipitation is common in the upper horizons of the laterite deposit, where it creates an indurated layer, protecting the deposit from erosion. The Bitincke lateritic nickel, Albania, comprises two distinct zones characterized by nickel silicate and iron oxide phases. The laterite formation and variations in thickness are controlled by the interaction between topography, faulting, and protolith fracture density.

The optimum temperatures and precipitation rates for the development of nickel laterite can be estimated by comparing paleoclimatic data for regions where appropriate ultramafic rocks are exposed. A compilation of paleoclimatic data provides the optimum periods of laterite

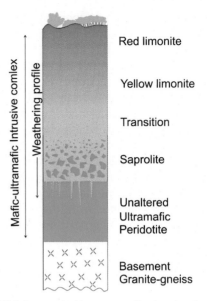

FIGURE 9.2   An ideal lateritic profile showing the formation of nickel–copper–cobalt–PGMs formed by a supergene enrichment process of physical and chemical weathering in humid tropical climates. The nickel–copper–cobalt metal enrichment is in the highest order in the saprolite/garnierite/serpentinite horizon and gradually diminishes to the up and down margins.

formation. The temperatures of formation can be obtained from goethite oxygen and hydrogen isotopes.

An ideal lateritic Ni–Cu–Co–PGE profile (Fig. 9.2) is composed of unaltered fresh mantle-derived mafic–ultramafic host rock that has been exposed to the surface. The overlying saprolite horizon comprises soft, thoroughly decomposed, and porous rock formed in the lower zone of the soil profile, often rich in clay. Saprolite is formed by the inplace deep chemical weathering of mafic–ultramafic rocks, especially common in humid and tropical climates. The deeply weathered profiles are widespread on the continental landmasses between latitudes 35°N and 35°S. The nickel–copper–cobalt metal of supergene enrichment is in the highest order in the saprolite horizon. This horizon may also contain garnierite and serpentinite (Box 9.1). The lateritic profile gradually changes upward

---

BOX 9.1

LATERITE PROFILE

Saprolite is soft, thoroughly decomposed and porous rock formed in the lower zone of the soil profile, often rich in clay. Saprolite is formed by the inplace, deep chemical weathering of mafic–ultramafic rocks, especially common in humid and tropical climates. The deeply weathered profiles are widespread on the continental landmasses between latitudes 35°N and 35°S. The nickel–copper–cobalt metal enrichment is in its highest order in the saprolite horizon.

Garnierite is a general name for a green nickel ore (NiO) to $(Ni,Mg)_6SiO_{10}(OH)_8$, which is found in pockets and veins within weathered and serpentinized ultramafic rocks. It forms by lateritic weathering of ultramafic rocks and occurs in many nickel laterite deposits in the world.

Serpentinite is an ultramafic rock composed of one or more serpentine group minerals (kaolin–serpentine). Minerals in this group are formed by serpentinization, a hydration and metamorphic transformation of mantle-derived ultramafic rock. Serpentinites are ideal host rocks for chromium, nickel, cobalt, PGMs, and manganese.

---

passing through a transition zone, yellow limonite, red limonite, and a thin soil cover in succession. The metal grade likewise gradually changes to higher than background level due to chemical weathering. The depth, width, and the metal enrichment (Table 9.1) of the complete profile and the subunits vary widely between a few m to hundreds of m depending on the composition of the primary magma, structural deformation, stability of paleoclimatic conditions, and surficial weathering.

The key controlling factors for the lateritic nickel–cobalt–PGE formation in mafic–ultramafic host rocks are:

1. The mantle-derived ultramafic host rocks must be exposed to the surface for extensive physical and chemical weathering for metal enrichment through the supergene process.
2. A stable paleoclimate is one of the major controls on formation of nickel laterite deposits. A comparison of climatic data conducive for nickel laterite formation revealed that more than 1000 mm annual precipitation must be received. The average temperature of cold months is between 15 and 27°C, and the average temperature of warm months varies between 22 and 31°C.

TABLE 9.1 Trend of Metal Distribution in Ideal Lateritic Nickel–Cobalt Deposit

Laterite Profile	% Ni	% Co	% Fe	% MgO
Top soil	Traces	Trace	5–10	Trace
Red limonite	0.3–0.5	<0.1	>50	<0.5
Yellow limonite	0.5–1.5	0.1–0.2	40–50	0.5–5.0
Transition zone	1.0–2.0		25–40	5–15
Saprolite/ garnierite/ serpentinite	1.5–3.0		10–25	15–35
Fresh ultramafic rock	0.1–0.3	0.01	<5	35–45

An assessment of the timing and duration of paleonickel laterite formation in ancient terrains can be postulated by compilation of paleoclimatic data and using the defined climatic window of nickel laterite deposit formation. Robert et al. (2012) reported that the ultramafic protolith formation of the Çaldağ nickel laterite deposit of western Turkey was exposed in the Early Eocene, and that the majority of the deposit formed during the Eocene.

3. Deeply weathered profiles are widespread on the continental landmasses between latitudes 35°N and 35°S.
4. Conditions for the formation of deeply weathered regolith include a topographically moderate relief flat enough to prevent erosion and to allow leaching of the products of chemical weathering. The other conditions are long periods of tectonic stability, during which tectonic activity and climate change are minimal that would otherwise cause surficial erosion.
5. The last condition is essentially prevailing humid tropical to temperate climate.

The majority of large lateritic nickel–copper–cobalt±PGE deposits are distributed in the humid tropical region along the eastern coasts of Brazil, Madagascar, Western Australia, India, Indonesia, Papua New Guinea, and New Caledonia.

9.2.4 Placer Model

The subsequent process of the surficial physical and chemical changes forming concentrations of ore material within the regolith is the erosion, transportation, and deposition of valuable minerals under suitable low-energy environmental conditions. This includes placer deposits and residual or eluvial deposits. Erosion and transportation are caused by weathering followed by wind and water action. Deposition is a sedimentary process including winnowing and density separation.

Alluvial placers are formed by the deposition of dense particles at a site where wind or water velocity remains below that required for further transport. The dense minerals of chromium, nickel, cobalt, gold, platinum–palladium, tin, tungsten, and diamond form placers where the topography or the stream gradients lie within the critical range for deposition of these minerals. Placers form on the inside bends of rivers and creeks, in natural hollows, at the break of slope in a stream, and at the base of an escarpment.

Heavy-mineral sand deposits essentially fall into two categories depending on the mode of deposition: alluvial or aeolian. The aeolian deposits are generally closely associated with marine beach placers, having been formed by the erosion, transport, and deposition of heavy minerals from adjacent marine beach placers by prevailing winds.

The discovery of platinum from large alluvial placer deposits in the Ural Mountains in Russia dates back to 1823 and placer mining operated starting in 1824. Since then, exploitation and commercial uses of the platinum metals continued and declined around the 1930s. Euhedral macrocrystals of Pt–Fe alloy has been reported from the Kondyor PGE placer, Khabarovskiy Kray, eastern Siberia, Russia, having a relatively constant composition of $Pt_{2.4–2.6}Fe$, tin, and antimony contents up to 0.3% and an unusually low content of all PGEs except Pt. A number of new platinum–palladium placers (occurrences) have been identified in British Columbia.

9.3 DEPOSIT TYPES

PGEs, nickel, and chromium deposits occur in diverse genetic environments and in association with one common basic characteristic: mantle-derived magmatic mafic–ultramafic source rock. The ore mineral association can be exclusively the PGMs, nickel–copper–cobalt, and chromium, or as multimetals in various ratios. The deposits can be grouped according to genetic model. The most significant examples illustrate the specific models as follows:

1. Igneous Intrusive Complex (PGE–Ni–Cr)
   Bushveld
   Stillwater
   The Great Dyke
   Northern Finland
   Sudbury

Pechanga
Voisey's Bay
Jinchuan
Sukinda–Nausahi
Lac des Iles
2. Lateritic Ni–Cu
   Indonesia
   South America
   New Caledonia
   Murrin Murrin Lateritic Ni Deposit
3. Placer PGE
   Russia

9.3.1 Bushveld Type, South Africa

The Bushveld Igneous Intrusive Complex contains the largest and richest chromite and PGEs deposit in the world. The complex characterizes the best-layered mafic–ultramafic intrusion in the world, with stratiform deposits of sulfide, chromitite, and magnetite. The most significant mineralized zones are the Merensky Reef and the UG2 chromitite located in the Upper Critical Zone. The ultramafic cumulates of the Bushveld complex crystallized from a magma type having a distinctive combination of high MgO (12–14%), high SiO_2 (54–56%), and low TiO_2 (0.3–0.4%).

The mafic–ultramafic magmatic events formed around the same period (2000 Ma) and are remarkably similar. Vast quantities of mantle-driven molten magma were intruded into the Earth's crust through long vertical feeder channels/cracks and formed a huge arcuate differentiated lopolithic chamber. Differential crystallization, crystal settling, and accumulation of molten magma with lowering of temperature resulted in the selective formation of particular minerals and rocks. The cyclic addition of new magmatic influxes within a vast magma chamber over time resulted in the formation of distinct structure that include diverse rock strata, two layered and stratified chromitites, and three PGM-bearing reefs (layers).

Mondal and Mathez (2007) proposed that the UG2 chromitite of the Bushveld Complex formed by injection of a new batch of magma containing suspended chromite and orthopyroxene crystals into an existing primitive magma chamber. The compositions of the two magmas are nearly similar, as both the magmas were crystallizing orthopyroxene and chromite of similar composition. The chromite from the newly injected magma readily separated from the orthopyroxene, rapidly accumulated, and spread out to form a dense monomineralic, and relatively impermeable, layer on the floor of the magma chamber.

The magmatic intrusion was emplaced as an early diabasic sill, outcrops of which are visible on the southeastern side of the complex. These are typically greenish in color, composed of clinopyroxene altered to hornblende and plagioclase, and are regarded as the earliest phase of the complex.

The orebodies within the complex include the UG2 (Upper Group 2) reef containing up to ~44% chromitite, and the platinum-bearing horizons Merensky Reef and Plat Reef. The Merensky Reef is composed of extensive chromitite and sulfide layers with thickness varying between 30 and 90 cm.

The Upper Critical Zone is the outcome of the resident magma in the chamber crystallizing orthopyroxene and plagioclase, followed by a succession of more primitive magma pulses. The primitive magma was injected along the cumulate–magma interface. These influxes were derived from a staging chamber in which previous bodies of magma en route to the complex had deposited sulfide (Naldrett et al., 2009).

Naldrett et al. (2011a,c) proposed that the PGEs in chromitites are contributed from two sources: (1) the chromite itself (IPGE), and (2) the sulfides that have largely been destroyed through reaction with nonstoichiometric chromite (PPGE and some IPGE). The LG1 to LG4 chromitites, characterized by low (Pt + Pd)/(Ru + Ir + Os) ratios, developed without sulfides.

From the LG5 upward to the UG2/3, sulfides accumulated continuously along with the chromite. The more mafic sequences in the interval between the top of the UG2 and the Merensky Reef are due to influxes of an unusually PGE-rich mafic magma across cumulates crystallizing from magma with both plagioclase and orthopyroxene as evident by variation in the Cu/Pd ratio of sulfide-poor rocks. The average PGE content of massive chromitites increases upward from the LG1 to the UG2/3/4. The marginal PGE deposits, such as, the Platreef and Sheba's Ridge, are zones along which magma escaped up the sides of the intrusive complex.

9.3.2 Sudbury Nickel Intrusion, Canada

The Sudbury basin/Structure/Complex, or the Sudbury Nickel Irruptive, is a major geological structure in Ontario, Canada. It is the second-largest (~10–15 km in diameter) known impact crater or astrobleme on Earth, as well as one of the oldest impacts into the ancient Nuna Supercontinent during the Paleoproterozoic era (1849 Ma). The debris from the impact scattered over an area of 1,600,000 sq. km. The subsequent geological processes have deformed the crater into the current smaller oval shape.

The Sudbury Igneous Complex (SIC) and footwall brecciated country rocks that include offset dikes and the Sublayer. The SIC is likely to be a stratified impact melt sheet composed from the base up of Sublayer norite, mafic norite, felsic norite, quartz gabbro, and granophyre. The theory of meteor impact including shatter cones and shock-deformed quartz crystals in the underlying rock was unanimously agreed upon in general by 1970.

The SIC has undergone severe weathering, deformation, and erosion over 1800 million years since its formation and that raises the debate whether to accept the theory of meteor impact as the origin and not the usual igneous process of mafic–ultramafic intrusion applicable to the majority of the Ni–Cu–PGE deposits. It is also argued that the region was volcanically active at around the same time as the impact, and some weathered volcanic structures can look like meteor-collision structures.

The geochemical analysis of concentration and distribution of the siderophile elements from the impact-melted rocks at Sudbury complex and the size of the area indicated the most likely cause of crater formation is by a comet and not an asteroid (Petrus et al., 2014).

9.3.3 Jinchuan Ni–Cu–PGE, China

The Jinchuan intrusion, located in the southwest margin of the Alashan Block, northwest China, hosts a world-class Ni–Cu–PGE sulfide deposit. The ultramafic rocks of the central and western segments comprise dunite, lherzolite, and olivine websterite from the center to margin of the intrusion. The orebodies are confined in dunite at the base of the intrusion. The massive sulfide ores occur mainly within and around the layered orebodies with length from a few meters to 300 m, and in thickness from a few cm to 30 m.

The Jinchuan intrusion was formed in a continental margin rift and that the ultramafic rocks are the cumulate phase of an originally tholeiitic magma.

The high-Mg primitive basaltic magma rich in olivine, chromite, and droplets of immiscible sulfide intruded up through a series of funnel-shaped feeder structures preserved along the tectonic rift-related faults in the middle and western part of the segments. The feeder funnels widened upward causing decrease in the velocity of the intruding magma. The suspended, higher-density material concentrated and became trapped in the conduit, which likely remained open while material near the periphery cooled in the central and adhered to the walls of the funnel. The magma intruded along the faults away from the main funnels forming a layered sequence. Some sulfide moved from the feeder funnels and deposited as stratabound zones of disseminated mineralization within the layered rocks.

A fresh cycle of magma was constantly injected into the system enriching the existing sulfides suspended in chalcophile metal melts and giving rise to the lateral zonation of the PGE contents in 100% sulfide. The sulfides in the central parts of the deposit upgraded to the greatest extent. The sulfides trapped in the early cooling, marginal parts of the funnel were exposed to leftover magma and became less enriched in sulfide. The PGE abundance of the rocks in the middle and western segments are more than three times higher than that of the eastern segment. The massive sulfide ores show the abundance of PGEs.

9.3.4 Sukinda–Nausahi Belt, India

The Sukinda–Nausahi Intrusive Complex is known for layered stratiform chromite deposits along the entire mafic–ultramafic belt with a large lateritic nickel deposit in the northwest and dominance of platinum–palladium in the northeast. The main chromite layers/lodes were formed by early differential crystallization, settling, and accumulation of chromites. The multiple chromite lodes are the result of repetitive cycles of magma influx (Fig. 9.1).

The large lateritic nickel formation in the southwest part of the Sukinda belt is due to deep weathering of the ultramafic rocks resulting in saprolite, yellow limonite, and red limonite (Fig. 9.3).

A detailed study of the Nausahi chromite deposit indicates that a part of the chromite might have crystallized later than the olivine, but simultaneous with pyroxene. The Nausahi chromite deposit is late with respect to olivine but early with respect to the magma as a whole.

Quartzite

Red limonite

Yellow limonite

Transition

Saprolite

Unaltered peridotite

FIGURE 9.3   A conceptual cum field model of lateritic profit at Sukinda layered ultramafic belt showing the natural sequence of unaltered peridotite at the base successively overlain by weathered saprolite, transition zone, yellow limonite, and red limonite. The large volume of nickel-bearing limonite is awaiting a metallurgical breakthrough.

The gravity separation of the chromium-rich late residual liquid suggests that the only way of differentiation was in the form of crystals that settle down along with crystals of olivine and pyroxene.

The Nausahi belt has witnessed the formation of a north–south-trending intense breccia zone that acted as a channel for the influx of PGE-bearing second-generation gabbronorite magma. The breccia effect is evidenced by sharply angled large fragments of massive chromite surrounded by platinum–palladium-bearing anorthosite.

Exploration Guide

Platinum-Nickel-Chromium Deposits
http://dx.doi.org/10.1016/B978-0-12-802041-8.00010-9

247

10.1 BACKGROUND

Demand and supply of the metals will control the market price trend and influence exploration and exploitation need. A long-term focused strategy and multidisciplinary expertise (Geology–Geochemistry–Geophysics) supported by Mining, Metallurgy, and Economics will lead to new and viable discoveries. The new geological environments must be understood and identified with respect to style of geological setting, age, and host-rock characteristics before a new venture. Innovative exploration concepts and improved integration of information from chemistry, mineralogy, and thermodynamics are the canvas of future viable targets.

10.2 DEPOSIT CHARACTERISTIC FEATURES

The platinum-group elements (PGEs), nickel, and chromite deposits coexist in nature and are primarily hosted by mafic–ultramafic plutonic–volcanic rocks. The majority of the chromite and lateritic nickel deposits are exposed at the surface and can easily be identified during geological mapping of target areas. The deposits of the Bushveld Complex, The Great Dyke, the Stillwater Complex, the Sudbury Complex, the Kola Peninsula, Noril'sk, Sukinda–Nausahi, the lateritic deposits of the Philippines, Indonesia, and New Caledonia exhibit some significant and characteristic common features that include:

1. Predominantly late Archean to early Proterozoic mafic–ultramafic complexes occurring as intrusions, extrusions, or both.
2. Cumulates with komatiitic, tholeiitic, or calc-alkalic magmatic affinities. Tholeiitic intrusions are the most common host rock for PGEs mineralization.
3. Favorable host rocks include mafic–ultramafic mantle-depleted units having high to extremely high magnesium content, and low silica, potassium, and aluminum.

4. Significant PGEs resources originate from enriched (primitive) mantle sources (Coldwell and Lac des Iles).
5. Impact of a large meteorite on the Earth's surface assimilating upper crustal rocks and mantle depleted in lithophile elements magnesium-rich magma (Sudbury Complex, Canada).
6. Proterozoic greenstone association.
7. Mesozoic ophiolite association.
8. Evidence of multicycles and extra-differentiated intrusive magmas.
9. Orebodies typically concentrate at the base of intrusions formed by coalescence and gravity settling of dense immiscible sulfide liquid droplets (Ni–PGEs) and early-crystallized components (chromium). Orebodies often occur within topographic lows in komatiitic flows (Sudbury) and primary embayment or riffle traps (Barnes, 2011).
10. Increase in Cu/Pd values above the reefs.
11. Chlorine-enriched minerals in the footwall.
12. Ore concentration and disseminations in layered ultramafic (peridotite–dunite–pyroxenite, and variations), and mafic (gabbro–norite–anorthosite) in the form of layers, reefs, lenses, pockets, seams, and stringers emplaced and structurally conformable with host rocks.
13. Type-I contact mineralization associates rocks containing elevated to normal abundances of PGEs. Type-II is internally disseminated. Type-III reef associates with rocks containing normal PGEs. Type-IV is hydrothermally mobilized in association with PGEs-enriched magma.
14. Large and fully developed lateritic nickel profile occurs in Arctic region with high precipitation, and contrasting hot summer and cool winter.
15. Size, metal content, and deposit style depend on volume of magma emplaced, repetitive cycles of emplacement, chemistry of source type, crustal contamination, and duration of total process.

16. Formation of PGE-dominated deposit requires high degree of partial melting to release melt to dissolve all sulfides of source, and a large volume of melt to produce large tonnage.

The common mineralization styles include:

1. Reef type: The Great Dyke, Stella, Pedra Branca, Skaergaard, Munni Munni, Weld Range, Panton Sill.
2. Contact/marginal type: Duluth Complex, Coldwell Complex.
3. Reef and contact types: Bushveld Complex, Stillwater, Federo–Pana Intrusion, Portimo Intrusion.
4. Stratiform: Bushveld, Seagull, Coldwell, Sukinda.
5. Hydrothermally deposited: Lac des Iles.
6. Tectonically mobilized: Trout Bay.

PGE–nickel–chromite deposits are divided into three distinct types to design exploration.

1. Ni–Cu-rich sulfide setting with PGEs enrichment: Sudbury or Noril'sk.
2. PGEs-dominated low or no sulfide in stratiform, reefs, layers within layered intrusions: Bushveld, Stillwater.
3. Massive and high-density mineral association: Sukinda–Nausahi chromite, Bushveld chromite, Sudbury nickel.

10.3 EXPLORATION

The exploration strategies can be simplified into two key themes:

1. Greenfield: Exploration for unconventional deposits, new area, new geological environment under Reconnaissance and Large Area Prospecting License
2. Brownfield: Exploration in and around working mines by extension or area known to host mineralization under Prospecting License and Mining Lease.

"**Reconnaissance**" is "Grassroots exploration" to identify new targets on regional scale. The preparations include literature survey; acquisition of geophysical data, if any; synthesis of all available data and concepts; and obtaining tenement permission from State/Provincial/Territorial Government. A comprehensive work program to execute in sequential manner during Reconnaissance is given in Table 10.1. The activities are planned and suitably modified based on results achieved.

10.3.1 Exploration Geology

Exploration geology is identification of a package of information comprising a complete

TABLE 10.1 Work Program During Reconnaissance

Year	Proposed Work Program
Year-1	1. Literature survey, reconnaissance License, regional geological check, mapping, and rock-chip sampling.
	2. Acquisition and interpretation of available airborne geophysical data from previous surveys.
	3. Identification of prospective geological packages/structures.
	4. Regional geochemical surveys: Soil/stream sediment sampling as required.
	5. Regional airborne geophysics, ground magnetic, and electromagnetic traverses as required.
Year-2	1. Integration and interpretation of geological, geophysical, and geochemical data to identify anomalies/targets (could be geological, geochemical, and geophysical).
	2. First-pass followup of anomalies/targets by detailed mapping, infill soil/rock-chip sampling, and ground geophysics.
	3. Prioritization of anomalies/targets for drill testing.
	4. Scout drilling of interesting targets.
Year-3	1. Second-pass followup and target definition.
	2. Reverse circulation/Diamond core drilling
	3. Downhole geophysics and drilling, if required.
	4. Reports/Recommendations.
	5. Prospecting license application—if encouraging results obtained.

sequence of activities. It ranges between searching for a new prospect (Reconnaissance) and evaluation of the property for economic mining (Scoping/Feasibility Study). It also includes augmentation of additional ore reserves in the mine and the whole of the mining district. There are various exploration techniques conducted by one or a combination. The programs are carried by multidisciplinary data generation. Exploration geology incorporates surface geological features that include:

1. Regional geological mapping: generalized surface geological map [e.g., Duluth Mafic Igneous Complex showing the Ni–Cu–PGEs deposits (Fig. 4.3)].
2. Deposit-level geological mapping at different scales: detail geologic map of Nausahi (Fig. 6.23).
3. Stratigraphic analysis to focus on identifying thickness of layered mafic-ultramafic units, base/bottom of lava flows, trough morphologies, sediment-free windows on basal contact, and areas with cumulate and flow dominated by multiple thin lava units. The region is significant for komatiite sequences to be drilled in areas of high magnetic anomalies.
4. Lithostratigraphic sequence with mineralized horizons: Munni Munni Complex (Fig. 7.1).
5. Significant and unique geological features: chromite lode exposed at the surface (Fig. 6.23), surface weathering with well-developed laterite e.g., laterite cover rock at Sukinda (Fig. 6.23).
6. Identification of host rocks.
7. Structure: layered mafic–ultramafic rocks at Namakkal (Fig. 6.23), and formation of PGEs-bearing gabbronorite breccia at Nausahi mafic–ultramafic Complex (Fig. 6.23).

The surface geological environments related to PGEs–Ni–Cr deposits are significantly associated with the flowing magmatic process and lithologic units:

1. Layered mafic–ultramafic intrusions
2. Flood basalt magmatism
3. Komatiitic magmatism
4. Greenstone belt magmatism
5. Alkali magmatism
6. Orogenic magmatism
7. Large igneous province
8. Impact meteoritic related magmatism

PGEs–chromium–nickel deposits identified as giant lopoliths/batholiths of Bushveld and The Great Dyke; medium-sized layered complexes of Stillwater and Munni Munni; narrow flat tube-shaped sill-like intrusions at Noril'sk; and plug-type breccia bodies at Lac des Iles. Largest deposits are usually associated with largest magmatic provinces. The geological exploration for different type of deposits include:

1. The best exploration targets are to locate large batholiths and lopoliths comprising mafic–ultramafic magmatic events emplaced into Late Archean–Early Proterozoic basement rocks. The area is precisely mapped with identifying the feeder zones, breccia units, and presence of sulfides. Soil and rock chip sampling, trenching follow to demarcate contacts. In case of encouraging surface signatures, drilling can delineate mineralization/orebody with shape, size, and grade variation across and along the strike. This is applicable to both reef- and contact-type deposits.
2. In case of stratabound reefs, the ultramafic–mafic contacts and transition within the series should be identified. The exploration should be focused within gabbroic horizons or below the first appearance of cumulus plagioclase as done at Great Dyke and Munni–Munni. The thin and persistent chromitite layers could generally contain PGEs. The magnetite segregation often indicates PGEs enrichment. The lithogeochemical profiles are highly effective in narrowing the search area by identifying intense changes in chalcophile metal ratios above and below a reef.
3. Geological exploration to sample every layered ultramafic/dunite for chromium and nickel, and gabbronorite breccia zone for PGEs.

10.3.2 Exploration Geochemistry

Exploration geochemistry fundamentally deals with enrichment or depletion of certain trace elements having genetic affinity to parent mineralization. The art of geochemistry is to identify "primary" and "secondary" haloes formed by natural dispersion of indicator elements. The haloes exist around the parent body or move distances by physical and chemical processes. The parameters of magmatic host environments include dispersion characteristics, crustal contamination, sampling medium, interval, depth, size fraction, analytical techniques, and anomaly enhancement.

Lithogeochemical studies of mafic–ultramafic rocks are useful in identification of favorable environments, direct identification of crustal contamination signatures, and sulfide liquid segregation. This information will lead to target selection and design for exploring mineralization. The rock chemistry is influenced by a variety of primary magmatic variables, overprinted by degrees of various elements due to metasomatic alteration.

10.3.2.1 Layered Mafic–Ultramafic Intrusive Complexes

The deposits hosted by layered mafic–ultramafic intrusions are known for large tonnage and rich multimetal grades throughout the world. The mineralization, in general, is stratabound and typically stratiform in nature dominated by the platinum group of metals (PGMs).

10.3.2.2 Komatiitic Volcanic Complexes

The komatiite-associated iron–nickel–copper–cobalt–platinum group of deposits form in a wide range of volcanic environments and overlie varied components of footwall rocks that include basalts (Kambalda, Australia), andesite (Alexo, Ontario), dacite (Silver Swan, Australia), rhyolites (Dee's Flow, Canada), sulfide-facies iron formations (Windarra, Australia), and sulfidic semipelites (Raglan, Canada).

Geochemically, Ni:Cr/Cu:Zn ratio in komatiitic hosted deposits, identify areas of enriched Ni, Cu, and depleted Cr and Zn. Cr is associated with fractionated, low-MgO rocks, and Zn is a typical sediment contaminant. If the ratio is around unity or greater than 1, the komatiite flow is considered fertile. Other useful geochemical trends include high MgO to identify the area with highest cumulate olivine, low-Zn flows, tracking Al content to identify contaminated lavas, and direct detection of anomalously enriched Ni. The mineralized layers display greater enriched signatures by crustal contamination than unmineralized ones. The presence of Ni-enriched sulfides is the best lithogeochemical indicator.

10.3.2.3 Lateritic Ni–Cu–Co–PGEs

Oxidation during the weathering of sulfide-rich deposits enhances the dispersion of elements in the surface environment, and the presence of Ni, Cu, Cr, and Co supplement the PGEs as potential indicator elements. Sulfide-bearing ores can preserve a large lateritic mass with distinct physical characteristics amenable to geochemical exploration. Extensive sampling of residual soils and gossans in search of PGE-bearing Cr–Ni–Cu deposits below thick lateritic cover has been useful in Western Australia. Laterites are usually developed over a long period of time on a land surface of low relief. The subsequent erosion and truncation of profiles or the PGE-enriched ferruginous layer are likely to be covered by transported overburden.

10.3.3 Exploration Geophysics

Geophysical methods are effectively used for exploration of chromium, magmatic nickel–copper sulfides ± PGEs, and regolith/lateritic covers. The methods are efficient for detection of mineralization (orebody/targets), mapping, and delineation to guide followup exploration drilling. Geophysical methods have been successfully used for the exploration of magmatic nickel deposits at Enonkoski (Laukunkangas), Pyhäsalmi, and Telkkälä Taipalssari in Finland;

TABLE 10.2 Appropriate Geophysical Methods for Platinum-Group Elements (PGEs)–Nickel–Chromium Exploration

Property	Method	Stratigraphy–Setting	Target Mineral
Acoustic	Seismic	Subhorizontal, layered stratigraphy, volcanic basalt flows.	Munni Munni PGEs in Australia, Kevitsa Ni–Cu–PGEs in Finland, goldfields of Witwatersrand basin, South Africa, and Bathurst mining camp, Canada.
Density	Gravity (measuring variation in the Earth's gravitational field)	High-density mafic–ultramafics (layered dunite–gabbro–norite intrusion).	Bushveld Cr, South Africa; Kola Peninsula PGEs, Russia; Sukinda–Nausahi Cr, India.
Susceptibility	Magnetic (measuring variation in the Earth's magnetic field)	Magnetic pyrrhotite, magnetic host rock such as serpentinized ultramafic rocks, mafic intrusive rocks, dikes, faults, and fractures.	Sudbury Ni–PGEs–Cu, Canada.
Polarization	Electrical (Spontaneous Potential (SP) and Induced Polarization (IP).	Disseminated sulfides	Stillwater PGEs North America.
Conductivity	Electromagnetic (EM)	Pyrrhotite, pentlandite, chalcopyrite	Ni–Cu sulfide deposits at several hundred meters depth beneath Voisey's Bay, Canada.

Kabanga, Mount Keith, Rocky's Reward, and Poseidon, in Australia; Vaisey Bay in Canada; and lateritic nickel at Sulawesi Island, Indonesia (Dentith and Mudge, 2014). There are five main geophysical methods in use (Table 10.2).

10.3.3.1 Seismic

The reflection seismic method has been a powerful and attractive geophysical tool for exploration of deep subhorizontal, layered mafic–ultramafic stratigraphy and volcanic basalt flows that provides a high-resolution image of the subsurface. The method identifies information about structural and lithological relationships that control mineral deposits (Malehmir et al., 2013). The reflection seismic survey unravels a better understanding of the ore-forming processes of two diverse mineral regions such as at Kevitsa Ni–Cu–PGEs deposit in Finland, and Bathurst Mining Camp, Canada. Two- (2D) and three-dimensional (3D) seismic data from the Kevitsa deposit clearly image the

3D geometry of the ore-bearing intrusion and provide information about its relationship to host rock units and nearby intrusions within a larger tectonic framework. Two-dimensional and 3D seismic data from the Brunswick No. 6 area in Bathurst Mining Camp suggest that the Brunswick horizon contains the bulk of the massive sulfide and associated iron deposits occurring within a reflective package extending down to ~6–7-km depth.

10.3.3.2 Gravity

The basic concept behind gravity survey is to investigate variation (gravity anomalies) in the Earth's gravitational field generated by differences of density between subsurface rocks. The variation in density is induced by the presence of a causative body, such as mafic–ultramafic (layered dunite–gabbro–norite) intrusion, heavy mineralization of chromite, faults, and folds within the surrounding subsurface. The size of the anomalies mainly depends on difference in density between host rocks and causative body (mineralization), their geometrical form, and depth of occurrence. The method is applicable to carry survey on the ground, in the air, and into the marine environment.

Gravity survey is often used in the initial stage of exploration to discover the contrasting higher-density mafic–ultramafic intrusion from the surrounding lower-density country rocks. The PGE-bearing mafic–ultramafic intrusive complex in the Fenno–Scandian Region is confined to low-density Archean basement rocks and high-density Proterozoic volcanic–sedimentary belts mainly controlled by large-scale transverse faults. The major mineralized intrusions include: Fedorova–Pana, Imandrovskaja, Moncha–Monchetundra, Mt. Generalskaja, Russia; Bushveld chromium, South Africa; and Sukinda–Nausahi chromium, India.

10.3.3.3 Magnetic

The magnetic method is the oldest and most widely used geophysical exploration tool. Magnetic surveying for mineral investigation with high-precision instruments can be operable in air (airborne), at sea (marine), and on land (ground). The airborne survey is attractive for scanning a large area during reconnaissance to detect targets for detailed ground survey during prospecting. The process is rapid and cost-effective. It is suitable for investigation of polymetallic nodules on the ocean floor. Ground magnetic survey is performed during prospecting and exploration over relatively small areas on previously defined targets by aircraft. The geophysical exploration can be formulated based on high/low-sulfide deposits.

High-sulfide ores, dominated mainly by magmatic and conductive pyrrhotite, represent easier geophysical targets. Economic concentrations of PGEs can be found distal to the main sulfide body within overall high-sulfide environments. The main sulfide body may contain an economic Ni–Cu–Au association ± disseminations of PGEs.

Low-sulfide settings with 2–5% sulfide ± PGEs will not make sufficient contrast in physical properties relative to the host rock to produce a measurable response. It is necessary to detect some other mineral, e.g., magnetite, and hope to establish an association with the host rock or structure serving as a pathway for concentrating the PGEs.

A mafic–ultramafic layered intrusion possesses the characteristic signature of high-magnetic remanence magnetization or residual magnetism such as at Stillwater, Fedorova–Pana, and Burakovsky. Magnetic survey is the most efficient tool for studying the internal structure of layered intrusions that include tectonic disruptions, magmatic layering, high-magnetic stratigraphic-marker horizons, and stratigraphic levels prospective for Ni–PGE reefs.

10.3.3.4 Electrical

Geoelectrical methods of mineral investigation depend on the properties of conductivity and resistivity of subsurface rock masses to pass electric current. The methods are exercised either by introduction of artificially generated current through the ground or making use of the naturally occurring electrical field within the earth. The current is driven through a pair of electrodes connected to the terminal of a transmitter, and the resulting distribution of potential in the ground is mapped by using another pair of electrodes connected to a sensitive voltmeter. The potential distribution and lines of electrical flux can be measured by the magnitude of current introduced and variation in receiving electrodes in the case of a homogeneous subsurface. The current deflects and distorts the normal potential in inhomogeneous conditions in case electrically conductive or resistive objects are present in the ground. The better conductive causative mineral bodies are massive and disseminated sulfide deposits. Two methods in use based on the type of current sources and their response to the subsurface rocks/mineralization under investigation are Induced Polarization (IP) and Self-Potential.

Induced Potential mechanism works on externally imposed voltage causing electrolytic flow in the pore fluid of rocks. The Self-Potential or Spontaneous Potential (SP) method based on natural potential differences results from electrochemical reactions in the subsurface. The method is unique as no current is artificially introduced into the ground.

Metallic sulfides and oxides are good conductors and respond better to this effect. The magnitude of electrode polarization depends both upon intensity of impressed voltage and concentration of conductive minerals. The massive and ubiquitous disseminated sulfide-bearing nickel–copper sulfides and chromite orebodies provide larger surface areas available for maximum ionic–electronic interchange and hence are extensively used by electrical surveying (Stillwater PGEs, United States).

10.3.3.5 Electromagnetic

Electromagnetic (EM) survey, both airborne and ground, is one of the most commonly used methods in mineral exploration. The technique is proficient in direct detection of conductive sulfide deposits, in which large conductivity contrasts exist between the orebodies and country/host rocks or thin overburden cover. The survey is based on responses of the ground to the propagation of EM fields composed of an alternating electric intensity and magnetizing force. A primary or inducing field is generated by passing an alternating current through a coil (a loop of wire called a transmitter coil) placed over the ground. The primary field spreads out in space, both above and below the ground, and is detected with minor reduction in amplitude by a suitable receiving coil in the case of a homogeneous subsurface. The magnetic component of an EM field penetrating through ground induces alternating currents or "eddy currents" to flow within the conductor in the presence of a conducting orebody. The eddy currents generate their own secondary EM field distorting the primary field. The receiver responds to the resultant of arriving primary and secondary fields so that the response differs in phase, amplitude, and direction from the response to the primary field. These differences between transmitted, and received EM field reveal the presence of a conductor and provide information on its geometry and electrical properties. The induction of current flow results from the magnetic component of the EM field. The total process of EM induction has been generalized in Fig. 10.1. The depth of penetration of an EM field depends upon its frequency and the electrical conductivity of the medium through which it is propagating.

The nickel sulfide deposits are considered effective superconductors in the geologic context for geophysical exploration. The deposits are explored by using EM techniques that measure the electric and magnetic fields generated in buried and concealed mineralization. Mapping of regional magnetic response and gravity is also of use in defining the komatiite sequences.

The EM survey is usually less effective for low-sulfide–PGE-bearing mineralization and shows resistivity contrast with the host rocks. There are examples of conductive PGE mineralization in sulfide deposits successfully discovered using EM, e.g., Ni–Cu sulfide deposits at several hundred m depth from Voisey's Bay, Canada, and platiniferous Moncha–Fedorova–Pana intrusion, Russia.

10.4 SAMPLING

Sampling is a key critical path in mineral exploration and performed by taking a small part of a geological entity such that the consistency of the part represents the whole. The sampling starts from the first day of a field visit to the last activity of project closure. It passes through Reconnaissance, Large Area Prospecting, Prospecting,

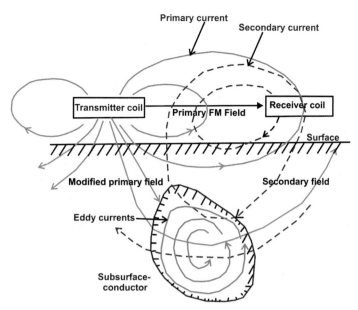

FIGURE 10.1 Conceptual diagram of electromagnetic (EM) induction processing system generating eddy currents in subsurface conductive mass (Haldar, 2013).

and grade assessment in mine production, mineral processing, smelting, and refining. There are various procedures of sample collection and several methods for analysis with minor variations during the phases of exploration. An authentic sample collection and precise analysis provide the economic survival of the project for a sustainable mine life.

10.4.1 Sampling Methods

Samples are collected by various means and methods as suitable and convenient to the situations without compromising quality and reproducibility. The collection tools are hammer and chisel, diamond core, reverse circulation, and air-core drills.

10.4.1.1 Soil–Stream Sediment

The soil and stream sediment samplings are the most common practices during Reconnaissance and Large Area Prospecting to identify target areas for detailed study. The samples are collected at random over the surface geological signature and in regular grid pattern as followup.

Soil is the unconsolidated weathering product and generally lies on or close to its source of formation as residual or transported soil. A soil survey is widely used in geochemical exploration and often yields successful results. Anomalous enrichment of elements from underlying mineralization occurs due to secondary dispersion in the overlying soil, weathered products, and leaching. Elemental dispersion spreads outward forming a larger exploration target than the actual size of the orebody. Soils display layering of individual characteristic horizons differing in mineralogy and trace element composition. Sampling of different horizons will present different results. The soil profile is classified into three broad horizons. The "A" horizon is composed of humus-charged top soil with minerals. The "B" zone represents an accumulation of clays enriched with trace elements.

The "C" horizon consists of bedrock fragments in various stages of degradation and gradually changes to hard parent rock. The indicator trace element concentration is generally higher in the "B" layer, and the most preferred sampling location (Fig. 10.2). Sampling of consolidated soil (Calcrete, Silcrete, Laterite, and Gossan) is carried out by the same procedure.

Stream sediment sampling is widely used during Reconnaissance and Prospecting survey from drainage basins. Sulfide minerals are unstable in weathering and will break down due to oxidation and other chemical reactions. The process will motivate secondary dispersion of both ore and indicator elements, and move in solid and solution form downstream to various distances within the drainage basin. The mobility of the different elements varies significantly, and eventually detrital fine-grained particles of rocks and minerals, clay, solutions, and organic and inorganic colloids enriched in ore and indicator elements are deposited downstream. The samples represent the best-possible composite of weathered and primary minerals of upstream catchments. It is an unconsolidated material in a state of mechanical transportation by streams, springs, and creeks. Sample density from 1 in 200 km² to close spaced following the course of a stream depends on the stage of exploration. The samples are collected in the dry season from natural sediment traps along stream courses below confluences. The choice of samples from first-, second-, and third-order streams will depend on the terrain, climate, and nature of weathering of the region.

10.4.1.2 Exploration Pit–Trench

Pitting is common practice during the initial stage of surface geochemical exploration to dig about $1 \times 1 \, m^2$ pits in a rectangular- or square-grid pattern covering the entire target area. The depth of a pit varies depending on the extent of weathering and the nature of the rocks. The material from each meter of depth is kept in separate low-height rectangular flat stacks to determine the variation in grade and other distinctive features. Pits showing the presence of mineralization are contoured to identify the strike and depth continuity of an orebody for drill testing.

Trenches are cut across the orebody (Fig. 10.3) to configure the probable mineralization boundary.

FIGURE 10.2   Soil sample collection during Reconnaissance/Orientation survey for Pt–Pd target search around abandoned Tagadur chromite–magnesite open-pit mine at Nuggihalli Schist belt, Karnataka, India (Finn Barrett, Consulting Geologist, Australia, in foreground).

FIGURE 10.3   Exploration trench sampling for platinum group of metals and chromium during Reconnaissance survey at Sitampundi layered intrusive complex, Tamil Nadu, India (Haldar, 2013).

The material recovered is stacked separately as a sample to identify variations. Trench walls are sampled by channel/chip for comparing with stack sampling.

10.4.1.3 Channel Chip

Channel sampling is appropriate for near-uniformly distributed mineralization in the form of veins, stringers, and disseminations. The sampling is done by cutting channels across mineralized body in fresh surface exposures or underground mine workings, such as face, walls, and roof. The area is cleaned to remove dusts, slimes, and soluble salts. A linear horizontal channel is cut between two marked lines at a uniform width and depth (Fig. 10.4). The length varies depending on variation in mineralization across the orebody. The tools are sharp drill-steel chisel and hammer or a pneumatic hammer with pointed or chisel bit/cutter.

Chip sampling is a matching technique for irregularly distributed or disseminated mineralization not easily recognized by the naked eye. It is a better alternative practice to collect samples by chipping off fragments of about 1–2-cm size across the entire surface exposure, underground

FIGURE 10.4 Channel sample cut by pneumatic cutter for exploration of platinum group of metals from the open pit bench face at Boula–Nausahi chromite mine, Orissa, India (Haldar, 2013).

mine face, wall, or roof in a regular grid point interval, say 25 cm × 25 cm. The area is cleaned before sample cutting. A sampler chips off fragments with the help of a hammer and a pointed chisel. Chips are collected in a clean box or satchel or on a canvas sheet spread on the floor. Chip sampling is preferred due to faster sampling and evaluation of a property.

10.4.1.4 Drilling

Sample collection from or near the surface is ideal during the Reconnaissance phase to identify targets. Mainstream sampling is conducted by drilling to collect samples from the subsurface on an interpreted vertical plane (cross-section: Figs. 10.5 and 10.6) across the mineralization/orebody at regular intervals along strike continuity. The section interval changes with complexity of mineralization.

The common drilling methods include Diamond Drilling (DD), Reverse Circulation (RC), and Air Core (AC).

Diamond drilling operates by diesel or electric power, is comparatively expensive, but most representative, with high sample quality, and reliable for precise orebody modeling, and estimation of reserves and resources with contained metals at 100% core recovery. The drilling capacity can be over 1000 m. Diamond drilling generates solid cylindrical continuous core (Fig. 10.7) under good conditions of machine, operators, and rocks. The drill core is cut into two equal halves, one-half being the mirror image of the other half.

Reverse circulation drilling uses hammer-bit cutting technology and air-force rig through the ground and brings rock-chip samples to the surface inside the inner tube. The system is low cost, faster, more efficient, and reasonably representative for quick estimation of tonnage and grade.

The Air-Core or Rotary Air Blast drilling is low cost, fast to a depth of ~125 m operates by air-force rig through the ground and rock-chip samples to the surface outside of the inner tube. Sampling is less efficient.

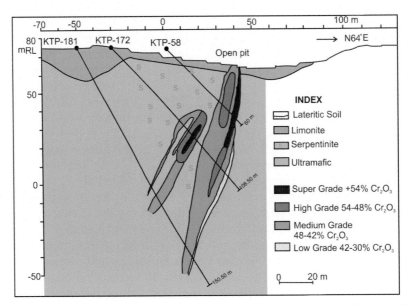

FIGURE 10.5 Drilling pattern and creation of cross-section for reserve estimation at Sukinda Layered Igneous Complex, Orissa, India.

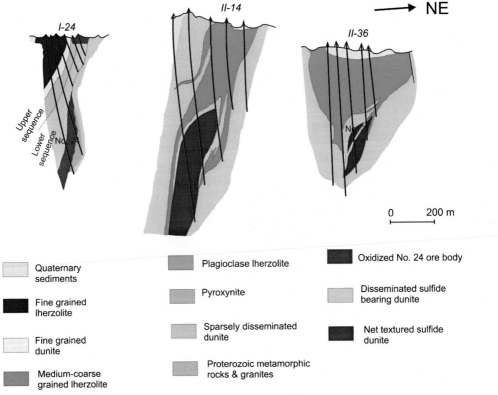

FIGURE 10.6 Simplified main geological exploration sections of the Jinchuan ultramafic intrusion and hosted Ni–Cu–PGEs sulfide deposit. I-24, II-14, and II-36 are the exploration section lines showing the lithological succession showing the net-textured, disseminated, and oxidized orebodies. *Song, X., Wang, Y., Chen, L., 2011. Magmatic Ni-Cu-(PGE) deposits in magma plumbing systems: features, formation and exploration. Geoscience Frontiers 2(3), 375–384. Production and hosting by Elsevier.*

FIGURE 10.7 Solid cylindrical core of 17-cm length. The core represents solid massive sulfide breccia and is composed of pyrrhotite, pentlandite, and chalcopyrite, rich in nickel, copper, cobalt, and PGEs.

FIGURE 10.8 Manual sorting of Run-of-Mine chromite ore with respect to size and grade for variation in market price, Boula–Nausahi, Orissa, India.

10.4.2 Run of Mine

The Run of Mine (ROM) is the ultimate quality of ore coming out of the mine or mine head. The run-of-mine ore is sorted manually (Fig.10.8) or screened with respect to size and grade for variation in market price. The ore of different size and grade is stacked separately for shipping. The stacked ore is sampled jointly for valuation. The fine or concentrated ore is sampled manually or by automatic sampler.

10.4.3 Accuracy and Due Diligence in Sampling

Possibilities exist of generating some inbuilt characteristic errors in every step of sample collection, processing, and chemical analysis.

A sampling due-diligence audit, that makes an authentic geological resource assessment, needs validation of six principal and critical components.

1. Sample representation, integrity, and security,
2. Accuracy of laboratory assays,
3. Insertion of blanks and standards at industry-accepted interval in the sample string within mineralization,
4. Adequacy of samples,
5. Implementation of Quality Assurance and Quality Control protocols,
6. Mineral resource continuity.

10.4.4 Quality Assurance and Quality Control

The most critical issue in sampling is to identify the bias likely to be associated during any step of collection, preparation, and analysis Quality Assurance and Quality Control (QA/QC) measures are essential in modern exploration programs for creating a reliable database which is free of any bias/error. The QA/QC pass database assures a trustworthy quantity (tonnes) and quality (% grade) within acceptable confidence of reserve and resource base of the deposit under feasibility study and mining investment. The sample audit eliminates the risk factors in estimation to a great extent. This is more relevant to upload the investigation data and reports in standard stock exchanges for commercial trading in a competitive manner. Some of the control measures include:

1. Several standard sampling methods in combination, such as the drill core, channel, and chip, may be conducted during exploration of a deposit. A sampling campaign should be piloted by creating a comparable database of drill core, mine face, muck, grab, channel, and chip samples over the same location and length of mineralization. The results can be compared

and statistically tested before accepting the best-suitable sampling technique for estimation of reserves and resources of the deposit.

2. In the case of drill-core sampling, usually one-half of the solid core is cut lengthwise by diamond cutter and processed for analysis. The analytical values are used for reserves–resources estimation without confirming the mirror image/equal representation of the other half in the system. The second half of the core should be processed, analyzed for a certain length of intersection, statistically verified, and accepted, modified, or rejected (Fig. 10.9).

3. Inherent human and process error can arise in a laboratory while analyzing a sample. Duplicate samples of known value are inserted at industry-accepted intervals of 10th or 20th position in a sample run and analyzed throughout the exploration phase at the same laboratory without disclosing identity of the sample. The results can be tested statistically for mean, variance, scatter plot, correlation coefficient "f," Paired "t" and Pool "t," before final acceptance, modification, or rejection.

4. Standard (certified reference material of known value) and blank (certified reference material of zero value) samples are inserted within the routine samples at industry-accepted intervals and dispatched to laboratory for quality-control purposes. These standard and blank samples can be inserted at the start, end, and every 10th or 20th position in the sample string. It is desirable to change the sequence of insertion of blanks and standards from time to time for quality assurance (similar to Fig. 10.9) in case of major differences. Samples are sent for repeat reanalysis. The QA/QC can be repeated in batches.

5. If the samples from the same deposit are analyzed at different laboratories, some bias is likely due to different laboratory personnel (analysts) and analytical procedures [Atomic

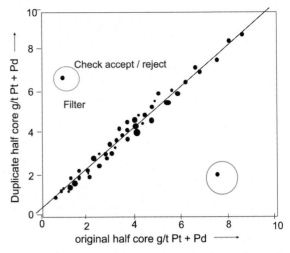

FIGURE 10.9   Plot of scatter diagram of first- and second-half core sample analytical values showing the comparative distribution of Pt + Pd and line of fit in regression. There may be samples with contrasting high or low values. The samples should be physically checked and reanalyzed before accepting or rejecting from total population. The correlation coefficient, after filtering, is high at 0.95 (+95% goodness of fit) (Haldar, 2013).

Absorption Spectrometry (AAS), Volumetric, X-Ray Diffraction (XRD), etc.]. A set of the same sample should be analyzed at all the concerned laboratories as well as in a Referee Lab of international repute. Results are statistically tested for equality.

6. The "data-error" between paired sets can be analyzed by various statistical tests. The simplest one will be a scatter plot of paired data. The scatter diagram will easily identify the presence of extreme erratic high or low sample values with respect to each other. The erratic sample pair must be sorted, identified, isolated, and their authenticities investigated along with probable sources of error. The samples can be verified and rejected if unsatisfactory and fit for the geological condition. The filtered data set will be suitable for QA/QC analysis.

The plot must show a remarkable degree of correlation at a high-confidence (r^2) level. If the check assay results exist in the acceptable range of standard deviation or within less than 5% variation from mean value at the 95% level of significance, then assay results are captured in the main assay database. The file turns into "Stable Data base [sic]." It is a continuous process with incoming addition assay input till the exploration ends. In the process, some of the results are unacceptable and the rejected values are not included in database. The total sample database can only be used for estimation of reserve and grade parameters if technically convinced by QA/QC protocols.

The confirmation of correlation between two complementary assay data strings can be performed by:

1. a percentile–percentile (PP) plot,
2. a quartile–quartile (QQ) plot, and
3. a cumulative frequency (CF) plot.

10.4.5 Sample Optimization

Sampling is a continuous process during exploration, mine planning, production scheduling, and grade control in a working mine. Drilling and the diamond drilling in particular is most authentic and all-out costly sampling tool. So the hardest question often raised in an exploration program is: "when to start drilling?" and even harder is "when to stop?"

The first step is governed by evidences identified from surface signature, supported by airborne and ground geophysical and geochemical anomalies indicating subsurface continuity of mineralization during Reconnaissance and Prospecting. The interpretative skill and experience of geologists act as prime factors at this juncture. Once the evidence potentially indicates existence of mineralization, then the decision for drilling is taken. The main purpose of drilling is to delineate the deposit and to establish continuity in strike and depth for resource estimation. A success story may begin with a bit of luck for estimation of reserves and resources.

A drilling program continues in a sequential manner to achieve the defined objectives. The drill interval and quantity will depend on complexity and value of commodity as in the case of the platinum group of precious metals. Drilling quantity should be adequate to establish 60% of the total reserve as Demonstrated (Measured + Indicated) category for feasibility study and investment decision. The drilling can be stopped for the time being at that stage until additional reserve is required.

The precision of width (tonnage) and grade (g/t or % metal) with increasing number of boreholes in an ongoing exploration program can be determined by applying statistical tools, and the exploration can be optimized to achieve a specific objective. The tools include statistical methods of Frequency, Probability, and Geostatistical model (Fig. 10.10). The curves become steady after drilling at a 100-m × 50-m spaced grid interval. This provides adequate confidence for definitive feasibility report preparation. The information is adequate and suggest for an investment decision for mine development. Further drilling improves the confidence marginally.

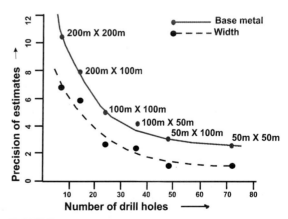

FIGURE 10.10 Precision of estimation and drill-hole spacing that indicates adequacy of sampling leading the investment decision for a large tonnage high-grade metal deposit (Haldar, 2013).

10.5 FUTURE EXPLORATION STRATEGY

10.5.1 Ni–Cu–PGEs Camp

Sulfide and silicate liquids are immiscible over a wide range of settings and will separate from magma when it becomes saturated in sulfur under reducing conditions. Ni, Cu, and PGEs have a strong tendency to partition into the sulfide liquid. The identification of the nickel sulfide deposit type is an important step to design a basic framework for understanding and developing successful discovery/exploration strategies. Ni-sulfide deposits are classified based on sulfide and MgO content of the parental magma. Regional/ District/ Deposit-scale exploration targeting can be focused according to the formational environment of the deposits.

Nickel sulfide deposits are associated with nuclei in Large Igneous Provinces (LIPs) at the regional scale. They represent a large volume and extensive area (region) of mantle-derived magmatic rocks emplaced rapidly (Noril'sk lavas in Siberia). The magmas are dominantly picritic to basaltic, or komatiites of Archean to Paleoproterozoic age.

Nickel sulfide deposits are closely associated with major deep-trapping structures that provide optimal conduits for magma flux from the mantle. The deposits form localized zones of concentrated intrusions or lava emplacement. They are characterized by larger volumes and differentiation of magmatic products (District/Camp: Kambalda in Australia). These camp-scale zones are localized by intersection of cross-structures with major mantle-tapping structures.

NiS orebodies are closely associated with magmatic conduit zones (Deposit-scale targeting), either volcanic or intrusive, and the identification of such zones is a priority for exploration. In general, the largest and most dynamic conduit zones are the most prospective (Flying Fox Ni deposit, Western Australia).

10.5.2 Magmatic Sulfide Association

Magmatic Ni–Cu and Cr deposits are formed by either layered/stratiform mafic–ultramafic intrusives or mafic volcanic basalts. Magmatic Ni–Cu–PGEs sulfide deposits occur as high-sulfur disseminated and massive, and low-sulfur disseminated types at Merensky Reef (Bushveld), Voisey's Bay, Jinchuan, and Duluth. The possible sources of sulfur may be solely of mantle origin and/or partly derived by hanging-wall/footwall contamination, and xenoliths from the country rocks. The formation of rich deposits requires a high-degree of partial melting of the mantle, crustal contamination, a dynamic system of sufficient metal availability, and suitable transfer apparatus for accumulation of immiscible sulfide liquids (Ripley, 2011). The early attainment of sulfide saturation is critical for the development of Ni-rich deposits.

A group of orebodies within one cluster or camp having common observed characteristic features can be identified for successful exploration. Exploration models for contact, footwall, and offset, massive and disseminated styles of sulfide mineralization will be different. Detailed exploration of marginally economic mineralized systems is often rewarding by critical quantitative analysis, eliminating bias at all levels. Exploration risk will be proportionately lower with additional and adequate investigation input. The risk of missing a new style of mineralization in a mature camp or mine environment or in a new geological environment is complex. A holistic geological model integrates the common characteristics of mineralization systems and maximizes the value from experiential geoscience observations. It forms the base for project selection and targeting of exploration using advanced geochemical and geophysical tools. Petrologic characteristics of host rocks can effectively be used to rank exploration modules in prospective mafic/ultramafic associations. It can also provide indication of productive source magma equilibrated with sulfide. Linkages between the abundance of metal in sulfides and host rock can focus efforts to define drill targets and better delineate orebodies at deposit scale.

10.5.3 Komatiitic Magmatism

Komatiite, mainly of Archean and Proterozoic age, derived from high degrees of partially melted mantle at extremely high temperatures and very low viscosity, are significant targets for magmatic sulfide mineralization. Volcanic rocks contain high MgO and display olivine–chromite cumulate texture. Sulfur saturation occurs when silicate magma exceeds its capacity to dissolve S and/or contaminated in excess by a large quantity of S to the magma from country rocks. Hydrothermal S is likely to be the most abundant source for komatiite. Komatiite is an ideal target for hosting magmatic PGEs-bearing Ni–Cu–Co sulfide deposits and is matured by metamorphic, tectonic, and hydrothermal processes.

10.5.4 Flood Basalt Magmatism

The geochemical signature of giant trap basalts at Noril'sk suggests that exploration may succeed in other trap basalts of the world. The key features of basaltic rocks indicate magmatic ore-formation events at Noril'sk intrusions in Devonian and lower Permian sediments. Minerals include disseminated and massive pyrrhotite–chalcopyrite association containing rich copper, nickel, platinum–palladium, and gold. The process of Ni sulfide formation imparts chemical signatures to parent magmas. The chemical signature of basalts at Noril'sk is empirically linked in space and time to ore formation. The vast spread of the Deccan Trap system in India is large, and regional lithogeochemistry offers a way to assess whether the magmas achieved S saturation and focus future exploration targeting (Lightfoot, 2011). The other future target areas are:

1. Continental: Midcontinental Rift Magmatism.
2. Oceanic: Triassic Magmatism.

10.5.5 Hydrothermal Ni-Sulfide System

Hydrothermal nickel sulfide systems are not well attended due to fewer occurrences, low tonnage, and low grade, incomplete available information, and lack of understanding. The occurrences are reported from Australia–Tasmania, the United States, Canada, Papua New Guinea, Russia, and Morocco (Bou Azzer).

The possible targets include obducted ophiolites in orogenic belts, close proximity of magmatic NiS deposits, mafic–ultramafic suites intruding or structurally linked to S sources and affected by a heat source, and low-MgO % mafic suites from LIPs (Álvarez et al., 2011).

10.5.6 Stream Sediment

Stream sediment sampling has been the most significant method for reconnaissance-scale geochemical exploration. Merensky Reef at Bushveld complex was discovered by mineralogical examination of stream sediments. Platinum was the principal indicator element. Enhancement of anomaly/background contrast of Pt is done by removal of coarser-grained fraction by sieving (−80 mesh), panning, and Density Media separation.

10.5.7 Glacial Region

Sampling of glacial tills and derived soils for distribution of these elements has been useful at Alaskan-type ultramafic–mafic complex, Tulameen, British Columbia.

10.5.8 Biogeochemical Survey

Practical application of biogeochemistry to geochemical exploration of noble metals has been limited by low abundance and below the detection level of existing analytical methods. Advancement of analysis by Inductively Coupled Plasma Mass Spectrometry (ICP–MS) at detection limit of 1–2 ppb for Au, Pt, and Pb on 1 g of ash from 50 g of dry twigs made a remarkable breakthrough. Sampling of organic trapping of noble metals along drainage course is popularly practiced.

10.6 EXPLORATION MODELING

Exploration models promote the concept of generating prospects using a series of decision loops based on successive multidisciplinary inputs from different sources. Each input contains new data to aid in the decision-making process. A model starts with study of existing literature and regional maps followed by various stages of Reconnaissance→Large Area Prospecting/Prospecting→Exploration→Phases—I, II, III→Estimation→Scoping/Feasibility study. A dynamic model presented by a logical flow diagram focusing objectives, time, and cost along with decision-making criteria to continue, modify, and temporarily abandon (stack in shelf) at any stage of activity. The exploration activity can be summarized in the following stages.

1. Regional geological activity: (Reconnaissance Permit)
   Mineral exploration starts with study of the demand–supply scenario over decades, existing literature, satellite maps, and imageries to support new investigation/investment, followed by application of a mathematical/forecasting model. A regional field check is essential to identify the presence of surface signatures, suitable stratigraphic packages, and favorable host rocks, structures like lineaments, shear, breccia, pathfinder elements, geochemical halo, and support from airborne geophysics. The final goal would be identification of a mineral province and/or belt.
2. District geological criteria: (Large Area Prospecting or Prospecting License)
   Activities involve detailed geological mapping, geochemical and ground geophysical survey, prioritization of targets, and diamond and reverse-circulation

drilling, and estimation of reserve and resources.
3. Local geological activity: (Mining Lease)
   The final stage of exploration comprises detailed close-spaced drilling to estimate reserves and grades with higher confidence supported by QC/QA analysis at all levels. Deposit is thoroughly understood when mineralization features are covered, along with stratigraphic control, host-rock assemblage, structure, lineament, and tectonic setting.

10.6.1 Modeling: A Holistic Dynamic Approach

Mineral exploration flow diagram is sequentially synthesized to evaluate property at the end of each stage for economic significance and opens two alternative paths suggesting either to "level pass" and continue successive exploration activity or to store conditionally on the shelf for the future (Fig. 10.11). The objectives of search and preparation requirements are defined at the beginning. It is proposed to analyze demand–supply scenario of a mineral or a group of minerals at the national and global levels for prioritizing long-term investment policy. The common key parameters are discussed with linkage if they exist. Understanding the stratigraphic horizon is essential to define a broad target area.

The modeling approach guides the search for a favorable host environment. A preliminary field check along with some spot geochemical samples from probable host rocks may indicate the significance of an area for submission of "Reconnaissance Permit," sequentially followed by Prospection and a Mining Lease.

10.6.2 Limitation

Geologic processes are events that occur on a geologic timescale lasting in millions of centuries, covering areas of thousands of square

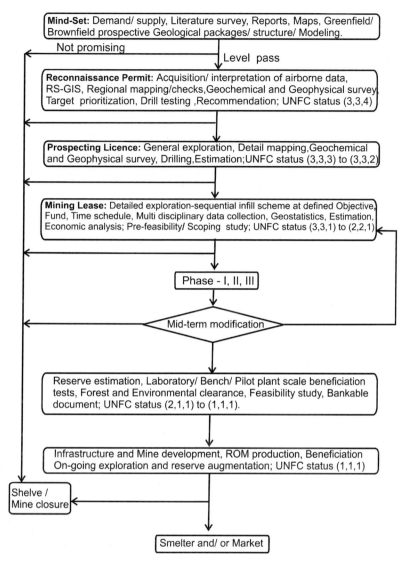

Mind-Set: Demand/ supply, Literature survey, Reports, Maps, Greenfield/ Brownfield prospective Geological packages/ structure/ Modeling.

Not promising

Level pass

Reconnaissance Permit: Acquisition/ interpretation of airborne data, RS-GIS, Regional mapping/checks,Geochemical and Geophysical survey, Target prioritization, Drill testing ,Recommendation; UNFC status (3,3,4)

Prospecting Licence: General exploration, Detail mapping,Geochemical and Geophysical survey, Drilling,Estimation;UNFC status (3,3,3) to (3,3,2)

Mining Lease: Detailed exploration-sequential infill scheme at defined Objective, Fund, Time schedule, Multi disciplinary data collection, Geostatistics, Estimation, Economic analysis; Pre-feasibility/ Scoping study; UNFC status (3,3,1) to (2,2,1)

Phase - I, II, III

Mid-term modification

Reserve estimation, Laboratory/ Bench/ Pilot plant scale beneficiation tests, Forest and Environmental clearance, Feasibility study, Bankable document; UNFC status (2,1,1) to (1,1,1).

Infrastructure and Mine development, ROM production, Beneficiation On-going exploration and reserve augmentation; UNFC status (1,1,1)

Shelve / Mine closure

Smelter and/ or Market

FIGURE 10.11   A holistic concept-and-approach flow diagram in mineral exploration modeling that is achieved by breaking the first fortune rock exposed on the Mother Earth to supply the finished goods to the common users in the society (Haldar, 2013).

kilometers, and measuring hundreds of meters in thickness. Compare this to everyday models from physics and engineering that operate at laboratory units and on the scale of one human lifetime. Geological concepts represent an abstraction of nature, and numerical models represent a tremendous simplification of a geological concept. Geologic models are conceptualized with the physical, chemical, and biological processes observed from stratigraphic sequences and syndepositional and postdepositional phenomena introduced by external and internal tectonic forces. Uncertainties are present at every level, and these limitations are to be appreciated. Outcomes need to be compared with observations. A geologic model can be considered predictive only when the applied parameters are within a reasonable range of values, supported by either measurements or some type of geologic reasoning, and only when the differences between the model outcome and observations are within acceptable ranges of uncertainty. There is no guarantee with respect to uniqueness of results in a particular situation, and hence these should be expressed in a probabilistic rather than in a single-value manner.

The exploration geologist should avoid purely model-driven exploration under any circumstances. One should remain open-minded and observation-driven to select and apply appropriate techniques in exploration. In most cases, effective geological mapping, prospecting, and structural analysis will provide the most valuable early-stage data for mineral targets. Nevertheless, companies that have traditionally explored for magmatic Ni–Cu–PGE–chromium deposits tend to be driven by geophysical techniques.

11.1 BACKGROUND

The platinum group of metals (PGMs), nickel, and chromium are in rising demand with global industrial growth. The platinum group of elements (PGEs) occurs as medium to large tonnages with usually low metal content (<1–5 g/t). Nickel occurs as medium to large tonnages with grades between 0.5% and 3% Ni. Chromite occurs as medium to large tonnages grading between 30% and 55% Cr_2O_3. South Africa, Zimbabwe, Russia,

the United States, Canada, and Australia are known for large PGEs resources. Sulfide/lateritic nickel deposits are located in Canada, Brazil, Australia, China, Nickel Rim of Indonesia, the Philippines, and New Caledonia. Historically, sufficiently high metallurgical-grade chromite reserves have been available at Bushveld, South Africa (+11,550 Mt), The Great Dyke, Zimbabwe (2574 Mt), Kazakhstan (320 Mt), Sukinda–Nausahi, India (190 Mt), Finland (161 Mt), Brazil (31 Mt), China, and other smaller producing countries (24 Mt).

11.2 RESOURCES AND RESERVES

Mineral resource and ore reserves are dynamic commodities, and continue to increase over time with additional exploration input, new discoveries, and/or new technology. Economically extractable ore reserves are considered a working inventory of the mining industry. Ore reserves and mineral resources form a continual chain of development and future supply of minerals (Boxes 11.1 and 11.2).

11.3 RESOURCE CLASSIFICATION SYSTEM

Mineral resources and ore reserves are estimations of tonnage and metal content of the deposit as outlined three-dimensionally (3D) with variation in sampling density, interpretation, and assumptions of continuity, shape, and grade. The sample database is checked and cross-checked by standard industry-acceptable norms and statistically assured before accepting for estimation.

BOX 11.1

RESOURCE AND RESERVE

Mineral resources are the quantity of natural concentrations of solid, liquid, and gaseous material in or on the Earth's crust. Volume (tonnage) is configured within the three-dimensional frame of a mineral body/deposit irrespective of the economic connotation. The mineral deposit must indicate the grade(s) of contained metal/nonmetal based on limited exploration input.

Ore reserve is an economically minable part of the orebody in which all-appropriate exploration, assessments, and feasibility studies have been carried out. All mining, metallurgy, commercial support, taxes, price, demand–supply trend, will and stability of government, and environmental aspects should have shown that the deposit can be mined at a profit after consideration of all factors impacting a mining operation.

BOX 11.2

QUALIFIED PERSON

A Qualified Person (QP) is a reputed professional with graduate or postgraduate degree in geosciences or mining engineering with sufficient experience (≥5 years) in mineral exploration, mineral project assessment, mine development, and mine operation. The QP may be preferably in good standing or affiliated with national and international professional associations or institutions. He or she is well informed with technical reports including Quality Assurance (QA), Quality Control (QC), data verification, discrepancy, and limitations, estimation procedure, quantity, grade, levels of confidence, categorization, economic status, order of magnitude, prefeasibility, and feasibility studies of the deposit. He/she is in a position to make statements and can vouche for the accuracy and completeness of the contained technical report. This is a matter of professional integrity and carries legal risk.

Mineral resource and ore reserve estimations are classified in standard and uniform codes by various countries for international acceptability in the open market system. Three commonly used classification system and reporting codes are the United States Geological Survey (USGS), the United Nations Framework Classification (UNFC), and the Joint Ore Reserves Committee (JORC).

11.3.1 USGS/USBM Classification Scheme

The U.S. Geological Survry (USGS) and the U.S. Bureau of Mines (USBM) system (1980) was developed in two combined axes of (1) increasing degree of geological assurances, and (2) increasing economic feasibility with division of identified reserves (Measured, Indicated, and Inferred) and unidentified resources (Hypothetical and Speculative), as shown in Fig. 11.1.

11.3.2 UNFC Scheme and Reporting Code

The United Nations Framework Classification (UNFC) system (2004) is formulated to give equal emphasis to all three criteria of

exploration, investment, and profitability of mineral deposits. The format provides (1) the stage of geological exploration and assessment, (2) the stage of feasibility appraisal, and (3) the degree of economic viability. The model is represented by multiple cubes (4 × 3 × 3 blocks) with geological (G), feasibility (F), and economic (E) axes (Fig. 11.2). The scheme is presented in 3D perspective with numerical codification facilitating the digital processing of information. The scheme is internationally understandable, communicable, and acceptable across national boundaries under economic globalization that makes it easy for the investor to make a correct decision.

11.3.3 JORC Classification Code

The Joint Ore Reserve Committee (JORC) Code (1971) was formulated by the Minerals Council of Australia, the Australian Institute of Mining and Metallurgy, the Australian Institute of Geoscientists, and the Australian Stock Exchange for public reporting of Exploration Results, Mineral Resources, and Ore Reserves (Fig. 11.3). The scheme emphasizes the basic principles of transparency, materiality, competency, and consideration of geological, economic,

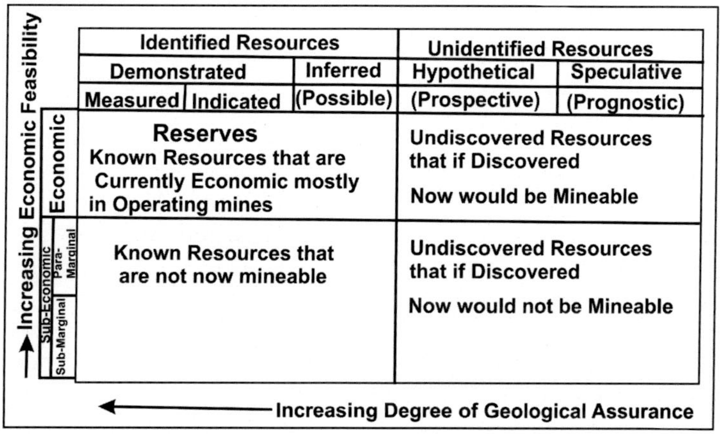

	Identified Resources			Unidentified Resources	
	Demonstrated		Inferred	Hypothetical	Speculative
	Measured	Indicated	(Possible)	(Prospective)	(Prognostic)
Economic	**Reserves** Known Resources that are Currently Economic mostly in Operating mines			Undiscovered Resources that if Discovered Now would be Mineable	
Sub-Economic (Para-Marginal / Sub-Marginal)	Known Resources that are not now mineable			Undiscovered Resources that if Discovered Now would not be Mineable	

Increasing Economic Feasibility →

← Increasing Degree of Geological Assurance

FIGURE 11.1 USGS resource classification scheme. *USGS*, US Geological Survey.

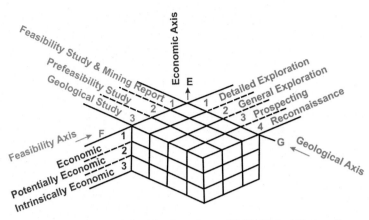

FIGURE 11.2   UNFC resource and reserve classification scheme. *UNFC*, United Nations Framework Classification.

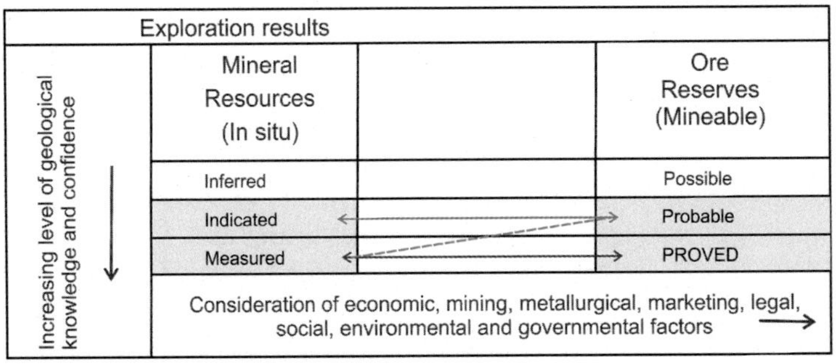

Exploration results			
Increasing level of geological knowledge and confidence	Mineral Resources (In situ)		Ore Reserves (Mineable)
	Inferred		Possible
	Indicated ←	→	Probable
	Measured ←	→	PROVED
	Consideration of economic, mining, metallurgical, marketing, legal, social, environmental and governmental factors →		

FIGURE 11.3   JORC Code developed by professionals of Australian Mining and Metallurgy. *JORC*, Joint Ore Reserves Committee

mining, metallurgical, marketing, legal, social, environmental, and governmental factors, and is effectively updated for comparable reporting standards introduced internationally.

11.4 RESERVE BASE

11.4.1 Platinum Group of Metals

The primary types of magmatic PGEs deposits can be divided into four broad groups:

1. PGMs-dominant deposits in association with sparsely dispersed Ni–Cu sulfides and/or Cr mineralization (Stillwater Complex at 25 g/t PGEs, 0.05% Ni, and 0.02% Cu).

2. Nickel–copper-dominated sulfide deposits with minor PGEs in komatiitic volcanic and intrusive rocks, rift, and continental flood basalts, or meteorite impacts (Sudbury Ni, Canada at 1% Ni, 1% Cu, and 1 g/t PGEs; Jinchuan Ni–Cu, China at 1.18% Ni, 0.63% Cu, and 0.26 g/t PGEs); Pechenga Ni–Cu, Russia at 1.18% Ni, 0.63% Cu, and 0.32 g/t PGEs; and Eagle Ni–Cu–PGEs, United States at 3.05% Ni, 2.51% Cu, and 1.17 g/t PGEs).

3. Deposits equally enriched with all metals to complement total economics of the project (Bushveld Complex, South Africa at 5.67 g/t PGEs, 45% Cr_2O_3, 0.13% Ni, and 0.06% Cu; Noril'sk, Russia at 9.50 g/t PGEs, 1.77% Ni,

and 3.57% Cu; and Raglan Complex, Canada at 3.76 g/t PGEs, 3.18% Ni, and 0.90% Cu).

4. The placer type of platinum deposits at low mining cost (Fifield in Western Australia at 0.61 g/t PGEs), and the Ural Mountains in Russia at low tenor.

The world reserve base of PGEs is estimated at 66,110 tonnes (~72,870 tons) with the Bushveld Complex contributing 95.30% (Table 11.1). The other countries having important reserves are Russia (1.66%), the United States (1.36%), Canada (0.47%), and all others (1.21%).

The grade-tonnage distribution of the known platinum group of deposits covering reserve base including past production and grade of total PGMs, plotted in log–log scale to accommodate extremely large and small values keeping the characteristics of the individual deposit, is plotted in Fig. 11.4.

Five extremely large deposits (>1000 Mt) exist in decreasing order of size include Bushveld (11,550 Mt), South Africa; Duluth (+3200 Mt), United States; The Great Dyke (2574 Mt) in Zimbabwe; Sudbury (1500 Mt), Canada; and Noril'sk (1309 Mt) in Russia. Three extremely low-tonnage deposits (≤10 Mt) include Lomalampi (3.06 Mt) in Finland; Eagle (5.33 Mt) in the United States; and Munali (10 Mt) in Zambia.

Four very high- to high-grade deposits (+5 g/t PGMs) include Stillwater (24.9 g/t), Noril'sk (9.5 g/t), Bushveld (5.7 g/t), and The Great Dyke (5.4 g/t). The remaining deposits contain grades between 0.19 and 2.90 g/t PGMs. These low-grade deposits are economically viable due to associated nickel, copper, and chromium in various magnitudes.

11.4.2 Nickel

Magmatic nickel sulfide is hosted by mafic–ultramafic intrusive/impact structures in Archean–Proterozoic basement rocks, volcanic flows, and ophiolites. The deposits are often layered and large, associated with copper, cobalt,

TABLE 11.1 World Reserves and Resources of PGM Estimated as of January 2015

Country	Reserve Base (Tonnes Metal)
South Africa	63,000
Russia	1100
United States	900
Canada	310
Other countries	800
World total reserve base	66,110
World total resources (including reserves)	~100,000

Adapted from U.S. Geological Survey, 2015. Mineral Commodity Summaries 2015. U.S. Geological Survey, p. 196. http://dx.doi.org/10.3133/70140094.

and ±PGEs, such as Sudbury, Jinchuan, Mount Keith, Leinster Camp, and Nebo–Babel. The lateritic deposits were formed during the Cretaceous and Tertiary periods. Effective weathering requires good geomorphology such as plateaus and adjoining gentle slopes, elongated flat ridges, high precipitation (+1000 mm/annum), and finally annual temperature averages ranging between 15 and 31°C with pronounced wet and dry seasons. Deposits are located spread over large surface areas such as Soroako, Gag Island, and Weda Bay in Indonesia; Goro and Koniambo in New Caledonia; and Kurumbukari in Papua New Guinea.

Magmatic Ni sulfide deposits contribute significant (>50,000) tonnes (~55,000 tons) to worldwide Cu production from Noril'sk, Sudbury, Voisey's Bay, and Jinchuan (Burrows and Lesher, 2012). The majority of Ni resources (~60%) are hosted by laterite deposits located in the tropics (Mudd and Jowitt, 2014). Three countries, namely, New Caledonia (22%), Philippines (16%), and Indonesia (15%) host ~53% of contained lateritic nickel resources. The remaining lateritic Ni resources are shared by Australia (12%), Central and South America (10%), Africa (7%), the Caribbean (7%), and others (11%) (Fig. 11.5).

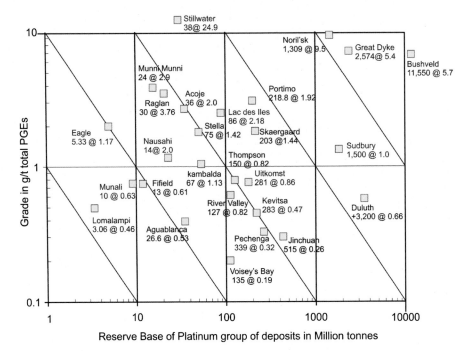

FIGURE 11.4 Grade-tonnage distribution of reserve base including past production of total platinum group of metals is plotted in log–log scale to accommodate extremely large and small values while keeping the characteristics of individual deposits. The Bushveld Complex is the largest deposit.

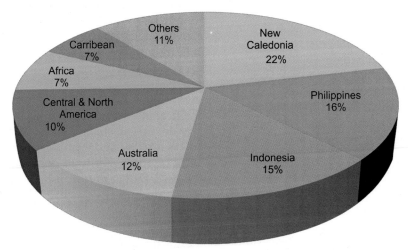

FIGURE 11.5 The comparative share of contained nickel in lateritic deposits in the world. The three countries, New Caledonia, the Philippines, and Indonesia, host ~56% of the contained lateritic nickel resources.

Identified land-based global (magmatic+ lateritic) nickel reserves are estimated at 81 million tonnes (~89 million tons) of contained metal as of January 2015. The shares of the reserve base are from Australia (23.5%), New Caledonia (14.8%), Brazil (11.2%), Russia (9.7%), Cuba (6.8%), Indonesia (5.6%), South Africa (4.6%), the Philippines (3.8%), China (3.7%), Canada (3.6%), Madagascar (2.0%), and the Dominican Republic (1.3%), making a total of ~92%. Remaining countries with small contribution share 8% of the reserve base. The identified land-based global resources of 130 Mt contain ~1% Ni (USGS, 2015).

Extensive deep-sea resources of nickel are in manganese crusts and polymetallic nodules, covering large areas of the abyssal ocean floor, particularly in the Pacific Ocean. The nickel resources on the seabed are likely to be many times those located on land. A statement of world nickel reserve base in decreasing order of abundance is given in Table 11.2 and Fig. 11.6 provide a global overview of nickel reserve base and distribution by countries/deposits.

The grade-tonnage distribution of known nickel deposits (magmatic+lateritic) covering reserve base including past production and grade of % Ni is plotted in log–log scale (Fig. 11.6).

The grade-tonnage distribution of nickel deposits shows a wide range and contrasting characteristics with respect to the magnitude (tonnage) and grade (% Ni). The reserves and grade vary between as low as 5.3 Mt at 3.05% Ni at Eagle, and as high as +3200 Mt at 0.13% Ni at Duluth, United States. Three very large-volume magmatic sulfide deposits exist, namely Duluth (+3200 Mt at 0.13% Ni, United States, +1500 Mt at 1.0% Ni at Sudbury, Canada, and 1300 Mt at 1.77% Ni at Noril'sk, Russia. Lateritic deposits are of medium tonnage (5–250 Mt) and medium grade (080–3.0% Ni).

11.4.3 Chromium

About 92% of the world's chromium reserve base is geographically concentrated in South Africa, Zimbabwe, Kazakhstan, Turkey, Finland,

TABLE 11.2 Worldwide Nickel Metal Reserves by Countries as of January 2015

Country	Reserve Base (Mt Ni Content)
Australia	19.00
New Caledonia	12.00
Brazil	9.10
Russia	7.90
Cuba	5.50
Indonesia	4.50
South Africa	3.70
Philippines	3.10
China	3.0
Canada	2.90
Madagascar	1.60
Colombia	1.10
Dominican Republic	0.93
United States	0.16
Other countries	6.5
World total reserves	81.00
World resources (including reserves)	~130.00

Adapted from U.S. Geological Survey, 2015. Mineral Commodity Summaries 2015. U.S. Geological Survey, p. 196. http://dx.doi.org/10.3133/70140094.

and India. The chromium resources of the United States are mainly in the Stillwater Complex. Other countries having sizeable resources are Brazil, China, Russia, and Albania. The average grade of deposits in Finland is low at 25% Cr_2O_3. The grade-tonnage distribution of known chromite deposits covering reserve base, including past production and grade of % Cr_2O_3, is plotted in log–log scale to accommodate extremely large and small tonnages (Fig. 11.7).

Total resources including reserves and past production of 10 deposits/groups of deposits indicate wide variation of resources ranging between 12.5 Mt at Bulqiza, Albania, and 11,550 million tonnes (~12,730 million tons) at

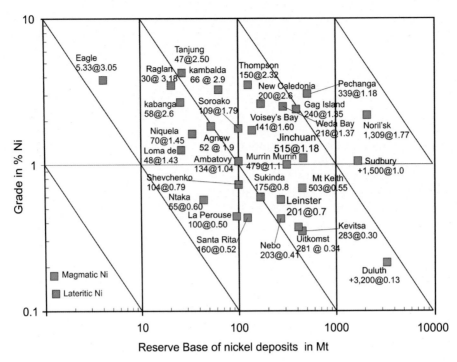

FIGURE 11.6   Worldwide grade-tonnage distribution of reserve base including past production and grade of nickel deposits plotted in log-log scale to highlight the population formed by magmatic sulfide (blue) and lateritic process (saffron). The lateritic deposits are comparatively of higher nickel content.

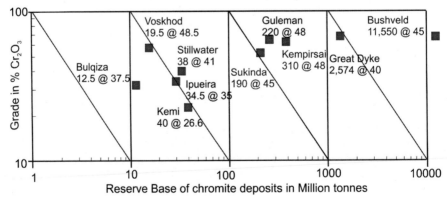

FIGURE 11.7   Grade-tonnage distribution of reserve base including past production and grade of Cr_2O_3 plotted in log-log scale. The chromite deposits are economically viable at 35–48% Cr_2O_3.

the Bushveld Complex. The % Cr_2O_3 ranges between 35 and 48 making the deposits economically viable. Low tonnage and low chromite grade is often boosted by naturally associated high-value metals like PGEs and nickel as main or byproducts.

11.5 RECOVERY MEASURES

11.5.1 Mining

Selection of mining method largely depends on the size, grade, and location of the deposit with respect to the surface. Universal mining methods are either open pit, underground, or in combination.

Open-pit mining is low cost, easily accessible, high productivity, and safer than underground mines. Open-pit mining is appropriate for near-surface, steeply dipping, and massive or large orebodies. Large-scale pit excavation would not cause significant environmental impact. Major components involve removing the overburden waste rocks/soils, digging the soft–ore, drilling and blasting the hard ore, removing by trucks, dumpers, or belt conveyor, and finally stockpiling prior to processing. Open-pit/surface mining is the first choice for any mining company.

Underground mining uses a variety of standard mining practices depending on the characteristics of the orebodies. The primary criterion would be the occurrence of the orebodies at depth where the ore-to-waste ratio will not permit economic open-pit mining. Access to the orebody is reasonably slow and costly with complexity of the deposit. Underground mining causes a comparatively high cost of production, ventilation, and lighting, and may be hazardous.

Opencast and underground mining are often combined in a single deposit that is exposed to the surface or near surface, but continues to greater depth. Both methods can take place simultaneously to access the shallow and deeper parts of the same orebody. An open-pit mine may be in operation while underground workings are in development. Open-pit/underground mining in combination is the best choice today for all the large metallic deposits.

11.5.2 Mine Production

Annual production of the platinum group, nickel, and chromite is reported by either mining of ore in millions of tonnes (e.g., chromite), or finished refined metals in tonnes (e.g., platinum, palladium, and/or nickel), or both in million tonnes and/or tonnes (e.g., copper, zinc, and/or lead).

11.5.2.1 Platinum Group of Metals

The in situ resources of PGE metals are limited to an uneven geographical distribution with more than 95% of metals concentrated in one single deposit, the Bushveld Complex. A statement of the global mine production of platinum and palladium metals contributed by the major producing countries during 2014 is given in Table 11.3. The share of the individual metals and in combination is depicted in Figs. 11.8–11.10.

11.5.2.2 Nickel

Abundance of identified nickel resources will satisfy the growing global demands for the next few decades. The decision to sanction a new project developed (or not) will be

TABLE 11.3 The Global Mine Production of Platinum and Palladium Metals by Major Producing Countries During 2014

Country	Platinum (t)	Palladium (t)	Total (Pt + Pd) (t)
South Africa	110	60	170
Russia	25	81	106
Zimbabwe	11	10	21
Canada	7	17	24
United States	4	12	16
Other countries	4	10	14
World total	161	190	351

Adapted from U.S. Geological Survey, 2015. Mineral Commodity Summaries 2015. U.S. Geological Survey, p. 196. http://dx.doi.org/10.3133/70140094.

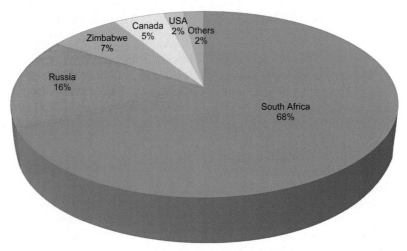

FIGURE 11.8   Global production share of platinum by major producing countries during 2014. The group of mines from South Arica led the production at 110 tonnes of platinum metal (68% global share) suppressing much beyond the remaining producing countries.

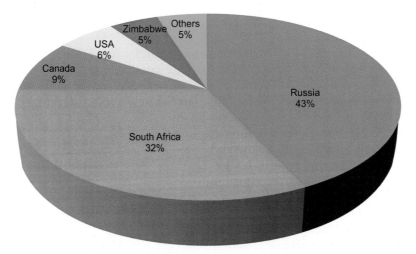

FIGURE 11.9   The global production share of palladium is by major producing countries during 2014. The group of mines from Noril"sk Region and Pechenga Belt produced the highest at 81 tonnes (43% of global share) in 2014, closely followed by South Africa at 60 tonnes (32% global share).

aimed at the economic, social, and environmental aspects. Global nickel production has grown steadily throughout the 20th century and has been matched by substantial growth in estimated Ni reserves and resources. Lateritic deposits account for >50% of global nickel production. The land-based resource is

expected to last in excess of 100 years at the present mining rate.

A statement of global mine production of nickel metal (magmatic sulfide and lateritic sources) contributed by all producing countries during 2014 is given in Table 11.4. The share of the nickel metal production by country is

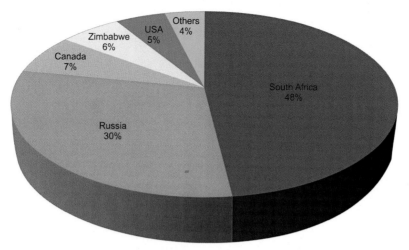

FIGURE 11.10 The global production share of platinum + palladium is by major producing countries during 2014. The combined metal production of 276 tonnes (48% global share) is achieved by South Africa. Russia at 30% share is meagrely supported by Zimbabwe, Canada, USA, and other countries.

TABLE 11.4 The Global Mine Production of Nickel Metal by Major Producing Countries During 2014

Country	Production in Tonnes of Ni Metal
Philippines	440,000
Russia	260,000
Indonesia	240,000
Canada	233,00
Australia	220,000
New Caledonia	165,000
Brazil	126,000
China	100,000
Colombia	75,000
Cuba	66,000
South Africa	54,700
Madagascar	37,800
United States	36,000
Dominican Republic	–
Other countries	410,000
World total	2,421,100

Adapted from U.S. Geological Survey, 2015. Mineral Commodity Summaries 2015. U.S. Geological Survey, p. 196. http://dx.doi.org/10.3133/70140094.

depicted in Fig. 11.11. Total mine production of refined nickel metal was ~2.43 Mt during 2014.

The Philippines, Russia, Indonesia, Canada, and Australia are the largest producers of nickel in the world totaling 58%. New Caledonia and Brazil together share 12% of the mine production. The remaining countries with smaller production are China, Colombia, Cuba, South Africa, Madagascar, the United States, Dominican Republic, and many smaller producers.

11.5.2.3 Chromium

The world-class superdeposit of the Bushveld Complex has been known for chromite ore mining since 1897 and has been the highest-producing chromite mine in the world for over a century. Chromite ore is mainly extracted from the Upper Group 2 (UG2), Lower Group (LG), and Middle Group (MG) chromitite seams. UG2 contains significant amounts of PGMs, and some platinum mines produce chromite as byproduct. Several primary chrome mines are specifically maintained to provide chromite feed to the ferrochrome industry. Most of South Africa's chromite mines are developed along the Eastern limb of the Bushveld Camp, in the Steelport Valley. South Africa produced an estimated

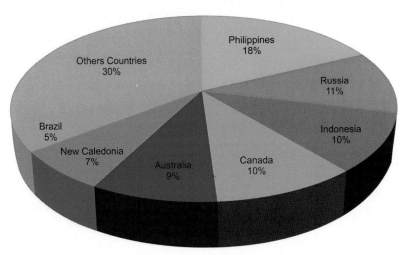

FIGURE 11.11   The main global players of refined nickel production in 2014 are Philippines (18%), Russia (11%), Indonesia (10%), Canada (10%), Australia (9%), New Caledonia (7%), Brazil (5%), and other countries (30%). The other Countries include China, Colombia. Cuba, South Africa, Madagascar, USA, Dominican Republic, and many small units.

15 Mt of chromite ore in 2014. Samancor Chrome is the second-largest ferrochrome producer in the world with total chromite resources exceeding 650 Mt.

Chromite ore production experienced stagnation between 1990 and 1999. From the year 2000 onward, market volumes witnessed steady growth exceeding 5% per annum from 15 to 29 Mt in 2014. This substantial increase is primarily due to rapid globally rising demand of stainless steel and chromium alloys. The other leading players producing more than 2 Mt/annum have been India, Kazakhstan, and Turkey.

A comparative statement of chromite ore + concentrate production during 2014 is given in Table 11.5.

Global chromite ore production during 2014 has been shared between South Africa (52%), Kazakhstan (14%), India (10%), and Turkey (8%); Brazil, Oman, Russia, and Zimbabwe together contributed 8%. The balance of 8% is shared by some 12 smaller producer countries with capacities anywhere less than 0.5 Mt/annum. The 12 smaller producing countries are Afghanistan, Albania, Australia, China, Finland, Greece, Iran,

TABLE 11.5   Statement of World Estimated Production of Chromite Ore and Concentrates in 2014.

Country	Production in Million Tonnes
South Africa	15.00
Kazakhstan	4.00
India	3.00
Turkey	2.40
Others	4.60
World: total	29.00

Adapted from U.S. Geological Survey, 2015. Mineral Commodity Summaries 2015. U.S. Geological Survey, p. 196. http://dx.doi.org/10.3133/70140094.

Madagascar, Pakistan, the Philippines, Sudan, and Viet Nam. Ninety-five percent of total ore and concentrates produced was metallurgical grade, 2% chemical grade, and the balance of 3% was refractory or foundry grade.

The share of chromite production by country is depicted in Fig. 11.12. Total mine production including concentrate of chromite was 29 Mt during 2014.

India was the number two leading chromite producer for many years. The annual

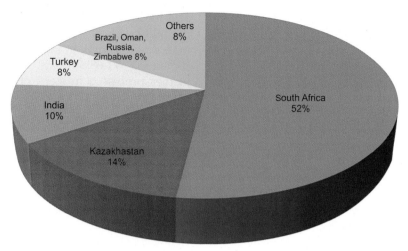

FIGURE 11.12 The major and minor global players chromite production during 2014 are South Africa (52%), Kazakhstan, India, and Turkey (32%), Brazil, Oman, Russia, and Zimbabwe (8%), and 12Twelve Smaller producers (8%).

production of +4 Mt has declined in recent years due to open-pit mines reaching ultimate pit bottom. Underground mining at deeper levels is becoming uneconomic because of lowering grade and thickness. Production was 4.8 Mt during 2007–2008 from 21 operating mines, mainly from Orissa (19), and minor quantities from Karnataka (2), Andhra Pradesh, Maharashtra, and Manipur. State Public Sector operates 8 mines, and 13 open-pit/underground mines by Private Sector Companies. Orissa continued to be the major producing State of chromite in India, accounting for 99.85% of the total production and the remaining 0.15% reported from Karnataka.

11.5.3 Metallurgy

Run-of-Mine (ROM) ore can be mafic–ultramafic-hosted PGEs (River Valley, Canada), nickel-dominated (Loma de Niquel, Venezuela), chromite-dominated (Kempirsai, Kazakhstan), chromite–PGEs (Bushveld, South Africa), nickel–PGEs (Sudbury, Canada), and laterite-hosted nickel deposits (Goro and Koniambo, New Caledonia). In any form, mined ore contains low metal grades (say, 0.0005% or 5 g/t PGEs)

unevenly mixed with the gangue minerals. Therefore, the ROM ores are physically and chemically processed to increase metal content passing through various stages of beneficiation, smelting, and refining to achieve user-friendly purity up to 99.99%. Metal concentration processes are generally performed at or close to the mining camp.

11.5.3.1 Beneficiation

Mineral beneficiation begins with crushing and grinding of mined ore for near-complete separation of ore and gangue minerals as well as between ore minerals. Each processing step is designed to increase the grade (concentration) of the valuable components of the original ore. Mined ore undergoes comminution by crushing and grinding, and gravity concentration by Dense Media Separation (DMS) removes the bulk of the rocks and gangue minerals. Installation of a DMS unit between the crusher and the grinder is extremely beneficial to eliminate large volumes of waste rocks from the ore. Consequently, the grinder, milling, and flotation unit will treat a significantly lower volume of higher-grade preconcentrate at a reduced operating cost with respect to energy, grinding media, and flotation reagents. The mineral pyrrhotite is often magnetically separated, collected, and treated to recover the minerals of PGEs and nickel. Sulfide minerals are further concentrated by froth flotation. The final concentrate upgrades the PGE content to 0.0150% (100–400 g/t) PGEs. The wet concentrate dewatered is thickened in large tanks, and filtered by disk or drum filters. The concentrate is dried in a spray drier or flash drier to reduce the energy requirement for smelting and the possible occurrence of explosions in the furnace. Dry concentrate is transferred pneumatically from the drier into the furnace for smelting.

11.5.3.2 Smelting

Smelting is necessary to maximize the separation of the gangue (oxide and silicate) minerals from the sulfide minerals associated with the noble metals. Flotation concentrates undergo smelting and converting in a furnace at a temperature exceeding 1300°C to produce a silicate melt (slag) and an immiscible sulfide melt (matte) due to density differences. Limestone is added to the furnace to reduce the melting temperature of the platinum and base-metal sulfides accumulated in the matte (green matte). The converted nickel–copper–PGE matte contains 0.20% PGEs.

11.5.3.3 Refining

The converted matte is milled by roll grinder/ball mill prior to treatment in the base-metal refinery adopting the hydrometallurgical pressure oxidation leaching process in diluted H_2SO_4

acid media. Copper and nickel are extracted by the sulfuric acid-leaching route. Leached residue makes up the high-grade PGM–Cu concentrate containing 30–65% PGEs. The concentrate is further processed in steps of electrolytic refining (electrolysis), pyrometallurgy, and selective-solvent extraction refining for final separation of the pure precious metals. The process of refining extracts the highest purity of metal, the final marketable purity at 99.90% for Rh, Ru, and Os, and 99.95% for Pt, Pd, and Au (Box 11.3 and Fig 11.13).

The nickel–copper-dominated sulfide concentrate is recovered through the extractive pyrometallurgical-process route by conventional roasting and reduction processes that yield a metal of greater than 75% nickel purity. Nickel with high purity can be used directly without further purification, depending on the composition of the impurities. Alternatively, the matte produced by the pyrometallurgical route is further refined by conventional hydrometallurgical pressure oxidation and electrolysis/selective solvent-extraction techniques. Copper is removed by adding hydrogen sulfide, leaving a metal concentrate of only nickel and cobalt. The selective solvent extraction will separate the cobalt and nickel to achieve the final nickel concentration up to 99.99% nickel metal.

11.6 DEMAND AND SUPPLY

The gross and net demand and supply of metals primarily depends on the availability of ore reserve base, the nature of orebodies,

---

BOX 11.3

AVERAGE PGE CONTENT

Ore: 0.0005% (5 g/t)
Flotation concentrate: 0.0150% (100–400 g/t)
Converter matte: 0.20%

PGM concentrate: 30–65%
Refined metals: 99.90% for Rh, Ru, Ir, and Os; 99.95% for Pt, Pd, and Au.

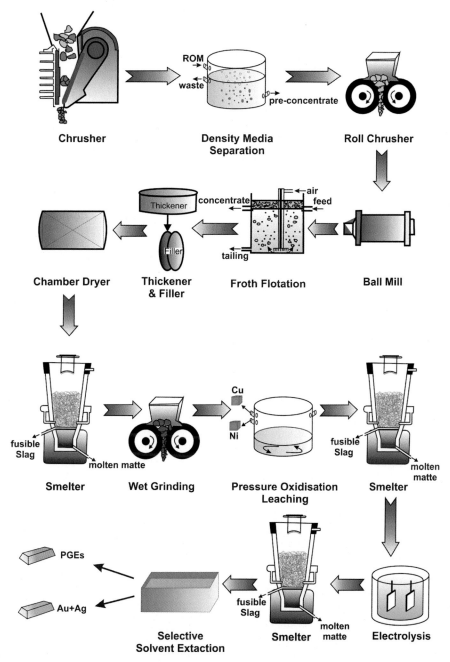

FIGURE 11.13 A complete flow diagram including crushing, grinding, density media separation, froth flotation, pyrometallurgical, and hydrometallurgical process route to achieve highest purity of metals.

BOX 11.4

WORKING DEFINITION

Gross demand figures represent the sum of manufacturer demand for a metal.

Recycling figures represent estimates of the quantity of metal recovered from open-loop recycling/recovery of metal from the end-of-life scrap.

Net demand figures are equivalent to the sum of gross demand less any metal recovery from open-loop scrap irrespective of industries and/

or applications. The gross and net demands are identical when no recycling from scrap is done.

Supply figures represent estimates of sales of primary metals at mine head, rather than the location of smelting, refining, and marketing.

The estimates are initially carried out by applications and regions and finally clubbed for global demand and supply.

---

mining technology, mineral processing, and the price of the metals. The demand and supply gap is minimized by adopting advanced technology in all aspects of the mineral industry (Box 11.4).

11.6.1 Platinum Group of Metals

PGMs are used in a wide range of technologies, including catalytic converters, electronics, and jewelry. Platinum market fundamentals are complex, and face very different issues than those of other strategic metals. The demand for jewelry, automobiles, and coin are all expected to rise, whereas the supply is from extremely limited locations. Very few significant platinum–palladium mining operations exist in the world. These two metals have the highest economic importance and are found in the largest quantities among the PGEs. The Bushveld Complex has the resources to supply world demand for platinum for the next century. However, miners' strikes complicate the economic stability of the noble metals, and the prices have skyrocketed due to strikes at South African mines.

South Africa is the largest producer of platinum from the Bushveld Complex since 1925, with palladium and rhodium as byproducts. Platinum was discovered in the rocks of The

Great Dyke, Zimbabwe, in 1918, but significant output from this extensive resource only began in the 1990s. The fresh supply of platinum in decreasing order is contributed by South Africa, Russia, North America (including Canada), and other countries. In addition, a bulk amount is received from the recycling of scrap which is about 25% of the fresh supply. Global demand and supply of platinum between 1990 and 2015 is provided in Fig. 11.14. The overall average demand of platinum is expected to rise by exceeding 3.5% per annum due to increasing need for catalytic converters, autocatalysts, jewelry, and coins.

Russia is the largest producer of palladium. The Stillwater Complex, United States, is a palladium-rich mine producing since 1987. Palladium demand is expected to rise by 3.5% per annum, and the supply is likely to drop in the future. This gap in demand and supply may result increase in the price of palladium over platinum. The global demand and supply trend of palladium between 1990 and 2015 is similar to that of platinum.

The average global supply of PGEs is shared by Bushveld (~78% Pt, and ~40% Pd), Noril'sk (~11% Pt, and ~44% Pd), Stillwater (~2% Pt, and ~6% Pd), and The Great Dyke (~3% Pt and ~2% Pd). The other four metals (rhodium, ruthenium, iridium, and osmium) are produced

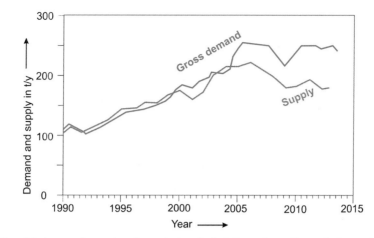

FIGURE 11.14 The global gross demand and supply trend of platinum is continuously increasing at a rate exceeding 3.5%. The gap between demand and supply is partially compensated by metal recovery from scraps.

as co-products of platinum and palladium. The annual production of these metals depends on production of the primary metals.

11.6.2 Nickel

Nickel occurs as oxides, sulfides, and silicates, with nickel ore mined, smelted, and/or refined in ~25 countries. Russia, Canada, New Caledonia, Australia, Indonesia, and Cuba are the main producers. Noril'sk began producing copper and nickel in 1935 with palladium and platinum as byproducts. Sudbury has been producing nickel and byproduct platinum since 1908. Nickel supplies are of four types: primary nickel, ferronickel, nickel in pig iron (NPI), and nickel scrap.

Global nickel supply is accelerating at a faster pace with growth rising 1.6% per annum in the 1980s, 2.9% in the 1990s, and finally, 3.5% since 2000 (Fig. 11.15). In 2011, nickel production was around 1.59 Mt—which was relatively small compared to 20 Mt copper and 45 Mt of aluminum. The emergence of a nickel supply from pig iron is a significant development in the mineral industry. Sky-high nickel prices in 2006 and 2007 forced China to find alternative means of getting nickel units, boosting NPI supply to around 275,000 tonnes (~303,000 tons) last year from 5000 tonnes (~5500 tons) in 2005.

The growth rate of nickel metal production between 2000 and 2015 is steady with steeply rising ranges between 1.25 and 1.49 Mt (2000–2005), ~1.50 Mt (2006–2010), and 2.00–2.60 Mt (2011–2014). The annual growth rate of nickel is closely related to crude-steel production to supply ferronickel and corrosion-resistant nickel plating. Production growth rate is limited due to the low tenor of mined ore (0.2–1.00% and above for Ni) and the types of ore (sulfide and laterite), despite the availability of large resources.

11.6.3 Chromium

The global ore reserve base of chromite stands at 9215 Mt, of which >90% is located in South Africa, Zimbabwe, Kazakhstan, and Turkey. Other countries having sizeable resources are Finland, India, Brazil, China, Russia, and Albania.

The average growth rate of global demand and supply of chromite in the 21st century is

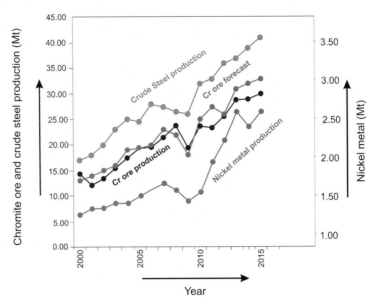

FIGURE 11.15   Global supply trend of primary nickel metal, chromite ore production, chromite demand forecast, and crude steel production in million tonnes/year between 2000 and 2014.

rising steady (Fig. 11.15). The average global demand growth rate for chromite between 2000 and 2015, based on the past consumption pattern, has been estimated at 5.8% per annum. The average supply growth rate for chromite for the same period, based on the actual mine production, stands at 6.0% per annum. Annual mine production steadily grew from 12.40 Mt of chromite ore in 2000 to 29.00 Mt in 2014. Ore production is closely commensurate with the demand growth rate of chromite ore, and annual production of crude steel for ferrochrome, and corrosion-resistant chrome plating/polishing. The average grade ranges between 40% and 52% Cr_2O_3. The grade upgradation is relatively easy by washing, screening, blending, gravity, and spiral separation, and partial flotation. Upgradation rarely requires pyrometallurgical or hydrometallurgical processes, or refining. The large chromite deposits occur as massive, layered, and stratiform types, and production can be increased as required under a congenial working environment.

India has been the second-largest leading supplier of chromite ore since the late 1960s. The country's production has declined in recent years as a majority of the open-pit mines reached ultimate bottom limit, and underground mining is less productive due to reduction in orebody thickness.

11.7 PGES–NICKEL–CHROMIUM: A GLOBAL REVIEW

An all-inclusive compilation of more than 60 platinum–palladium–nickel–chromium deposits of large and small tonnages, high and low grade from six continents, has been made in Table 11.6. The salient features include deposit/country, primary commodity, discovery, surface area/depth, stratigraphy, host rocks, mineralization, distinctive features, resources (Mt), contained metals (g/t PGEs, %Ni, %Cu, and %Cr_2O_3), and age of formation (Ma).

TABLE 11.6 Comparative Statement of All-inclusive Primary Characteristic Features of Global PGE–Nickel–Chromium Deposits Covering Six Continents

Deposit/Country	Primary Commodity	Discovery	Area (sq.km)/ Depth (km)	Stratigraphy	Host Rocks	Mineralization	Distinctive Features	Resources (Mt)/g/t PGE, %Ni, %Cu, %Cr$_2$O$_3$	Age (Ma)
Bushveld igneous complex, South Africa	Chromium, Platinum group	1897	65,000/7–9	Bushveld igneous complex	Peridotite, chromitite, gabbro, norite	Cr, Pt, Pd, Ni, Cu	Repeated magmatic layering	**11,550/**5.67, –, –, **40–50.**	2060
Uitkomst complex, South Africa	Platinum group, Nickel	1990	9/	Mafic–ultramafic intrusion	Gabbronorite, pyroxenite, harzburgite.	PGEs, Ni, Cu, Co, Cr	Shallow depth	**281/**0.84, **0.34,** 0.12, –	2044
Stella intrusion, South Africa	Platinum group	–	18/	Kraaipan-greenstone	Cumulate gabbro, magnetite	PGEs	Layered intrusion	**75/**1.42, –, –, –	3033
Kabanga, Tanzania	Nickel	1914–1976	4/–	Mafic–ultramafic intrusion	Gabbronorite, orthopyroxenite	Ni sulfide	High-magnesium magma	**58/** , 2.60, –, –	1780–1370
Ntaka hill, Tanzania	Nickel, Copper	–	15/-	Mafic–ultramafic intrusion	Pyroxenite, websterite, peridotite, gabbro	Ni, Cu, Co	High-Mg magmatic melt	**55/**–, 0.60, 0.14, –	640–596
Munali Ni–PGEs, Zambia	Nickel, Platinum group, Copper	1969	1.25/-	Mafic–ultramafic intrusive	Gabbro core,	PGEs, Ni, Cu, Co	Gabbro core, brecciated ultramafic	**10/**0.63, 1.20, 0.20, –	870–820(855)
The great Dyke, Zimbabwe	Platinum group, Chromium	1867–1918	4400/4.2	Ultramafic– mafic intrusive.	Pyroxenites, gabbro, norite	Pt, Pd, Cr, Au, Ag, Cu, Ni, Co	Magmatic layering	**2574/**5.42, –, –, **40**	2579±7
Ambatovy Ni–Cu, Madagascar	Nickel	1960	18/20–100m	Mafic–ultramafic intrusive	Saprolite, ferralite (laterite)	Ni, Co	Large lateritic Ni-deposit	**134/** , 1.04, –, –	–
Stillwater, United States	Platinum group, Chromium	1880–1973	180/6.5	Mafic–ultramafic intrusion	Dunite, gabbro	Pt, Pd, Cr, Cu, Ni	Layered complex	**38/**25, 0.05, 0.02, 35–47	2705±4
Duluth, United States	Platinum group, Copper	1948	–	Magmatic Intrusion	Troctolite, gabbroic cumulate	Ni, Cu, PGEs	Layered intrusion	**+3200/**0.66, 0.13, 0.46, –	1100

Continued

TABLE 11.6 Comparative Statement of All-inclusive Primary Characteristic Features of Global PGE–Nickel–Chromium Deposits Covering Six Continents—cont'd

Deposit/ Country	Primary Commodity	Discovery	Area (sq.km)/ Depth (km)	Stratigraphy	Host Rocks	Mineralization	Distinctive Features	Resources (Mt)/ g/t PGE, %Ni, %Cu, %Cr$_2$O$_3$	Age (Ma)
Eagle, United States	Nickel Copper Platinum group, Copper	2002	480×100m/ 340m	Mafic–ultramafic intrusion	Peridotite, melatroctolite, melagabbro, pyroxenite	Ni, Cu, PGEs	Massive and disseminated mineralization	**5.33/**1.17, **3.05,** 2.51, –	1100
La Perouse intrusion, Alaska, United States	Nickel, Copper, Platinum	1958	–	Mafic–ultramafic pluton	Olivine gabbro, norite	Ni, Cu, PGEs	Interlayered gabbro-norite	**100/**0.18, 0.50, 0.30, –	30
Sudbury complex, Canada	Nickel, Copper, Platinum	1856–1883	1860/15	Sudbury igneous complex	Norite, gabbro	Ni, Cu, PGEs,	Layered intrusion	**+1500/**1.0, 1.0, 1.0, –	1850
Lac des Iles, Canada	Platinum group	1958–1963	300/–	Ultramafic intrusives	Pyroxenite, websterite, peridotite, gabbro	Pd, Pt, Au, Cu, Ni	Layered ultramafic	86/2.18, 0.07, 0.07, –	2738±27
Voisey's Bay, Canada	Nickel, Copper	1993	–	Mafic intrusion in upper Eastern Deeps and lower Reid Brook chambers	Troctolite, olivine gabbro	Ni, Cu, Co	Discovery Hill gossan cap, massive, disseminated, layered	**141/** – 1.60, 1.25, –	1334
Thompson, Canada	Nickel, Platinum group	1956	300 km/1.2	Mafic–ultramafic intrusion	Serpentinite, dunite, pyroxenite, chemical sediments	Ni, PGEs, Cu, Co	Stratabound, stratiform, komatiite	**150/**0.83, **2.32,** 0.16, –	1900
River valley, Canada	Platinum, group	1990	0.0045/0.2	Layered mafic intrusion	Gabbronorite, leucogabbro, anorthosite	PGEs, Ni, Cu	Tholeiitic, plagioclase-rich magmas	**127/**0.82, 0.02, 0.06, –	2480
Raglan, Canada	Platinum, group, Nickel, Copper	1898	–	Mafic–ultramafic intrusion	Peridotite, massive gabbro	Ni, Cu, Co, PGEs	Tholeiitic, komatiitic basalt	30/3.76, 3.18, 0.90, –	2004–1920

Name/location	Commodity	Year	Dimensions/profile	Deposit type	Rock types	Metals	Description	Tonnage/grade	Age
Dumont sill, Abitibi, Canada	Nickel	–	–	Mafic–ultramafic differentiated intrusion	Peridotite, dunite, clinopyroxene, gabbro	Ni±PGEs	Layered enriched nickel sulfide	**180**/–, 0.32, –, –	–
Skaergaard, Greenland	Platinum, group, Gold	1961	10/1.1	Plutonic and volcanic magmatic complex	Gabbro, anorthosite, syenite, and granophyre.	Pd, Pt, Au	Layering significant	**203**/1.44, –, –, –	55
Santa Rita, Brazil	Nickel, Platinum, group	1976–1990s	–	Mafic–ultramafic intrusion	Gabbronorite, dunite, harzburgite, pyroxenite, websterite	Ni, Cu, Co, PGEs	Stratiform layer containing magmatic Fe–Ni–Cu–Co sulfides	**+160**/–, 0.52, 0.13,–	–
Niquelândia laterite, Brazil	Nickel	–	20.36/	Mafic–ultramafic complex	Gabbronorite, peridotite, pyroxenite, dunite, laterite, saprolite	Ni, Cu, Co sulfides	Lateritic weathering profiles	**70**/1.45, –, –	Proterozoic
Ipueira–Medrado sill, Brazil	Chromium	–	–	Mafic–ultramafic intrusion	Komatiitic orthopyroxenite sill	Cr	Layered stratiform chromite	**34.5**/–, –, –, 30–40	Paleoproterozoic
Americano do Brasil, Brazil	Nickel, Copper	1969	–	Mafic–ultramafic intrusion	Interlayered dunite, peridotite, websterite, gabbronorite	Ni, Cu	Layered intrusion	**3.1**/–, 1.12, 1.02, –	Neoproterozoic (~600)
Cerro Matoso Nickel deposit, Columbia	Nickel	1980s	Elongated low-lying hill	Peridotite-protolith	Blanket saprolite, yellow-red laterite	Ni, Cu, Co	Significant weathered profile	**108**/–, 0.57, –, –	–
Loma de Niquel, Venezuela	Nickel	–	6.2/10 m lateritic profile	Ophiolite complex	Basalt, gabbro, peridotite, serpentines, serpentinized harzburgite	Ni, Co	Lateritic blanket type	**48**/–, 1.43, –, –	Lower Cretaceous

Continued

TABLE 11.6 Comparative Statement of All-inclusive Primary Characteristic Features of Global PGE–Nickel–Chromium Deposits Covering Six Continents—cont'd

Deposit/Country	Primary Commodity	Discovery	Area (sq.km)/Depth (km)	Stratigraphy	Host Rocks	Mineralization	Distinctive Features	Resources (Mt)/ g/t PGE, %Ni, %Cu, %Cr$_2$O$_3$	Age (Ma)
Jinchuan, China	Nickel, Copper, Platinum group	1958	3/1.1	Ultramafic intrusion	Dunite, pyroxenite, lherzolite	Ni, Cu, Pt, Pd, Os	Super large magmatic deposit	515/0.26, 1.18, 0.63, –	870–820
Huangshandong, China	Nickel, Copper	–	3/1	Mafic–ultramafic intrusion	Gabbronorite, lherzolite, diorite	Ni, Cu	Massive and layered sequence	135/, 0.52, 0.27, –	270
Kalatongke, China	Nickel, Copper	1970s	Small/0.6	Mafic–ultramafic intrusion	Mafic ultramafic	Ni, Cu, PGEs	Small elongated body	33/–, 0.80, 1,13, –	283±3
Xiarihamu, China	Nickel	–	East-west trend	Mafic–ultramafic	Gabbronorite, websterite, olivine orthopyroxenite	Ni, Cu, Co, PGEs	Two large ore pods: disseminated (west) and multiple sub-layers (east)	157/–, 0.65, 0.14, –	406Ma
Sukinda, India	Chromium, Nickel	1940s–1949	28/+100m	Sukinda igneous intrusion	Pyroxenite, dunite, chromitite,	Cr, Ni ± PGEs	Rhythmic layering	190/–, 0.80, –, 40–50	3200–2700
Boula–Nausahi, India	Chromium, Platinum group	1942	~5/	Intrusive complex	-Do- gabbro, anorthosite	Cr, PGEs	-Do- massive	15/1–2, –, –, 45	3200–2700
Nuggihalli schist belt, India	Chromium	1970s	180/–	Intrusives	Peridotite, serpentinite	Cr, Ti, Au	Layered	–/1.34, –, –, 40	~3000
Hanumalapura, India	Platinum group	–	2/–	Intrusives	Pyroxenite, gabbro, anorthosite,	Cr, V, Ti, PGEs	Layered	0.84/1.79, –, –, –	3000–2500
Sittampundi, India	Chromium, Platinum group	–	50/–	Intrusives	Chromitite, anorthosite	Cr, PGEs	Layered and Potential target	–/–2, –, –, na	Proterozoic
Soroako, Indonesia	Nickel	–	10,000/–	Ultramafic ophiolite	Harzburgite, peridotite, laterite, saprolite, limonite	Ni	Lateritic profile	109.4/, 1.79, –, –	Cretaceous

Locality	Commodity	Year		Setting	Rock types	Metals	Ore form	Resource	Age
Gag Island, Indonesia	Nickel	–	–	Ophiolitic serpentinite and mélange,	Ni-bearing laterite and saprolite	Ni, Co	Laterite profile	**240**/ 1.35, –, –	Cretaceous
Weda Bay, Indonesia	Nickel	1990s	–	Ophiolitic serpentinite	Ni-hydrosilicate in saprolite	Ni, Co	Laterite	**216**/ 1.37, –, –	Cretaceous
Tanjung–Buli, Indonesia	Nickel	1970s	390/	Mafic–ultramafic complex	Dunite, pyroxenite, serpentinite, limonite, gabbro,	Ni, Fe	Laterite, saprolite	**47**/ 2.50, –, –	Cretaceous
Kempirsai massif, Kazakhstan	Chromium	1936	–	Large ophiolitic complex	Gabbro, dunite, wehrlite, pyroxenite,	Cr	Massive, pods	**310**/ –, –, **48**	Early Paleozoic
Voskhod ophiolite, Kazakhstan	Chromium	1963	/450	Ophiolite	Dunite, harzburgite	Cr	Massive, disseminated	**19.5**/ –, –, **48.5**	Early Paleozoic
Shevchenko laterite, Kazakhstan	Nickel	–	Laterite profile	Ophiolite, serpentinite	Saprolite, red and yellow limonite	Ni	Laterite column	**104.4**/ 0.79, –, –	Tertiary
Acoje ophiolite, Philippines	Nickel, Platinum group			Mafic–ultramafic ophiolite	Harzburgite, dunite, serpentinite, gabbro, norite, pegmatite	Ni, Co, PGE	Basal platiniferous Cr + Ni. Upper lateritic nickel.	**36**/–2, 1.1, –, –	Phanerozoic
Noril'sk–Talnakh, Russia	Platinum–Palladium, Copper, Nickel	1935	/3.50	Mafic intrusion and trap	Picritic gabbro, gabbro, dolerite	Pt, Pd, Ni, Cu, Au, massive and disseminated	Flood basalts with flow structure in sedimentary rocks.	**1309**/0.50, 1.77, 3.57, –	250
Pechenga belt, Russia	Nickel, Copper Cobalt, Platinum group	~1930	–	Mafic–ultramafic intrusion	Wehrlite through pyroxenite to gabbro–wehrlite	Ni, Co, PGEs	Layered	**339**/0.32, 1.18, 0.63, –	2500–2400
Burakovsky, Russia	Chromium, Platinum group	–	700/4-6	Lopolithic mafic–ultramafic intrusion	Dunite, peridotite, pyroxenite, wehrlite, gabbro, norite	Cr, PGEs	Large layered	**Na**/ –, –, **49–52**	2430

Continued

TABLE 11.6 Comparative Statement of All-inclusive Primary Characteristic Features of Global PGE–Nickel–Chromium Deposits Covering Six Continents—cont'd

Deposit/ Country	Primary Commodity	Discovery	Area (sq.km)/ Depth (km)	Stratigraphy	Host Rocks	Mineralization	Distinctive Features	Resources (Mt)/ g/t PGE, %Ni, %Cu, %Cr$_2$O$_3$	Age (Ma)
Guleman, Turkey	Chromium	1930s	>200/–	Allochthonous ophiolite	Peridotite	Cr	Layered	220/–, –, –, 48	Cretaceous-Eocene
Munni Munni, Australia	Platinum group	1980s	225/5	Mafic–ultramafic intrusives	Wehrlite, clino-pyroxenite, dunite, gabbro, norite	Pt, Pd, Au, Cu, Rh, Ni	Layered	24/2.9, –, –, –	2927 ±13
Fifield, Australia	Platinum group	1880	336/–	Ultramafic	Dunite, pyroxenite	Pt	Alluvial	13/0.61, –, –, –	–
Kambalda, Australia	Nickel, Platinum group	1968	–	Mafic–ultramafic complex	Komatiite–basalt sequence	Ni, Cu, PGEs	Komatiite flow units in stratigraphic contact	67/1.13, 2.90, 0.21, –	2710–2700
Mount Keith, Australia	Nickel	1968	–	Ultramafic complex	Adcumulates, gabbro, basalt	Ni, Cu, Co	Mineralized komatiite at base	503/–, 0.55, –, –	2700
Leinster camp, Australia	Nickel	–	–	Ultramafic complex	Layered and banded komatiite	Ni, Cu, PGEs	Basal sulfide enrichment	201/–, 0.74, –, –	2700
Nebo–Babel large igneous province, Australia	Nickel, Copper	2000–	7.5/–	Mafic complex	Gabbronorite	Ni, Cu	Massive and disseminated mineralization	203/–, 0.41, 0.42, –	1068 ±4.3
Goro laterite, New Caledonia	Nickel	–	–	Ultramafic complex	Saprolite, yellow and red limonite	Ni, Co	Lateritic profile	179/–, 1.48, –, –	Miocene
Koniambo laterite, New Caledonia	Nickel	–	21/–	Ultramafic complex	Saprolite, yellow and red limonite	Ni, Co, Fe	Lateritic profile	153/–, 2.47, –, –	Miocene
Kurumbukari laterite, Papua New Guinea	Nickel	–	–	Dunite, gabbro, pyroxenite	Saprolite, yellow and red limonite	Ni, Co	Laterite profile of ~15-m thickness	143/–, 1.01, –, –	Miocene
Bulqiza ultramafic, Albania	Chromium	–	4.5/1.6	Ophiolite sequence	Harzburgite, dunite cumulates	Cr	Massive, disseminated	12.5/–, –, –, 37.50	Triassic–Jurassic

Deposit	Commodity	Discovery	Size/grade	Deposit type	Rock types	Metals	Morphology	Grade	Age
Bitincke laterite, Albania	Nickel	–	–/<20 m	Ultramafic ophiolite	Peridotite, serpentine, iron oxide	Ni, Co	Laterite	**35.6**/–, 1.2, –, –	Late Jurassic
Kemi intrusion, Finland	Chromium	1959	22/0.04	Mafic ultramafic intrusion	Chromite, pyroxenite, gabbro cumulates	Cr	Layered	**40**/–, –, 26.6	2400
Lomalampi complex, Finland	Platinum group, Nickel	2004	–	Volcanic brecciated lava and tuffs	Komatiite olivine cumulate	Pt, Pd, Au, Ni, Cu, Co	Brecciated	**3.06**/0.46, 0.17, 0.06, –	2060
Kevitsa complex, Finland	Platinum, Nickel, Copper	1987	–	Mafic ultramafic komatiite complex	Dunite, websterite, pyroxenite, gabbro	Pt, Pd, Ni, Cu, Au, Co	Volcanic basalt flows	**283**/0.47, 0.30, 0.42, –	2060
Portimo intrusion, Finland	Platinum group	–	–	Ultramafic intrusion	Olivine-rich cumulate layers	Pt, Pd, Ni, Cu	Layered and reef type	**218.6**/1.92, 0.08, 0.18, –	2440
Aguablanca, Spain	Platinum, Nickel, Copper	1993	–	Mafic intrusion	Gabbro	Pt, Pd, Ni, Cu, Co,	Massive, banded in breccia zone	**26.56**/0.53, 0.41, 0.35, –	338–334

Bibliography

Alapieti, T.T., Kujan'Ad, J., Lahtinen, J.J., Papunen, H., 1989. The Kemi stratiform chromitite deposit, northern Finland. Economic Geology 84, 1057–1077.

Alapieti, T.T., Hakkoaho, T.A.A., Devaraju, T.C., Jayaraj, K.R., 1994. Chromite hosted PGE mineralization in the Channagiri area, Karnataka State, India. In: VIIth International Platinum Symposium. 1–4 August, Moscow (Abstract), pp. 3–4.

Alapieti, T.T., Devaraju, T.C., Kaukonen, R.J., 2002. Silicate-chromite type PGE mineralization in the Hanumalapur complex, Karnataka State, India. In: Boundreau, A. (Ed.), 9th International Platinum Symposium. 21–25 July, 2002, Billings, Montana, USA, Extended Abstracts, pp. 5–8.

Alapieti, T.T., Devaraju, T.C., Kaukonen, R.J., 2008. PGE mineralization in the late archaeaniron-rich mafic-ultramafic Hanamalapur complex, Karnataka, India. Mineralogy and Petrology 92 (1–2), 99–128.

Alonsoa, M.F., Cuttenb, H., Waelec, B.D., Tacka, L., Tahona, A., Baudeta, D., Barritt, S.D., 2012. The Mesoproterozoic Karagwe-Ankole Belt (formerly the NE Kibara Belt): the result of prolonged extensional intracratonic basin development punctuated by two short-lived far-field compressional events. Precambrian Research 216–219, 63–86. Elsevier www.elsevier.com/locate/precamres.

Altona Mainlining, 2013. Annual Report of the Directors to Share Holders and Financial Statement for 2012–2013, p. 116. www.altonamining.com.

Álvarez, G., Porwal, I., Beresford, A., McCuaig, S.W., Maier, T.C., February 23, 2011. Exploration targeting criteria for hydrothermal Ni systems. In: Advances in Ore Systems and Exploration Success. Perth, p. 6.

Arai, S., 1994. Podiform chromitites of the Tari-Misaka Ultramarine complex, South-Western Japan, as mantle-melt interaction products. Economic Geology 89, 1279–1288.

Arndt, N.T., 2011. Insights into the geologic setting and origin of Ni-Cu-PGE sulfide deposits of the Norilsk-Talnakh region, Siberia. Economic Geology 17, 199–215.

Bagai, Z., Armstrong, R.A., Kampunzu, A.B., 2002. U-Pb single zircon geochronology of granitoids in the Vumba granite-greenstone terrain (NE Botswana): implications for the evolution of the Archean Zimbabwe craton. Precambrian Research 118, 149–168.

Barnes, S.J., Gomwe, T.S., 2011. The Pd deposits of the Lac Des Iles Complex, North-Western Ontario. Economic Geology 17, 351–370.

Barnes, S.J., Hoatson, D.M., 1994. The Munni Munni complex, western Australia: stratigraphy, structure and petrogenesis. Journal of Petrology 35 (3), 715–751.

Barnes, S., Jones, S., 2013. Deformed chromitite layers in the Coobina intrusion, Pilbara craton, western Australia. Economic Geology 108, 337–354.

Barnes, S.J., Maier, W.D., 2002. Platinum-group element distributions in the Rostenberg layered suite of the Bushveld complex South Africa. In: Cabri, L.J. (Ed.), The Geology Geochemistry Mineralogy and Mineral Beneficiation of Platinum-group Elements. Can Inst Mine Met Spec, vol. 54, pp. 431–458 Ottawa Ontario.

Barnes, S.J., Maier, W.D., Ashwal, L.D., 2004. Platinum-group element distribution in the main zone and upper zone of the Bushveld complex, South Africa. Chemical Geology 293–317.

Barnes, S.J., Osborne, G.A., Cook, D., Barnes, L., Maier, W.D., Godel, B., 2011a. The Santa Rita nickel sulfide deposit in the Fazenda Mirabela intrusion, Bahia, Brazil: geology, sulfide geochemistry, and genesis. Economic Geology 106, 1083–1110.

Barnes, S.J., Fiorentini, M.L., Fardon, M.C., 2011b. Platinum group element and nickel sulfide ore tenors of the Mount Keith nickel deposit, Yilgarn Craton, Australia. Mineral Deposita 1–22. http://dx.doi.org/10.1007/s00126-011-0348-5.

Barnes, S.J., Heggie, G.J., Fiorentini, M.L., 2013. Spatial variation in platinum group element concentrations in ore-bearing komatiite at the long-victor deposit, Kambalda Dome, western Australia: enlarging the footprint of nickel sulfide orebodies. Economic Geology 108 (5), 913–933.

Barnes, S.J., 1986. The effect of trapped liquid crystallization on cumulus mineral compositions in layered intrusions. Contribution to Mineralogy and Petrology 93, 524–531 Springer-Verlag.

Barnes, S., 2011. Advances in Nickel Sulfide Targeting, Advances in Ore Systems and Exploration Success. CSIRO Earth Science and Resource Engineering, pp. 1–22. www.csiro.au.

Barnes, S.J., Pagé, P., Prichard, H.M., Zientek, M.L., Fisher, P.C., 2015. Chalcophile and platinum-group element distribution in the ultramafic series of the Stillwater Complex, MT, USA—implications for processes enriching chromite layers in Os, Ir, Ru, and Rh. Mineral Deposita 1–23. http://dx.doi.org/10.1007/s00126-015-0587-y.

Begg, G.C., Hronsky, J.A.M., Arndt, N.T., Griffin, W.L., O'Reilly, A.Y., Haywapd, N., 2010. Lithospheric, cratonic, and geodynamic setting of Ni-Cu-PGE sulfide deposits. Economic Geology 105, 1057–1070.

Bekker, A., Grokhovskaya, T.L., Hiebert, R., Sharkov, E.V., Bui, T.H., Stadnek, K.R., Chashchin, V.V., Wing, B.A., 2015. Multiple sulfur isotope and mineralogical constraints on the genesis of Ni-Cu-PGE magmatic sulfide mineralization of the Monchegorsk Igneous Complex, Kola Peninsula, Russia. Mineral Deposita 19. http://dx.doi.org/10.1007/s00126-015-0604-1.

Belinda, G., Sarah, J.B., Wolfgang, D.M., 2007. Platinum-group elements in sulphide minerals, platinum-group minerals, and whole-rocks of the Merensky Reef (Bushveld Complex, South Africa): implications for the formation of the reef. Journal of Petrology 48 (8), 1569–1604.

Benkó, Z., Mogessie, A., Molnár, F., Severson, M.J., Hauck, S.A., Raic, S., 2015. Partial melting processes and Cu-Ni-PGE mineralization in the footwall of the South Kawishiwi intrusion at the Spruce road deposit, Duluth complex, Minnesota. Economic Geology 110, 1269–1293.

Bohidar, S., Mohapatra, S., Sokla, L.B., 2009. Nickel recovery from chromite overburden of Sukinda using fungal strains. International Journal of Integration Biology 5 (2), 103–108.

Boudreau, A., Djon, L., Tchalikian, A., Corkery, J., 2014. The Lac Des Iles palladium deposit, Ontario, Canada part I. The effect of variable alteration on the offset zone. Miner Deposita 1–30. http://dx.doi.org/10.1007/s00126-014-0510-y.

Boyd, R., Mathiesen, C.O., 1979. The nickel mineralization of the Råna mafic intrusion, Norland, Norway. Canadian Mineralogist 17, 287–298.

Bremner, H.T.M., Wayne, G., Savage, J.R., 1996. Trends in Deep Drilling in the Sudbury Basin, Short Course on Technologies, and Case Histories for the Modern Explorationist, Toronto, pp. 53–75.

Brugmann, G.E., Naldrett, A.J., 1988. Platinum Group Elements abundances in Mafic and Ultramafic Rocks: Preliminary Studies at the Lac Des Lies Complex. Ontario Geological Survey, District of Thunder Bay, Ontario, p. 43.

Bulle, F., Layne, G.D., 2015a. Trace element variations in olivine from the eastern deeps intrusion at Voisey's Bay, Labrador, as a monitor of assimilation and sulfide saturation processes. Economic Geology 110, 713–731.

Bulle, F., Layne, G.D., 2015b. Multi-element variations in olivine as geochemical signatures of Ni-Cu sulfide mineralization in mafic magma systems – examples from Voisey's Bay and Pants Lake Intrusions, Labrador, Canada. Mineral Deposita 1–21. http://dx.doi.org/10.1007/s00126-015-0591-2.

Burrows, D.R., Lesher, C.M., 2012. Copper-Rich Magmatic Ni-Cu-PGE Deposits. Economic Geology Special Publication 16, pp. 515–552 (Chapter 20).

Capistrant, P.L., Hitzman, M.W., Wood, D., Kelly, N.M., Williams, G., Zimba, M., Kuiper, Y., Jack, D., Stein, H., 2015. Geology of the enterprise hydrothermal nickel deposit, North-western province, Zambia. Economic Geology 110, 9–38.

Cathelineau, M., Quesnel, B., Gautier, P., Boulvais, P., Couteau, C., Drouillet, M., 2016. Nickel dispersion and enrichment at the bottom of the regolith: formation of pimelite target-like ores in rock block joints (Koniambo Ni deposit, New Caledonia). Mineral Deposita 51, 271–282. http://dx.doi.org/10.1007/s00126-015-0607-y.

Cawthom, R.G., Merkle, R.K., Viljoen, M.J., 2002. Platinum-group element deposits in the Bushveld complex South Africa. In: Cabri, L.J. (Ed.), The Geology Geochemistry Mineralogy and Mineral Beneficiation of Platinum-Group ElementsCan Inst Min Met. Spec, vol. 54, pp. 389–429 Ottawa Ontario.

Cawthorn, R.G., 2005a. Contrasting sulfide contents of the Bushveld and Sudbury igneous complexes. Mineralium Deposita 40, 1–12. http://dx.doi.org/10.1007/s00126-005-0465-0.

Cawthorn, R.G., 2005b. Stratiform platinum-group element deposits. In: Mungall, J.E. (Ed.), Exploration for Platinum-group Element Deposits, Mineralogical Association of CanadaShort Courses Series, vol. 35, pp. 57–73.

Cawthorn, R.G., Maske, S., Wet, M.D., 1988. Contrasting magma types in the mount Ayliff intrusion (Insizwa complex), Transkei: evidence from Ilmenite compositions. Canadian Mineralogist 26, 145–160.

Chakraborty, K.L., Chakraborty, T.L., Mazumder, T., 1980. Stratigraphy and structure of the precambrian banded iron formation and chromite bearing ultramafic rocks of Sukinda valley, Orissa. Journal of the Geological Society of India 21, 398–404.

Chen, L.M., Song, X.Y., Keays, R.R., Tian, Y.L., Wang, Y.S., Deng, Y.F., Xiao, J.F., 2013. Segregation and fractionation of magmatic Ni-Cu-PGE sulfides in the western Jinchuan intrusion, North-western China: insights from platinum group element geochemistry. Economic Geology 108, 1793–1811.

Colin, F., Nahon, D., Trescases, J.J., Melfi, A.J., 1990. Lateritic weathering of pyroxenites at Niquelandia, Goias, Brazil; the supergene behavior of nickel. Economic Geology 85, 1010–1023.

Crocket, J.H., Paul, D.K., 2007. Platinum-group elements in igneous rocks of the Kutch rift basin, NW India: implications for relationships with the Deccan volcanic province. Elsevier. CHEMGE-15174 Chemical Geology xx, 17 xxx–xxx.

Dare, S.A.S., Barnes, S.J., Prichard, H.M., Fisher, P.C., 2014. Mineralogy and geochemistry of Cu-Rich ores from the McCreedy East Ni-Cu-PGE deposit (Sudbury, Canada): implications for the behavior of platinum group and chalcophile elements at the end of crystallization of a sulfide liquid. Economic Geology 109, 343–366.

De Waal, S.A., Maier, W., 2001. Parental magma and emplacement of the Stratiform Uitkomst complex, South Africa. The Canadian Mineralogist 39, 557–571.

Deng, Y.F., Song, X.Y., Chen, L.M., Zhou, T., Piraino, F., Yuan, F., Xie, W., Zhang, D., 2014. Geochemistry of the Huangshandong Ni–Cu Deposit in Northwestern China: Implications for the Formation of Magmatic Sulfide Mineralization in Orogenic Belts. Ore Geology Review, vol. 56. Elsevier, pp. 181–198.

Dentith, M., Mudge, S.T., 2014. Geophysics for the Mineral Exploration Geoscientist. Cambridge University Press, p. 454.

Devaraju, T.C., 2009. Field tour guide to the Hanumalapur PGE prospect, Davangere district, western Dharwad craton, Southern India. In: International Symposium on Magmatic Ore Deposits, Bhubaneswar, pp. 22–26.

Ding, X., Ripley, E.M., Li, C., 2011. PGE geochemistry of the Eagle Ni–Cu–(PGE) deposit, upper Michigan: constraints on ore genesis in a dynamic magma conduit. Mineral Deposita 1–16. http://dx.doi.org/10.1007/s00126-011-0350-y.

Distler, V.V., Marina, A.Y., 2005. Polymetallic platinum-group elements (PGE)-Au mineralization, of the Sukhoi Log deposit, Russia: exploration for platinum-group element deposits, mineralogical association of Canada. Short Courses Series 35, 475–485.

Donoghue, K.A., Ripley, E.M., Li, C., 2014. Sulfur isotope and mineralogical studies of Ni–Cu sulfide mineralization in the Bovine igneous complex intrusion, Baraga basin, northern Michigan. Economic Geology 109, 325–341.

Doreen, E.A., 2008. Mineral Deposits of Canada, District Metallogeny Ni-Cu-PGE: Metallogeny of the Sudbury Mining Camp. Geological Survey of Canada, Ontario. www.gsc.nrcan.gc.ca.

Duan, J., Li, C., Qian1, Z., Jiao, J., Ripley, E.M., Feng, Y., 2015. Multiple S isotopes, zircon Hf isotopes, whole-rock Sr-Nd isotopes, and spatial variations of PGE tenors in the Jinchuan Ni-Cu-PGE Deposit, NW China. Mineral Deposita 18. http://dx.doi.org/10.1007/s00126-015-0626-8.

Duuring, P., Bleeker, W., Beresford, S.W., Hayward, N., 2010. Towards a volcanic–structural balance: relative importance of volcanism, folding, and remobilisation of nickel sulphides at the Perseverance Ni–Cu–(PGE) deposit, Western Australia. Mineral Deposita 45, 281–311. http://dx.doi.org/10.1007/s00126-009-0274-y.

Eckstrand, R., 2005. Ni-Cu-Cr-PGE mineralization types: distribution and classification. In: Mungall, J.E. (Ed.), Exploration for Platinum-Group Element Deposits, Mineralogical Association of Canada. Short Courses Series, vol. 35, pp. 487–494.

Eliopoulos, D.G., Economou-Eliopoulos, M., Apostolikas, A., Golightly, J.P., 2012. Geochemical features of nickel-laterite deposits from the Balkan Peninsula and Gordes, Turkey: the genetic and environmental significance of arsenic. Ore Geology Review 48, 413–427 Elsevier.

Ernowo, Oktaviani, P., 2010. Review of chromite deposits of Indonesia. Bulletin Sumber Daya Geologi 5 (1), 10–19.

Ertel, W., Dingwell, D.B., Sylvester, P.J., 2008. Siderophile elements in silicate melts – a review of the mechanically assisted equilibration technique and the nanonugget issue. Chemical Geology 248, 119–139 Elsevier, ScienceDirect.

Evans, D.M., Barrett, F.M., Prichard, H.M., Fisher, P.C., 2011. Platinum – palladium – gold mineralization in the Nkenja mafic–ultramafic body, Ubendian metamorphic belt, Tanzania. Mineral Deposita 1–22. http://dx.doi.org/10.1007/s00126-011-0353-8 Springer.

Evans, D.M., 2012. Geodynamic setting of Neoproterozoic nickel sulphide deposits in eastern Africa. Applied Earth Science (Transactions Institution of Mining and Metallurgy Section B) 120 (4), 175–186.

Fiorentini, M.L., Rosengren, N., Beresford, S.W., Grguric, B., Barley, M.E., 2007. Controls on the emplacement and genesis of the MKD5 and Sarah's find Ni–Cu–PGE deposits, Mount Keith, Agnew–Wiluna greenstone belt, western Australia. Mineral Deposita 42, 1–31. http://dx.doi.org/10.1007/s00126-007-0140-8.

Fleet, M.E., Angel, N., Pan, Y., 1993. Oriented chlorite lamellae in chromite from the Pedra Branca mafic-ultramafic complex, Ceari, Brazil. American Mineralogist 78, 68–74.

Gao, J.F., Zhou, M.F., Lightfoot, P.C., Wang, C.Y., Qi, L., Sun, M., 2013. Sulfide saturation and magma emplacement in the formation of the Permian Huangshandong Ni-Cu sulfide deposit, Xinjiang, Northwestern China. Economic Geology 108, 1833–1848.

Ghorfi, M.El, Melcher, F., Oberthur, T., Boukhari, A.E., Maacha, L., Maddi, A., Mhaili, M., 2008. Platinum group of minerals in podiform chromitites of the Bou Azzer ophiolite, Anti Atlas, Central Morocco. Mineralogy and Petrology 92, 59–80.

Godel, B., Barnes, S.J., Maie, W.D., 2010. Parental Magma Composition 1 Inferred from in Situ Trace Elements in 2 Cumulus and Intercumulus Silicate Minerals: Example from the Lower 3 and Lower Critical Zones of the Bushveld Complex (South-Africa). Elsevier Litho, p. 67.

Godel, B., Barnes, S.J., Maier, W.D., 2007. Platinum-group elements in sulphide minerals, platinum-group minerals, and Whole-rocks of the Merensky reef (Bushveld complex, South Africa): implications for the formation of the reef. Journal of Petrology 48 (8), 1569–1604.

Godel, B., Maieri, W.D., Barnes, S.J., 2008. Platinum-group elements in the Merensky and J-M reefs: a review of recent studies. Journal Geological Society of India 72, 595–609.

Godel, B., Seat, Z., Maier, W.G., Barn, S.J., 2011. The Nebo-Babel Ni-Cu-PGE sulfide deposit (West Musgrave block, Australia): Pt. 2. Constraints on parental magma and processes, with implications for mineral exploration. Economic Geology 106, 557–584.

Good, D.J., Epstein, R., McLean, K., Linnen, R.L., Samson, I.M., 2015. Evolution of the main zone at the Marathon Cu-PGE sulfide deposit, Midcontinent rift, Canada: spatial relationships in a magma conduit setting. Economic Geology 110, 983–1008.

Golder Associates, 2009. Technical Report on the Aguablanca Ni-cu Deposit, Extremadura Region, Spain Report No. 08511150292 , pp. 1–169. www.lundinmining.com.

Greenoush, J.D., Kamo, S.L., Theny, L., Crowe, S.A., Fipke, C., 2011. High-precision U-Pb and geochemistry of the mineralized (Ni-Cu-Co) Suwar intrusion, Yemen. Canadian Journal of Earth Sciences 48 (2), 495–514.

Gresham, J.J., Loftus-Hills, G.D., 2008. The geology of the Kambalda nickel field, Western Australia. Economic Geology 76 (6), 1373–1416.

Grguric, B.A., Seat, Z., Karpuzov, A.A., Simonov, O.N., 2013. The West Jordan deposit, a newly-discovered type 2 dunite-hosted nickel sulphide system in the northern Agnew–Wiluna belt, Western Australia. Ore Geology Review (52), 79–92 Elsevier.

Haldar, S.K., Tišljar, J., 2014. Introduction to Mineralogy and Petrology. Elsevier Publication, p. 356.

Haldar, S.K., 1967. Geology of the chromite deposits occurring in the area around Sukarangi, Cuttack Dist. Orissa. Quarterly Journal of the Geological, Mining, and Metallurgical Society of India xxxix (1), 59–61.

Haldar, S.K., 2007. Exploration Modeling of Base Metal Deposits. Elsevier Publication, p. 227.

Haldar, S.K., 2011. Platinum-Nickel-Chromium: resource evaluation and future potential targets. New Paradigms of Exploration and Sustainable Mineral Development: Vision 2050, 67–81.

Haldar, S.K., 2013. Mineral Exploration – Principles and Applications. Elsevier Publication, p. 374.

Harris, C., Chaumba, J.B., 2001. Crustal contamination and fluid-rock interaction during the formation of the platreef, northern limb of the Bushveld complex, South Africa. Journal of Petrology 42 (7), 1321–1347.

Hiebert, R.S., Bekker, A., Wing, B.A., Rouxe, O.J., 2013. The role of Paragneiss assimilation in the origin of the Voisey's bay Ni-Cu sulfide deposit, Labrador: multiple S and Fe isotope evidence. Economic Geology 108, 1459–1469.

Hill, R.E.T., Barnes, S.J., Gole, M.J., Dowling, S.E., 1995. The volcanology of komatiites as deduced from field relationships in the Norseman–Wiluna greenstone belt, Western Australia. Lithos 34, 159–188.

Holwell, D.A., Keays, R.R., 2014. The formation of low-volume, high-tenor magmatic PGE-Au sulfide mineralization in closed systems: evidence from precious and base metal geochemistry of the Platinova reef, Skaergaard intrusion, East Greenland. Economic Geology 109, 387–406.

Holwell, D.A., Keays, R.R., Firth, E.A., Findlay, J., 2006. Geochemistry and mineralogy of platinum group element mineralization in the river valley intrusion, Ontario, Canada: a model for early-stage sulfur saturation and multistage emplacement and the implications for "Contact-Type" Ni-Cu-PGE sulfide mineralization. Economic Geology 109, 689–712.

Holwell, D.A., Abraham-James, T., Keays, R.R., Boyce, A.J., 2011. The nature and genesis of marginal Cu–PGE–Au sulfide mineralization in Paleogene Macrodykes of the Kangerlussuaq region, East Greenland. Mineral Deposita 1–19. http://dx.doi.org/10.1007/s00126-010-0325-4.

Houlé, M.G., Lesher, C.M., 2011. Komatiite-associated Ni-Cu-(PGE) deposits, Abitibi greenstone belt, superior province, Canada. Economic Geology 17, 89–121.

Houlé, M.G., Lesher, C.M., Davis, P.C., 2011. Thermomechanical erosion at the Alexo Mine, Abitibi greenstone belt, Ontario: implications for the genesis of komatiite-associated Ni-Cu-(PGE) mineralization. Mineral Deposita 1–24. http://dx.doi.org/10.1007/s00126-011-0371-6.

Hutchinson, D., McDonald, I., 2008. Laser ablation ICP-MS study of platinum-group elements in sulphides from the Platreef at Turfspruit, northern limb of the Bushveld Complex, South Africa. Mineral Deposita 695–711.

IBM, 2003. Indian Mineral Year Book, pp. 1–23 Chromite.

Indian Bureau of Mines, 2013. Monograph on Chromite, p. 162.

Johnson, R.S., 1986. The Phoenix and Selkirk nickel–copper sulphide ore deposits, Tati Greenstone Belt, eastern Botswana. In: Anhaeusser, C.R., Maske, S. (Eds.), Mineral deposits of Southern Africa. Geological Society of South Africa, pp. 243–248.

Junge, M., Oberthür, T., Melcher, F., 2014. Cryptic variation of chromite chemistry, platinum group element, and platinum group mineral distribution in the UG-2 chromitite: an example from the Karee Mine, western Bushveld complex, South Africa. Economic Geology 109, 795–810.

Karykowski, B.T., Polito, P.A., Maier, W.D., Gutzmer, J., 2015. Origin of Cu-Ni-PGE mineralization at the Manchego Prospect, West Musgrave province, western Australia. Economic Geology 110, 2063–2085.

Keays, R.R., Lightfoot, P.C., 2010. Crustal sulfur is required to form magmatic Ni–Cu sulfide deposits: evidence from chalcophile element signatures of Siberian and Deccan Trap basalts. Mineral Deposita 241–257. http://dx.doi.org/10.1007/s00126-009-0271-1.

Keays, R.R., Lightfoot, P.C., Hamlyn, P.R., 2011. Sulfide saturation history of the Stillwater Complex, Montana: chemostratigraphic variation in platinum group elements. Mineral Deposita 1–23. http://dx.doi.org/10.1007/s00126-011-0346-7.

Kinnaird, J.A., 2014. The Bushveld Large Igneous Province. kinnairdj@geosciences.wits.ac.za and www.largeigneousprovinces.org/sites/default/files/BushveldLIP.pdf 30 p.

Kirste, J., 2009. Reconstruction of Contact Metamorphism of the Uitkomst Complex, Near Badplaas, Mpumalanga Province, South Africa, Based on Mineralogical and Petrological Investigations of the Contact Aureole. University of the Free State, Bloemfontein, South Africa, p. 246.

Konnunaho, J.P., Hanski, F.J., Bekker, A., Halkoaho, T.A.A., Hiebert, R.S., Wing, B.A., 2013. The Archean komatiite-hosted, PGE-bearing Ni–Cu sulfide deposit at Vaara, eastern Finland: evidence for assimilation of external sulfur and post-depositional desulfurization. Mineral Deposita 1–23. http://dx.doi.org/10.1007/s00126-013-0469-0.

Leshers, C.M., Burnham, O.M., Keays, R.R., 2001. Trace element geochemistry and Petrogenesis of Barren and ore-associated Komatiites. The Canadian Mineralogist 39, 673–693.

Lewins, J.D., Hunns, S., Badenhorst, J., 2008. The Kalahari platinum project. In: Third International Platinum Conference 'Platinum in Transformation', the Southern African Institute of Mining and Metallurgy, Paper 48, pp. 355–366.

Li, C., Ripley, E.M., 2011. The giant Jinchuan Ni-Cu-(PGE) deposit: tectonic setting, magma evolution, ore genesis, and exploration implications. Economic Geology 17, 163–180.

Li, C., Ripley, E.M., Oberthür, T., Miller Jr., J.D., Joslin, G.D., 2007. Textural, mineralogical and stable isotope studies of hydrothermal alteration in the main sulfide zone of the Great Dyke, Zimbabwe and the precious metals zone of the Sonju Lake Intrusion, Minnesota, USA. Mineral Deposita 1–14. http://dx.doi.org/10.1007/s00126-007-0159-x Springer-Verlag.

Li, C., Ripley, E.M., Naldrett, A.J., Schmitt, A.K., Moore, C.H., 2009. Magmatic Anhydrite-Sulfide Assemblages in the Plumbing System of the Siberian Traps. Geological Society of America, pp. 259–262. http://dx.doi.org/10.1130/G25355A.1.

Lightfoot, P.C., Naldrett, A.J., 1983. The geology of the Tabankulu section of the Insizwa complex, Transkei, South Africa, with reference to the nickel sulphide potential. Transactions of the Geological Society of South Africa 86, 169–187.

Lightfoot, P.C., Keays, R.R., Lamswood, D.E., Wheeler, R., 2011. S saturation history of Nain Plutonic Suite mafic intrusions: origin of the Voisey's Bay Ni–Cu–Co sulfide deposit, Labrador, Canada. Mineral Deposita 1–28. http://dx.doi.org/10.1007/s00126-011-0347-6.

Lightfoot, P.C., 2011. Geochemical signatures of giant ore systems in Siberian Trap basalts at Noril'sk: implications for exploration in other flood basalts. In: Advances in Ore Systems and Exploration Success. Perth. 21–24 Feb, p. 10. peter.lightfoot@vale.com.

Liu, Y., Mungall, J.E., Doreen, E.A., 2016. Hydrothermal redistribution and local enrichment of platinum group elements in the tootoo and mequillon magmatic sulfide deposits, South Raglan Trend, Cape Smith belt, new Quebec Orogen. Economic Geology 111, 467–485.

Lord, R.A., Prichard, H.M., SÁ, J.H.S., Neary, C.R., 2004. Chromite geochemistry and PGE fractionation in the Campo Formoso complex and Ipueira-Medrado Sill, Bahia State, Brazil. Economic Geology 99, 339–363.

Lundgaard, K.L., Tegner, C., Cawthorn, R.G., Kruger, F.J., Wilson, J.R., 2006. Trapped intercumulus liquid in the main zone of the eastern Bushveld complex, South Africa. Mineralogy and Petrology (151), 352–369.

Lundstrom, C.C., Gajos, N., 2014. Formation of the PGE Reef horizon in the Sonju lake layered mafic intrusion by thermal migration zone refining. Economic Geology 109, 1257–1269.

Maier, W.D., Barnes, S.-J., 2010. The Kabanga Ni sulfide deposits, Tanzania: II. Chalcophile and siderophile element geochemistry. Mineral Deposita 45, 443–460. http://dx.doi.org/10.1007/s00126-010-0283-xSpringer-Verlag.

Maier, W.D., Barnes, S.J., Gartz, V., Andrews, G., 2003. Pt-Pd reefs in magnetites of the Stella layered intrusion, South Africa: a world of new exploration opportunities for platinum group elements. Geology 31, 885–888 (geology.gsapubs.org).

Maier, W.D., Gomwe, T., Barens, S.J., Theart, H., 2004. Platinum group elements in the Uitkomst complex, South Africa. Economic Geology 99, 499–516.

Maier, W.D., Barnes, S.J., Chinyepi, G., Barton Jr., J.M., Eglington, B., Setshedi, I., 2007. The composition of magmatic Ni–Cu–(PGE) sulfide deposits in the Tati and Selebi-Phikwe belts of eastern Botswana. Mineral Deposita 1–24. http://dx.doi.org/10.1007/s00126-007-0143-5 Springer-Verlag.

Maier, W.D., Barnes, S.-J., Bandyayera, Livesey, D.T., Li, C., Ripley, E., 2008. Early Kibaran rift-related mafic–ultramafic magmatism in western Tanzania and Burundi: Petrogenesis and ore potential of the Kapalagulu and Musongati layered intrusions. Lithos 101, 24–53. ScienceDirect. Elsevier www.sciencedirect.com.

Maier, W.D., Barnes, S.-J., Sarkar, A., Ripley, Ed, Li, C., Livesey, T., 2010. The Kabanga Ni sulfide deposit, Tanzania: I. Geology, petrography, silicate rock geochemistry, and sulphur and oxygen isotopes. Mineral Deposita 45, 419–441. http://dx.doi.org/10.1007/s00126-010-0280-0 Springer-Verlag.

Maier, W.D., Barnes, S.-J., Ripley, M., 2011. The Kabanga Ni sulfide deposits, Tanzania: a review of ore-forming processes. Economic Geology 17, 217–234.

Maier, W.D., Marsh, J.S., Barnes, S.J., Dodd, D.C., 2013a. The distribution of platinum group elements in the Insizwa Lobe, mount Ayliff complex, South Africa: implications for Ni-Cu-PGE sulfide exploration in the Karoo igneous province. Economic Geology 97 (6), 1293–1306.

Maier, W.D., Rasmussen, B., Fletcher, I.R., Li, C., Barnes, S.-J., Huhma, H., 2013b. The Kunene Anorthosite complex, Namibia, and its satellite intrusions: geochemistry, geochronology, and economic potential. Economic Geology 108, 953–986.

Malehmir, A., Koivisto, E., Manzi, M., Cheraghi, S., Durrheim, R.J., Bellefleur, G., Wijns, C., Hein, K.A.A., King, N., 2013. A review of reflection seismic investigations in three major metallogenic regions: the Kevitsa Ni–Cu–PGE district (Finland), Witwatersrand goldfields (South Africa), and the Bathurst Mining Camp (Canada). Ore Geology Review 19. www.elsevier.com/locate/oregeorev xxx xxx–xxx.

Manor, M.J., Scoates, J.S., Nixon, G.T., Ames, D.E., 2016. The giant Mascot Ni-Cu-PGE deposit, British Columbia: mineralized conduits in a convergent Margin Tectonic setting. Economic Geology 111, 57–87.

Markwitz, V., Maier, W.D., González-Álvarez, I., McCuaig, T.C., Porwal, A., 2010. Magmatic nickel sulfide mineralization in Zimbabwe: review of deposits and development of exploration criteria for prospectivity analysis. Ore Geology Reviews 38, 139–155 Elsevier.

Marques, J.C., Filho, C.F.F., Carlson, R.W., Pimentel, M.M., 2002. Re-Os and Sm-Nd isotope and trace element constraints on the origin of the chromite deposits of the Ipueira-Medrado Sill, Bahia, Brazil. Journal of Petrology 44 (4), 659–678.

Matthews, D.L., Lesher, C.M., Liwang, J., Halden, N., Burnham, M., Huibert, L., Peck, D.C., Keays, R.R., 2011. Mineralogy, geochemistry, and genesis of komatiite-associated Ni-Cu-PGE mineralization in the Thompson nickel belt, Manitoba. Economic Geology 17, 123–143.

McCracken, T., Kanhai, T., Bridson, P., McBride, W.R., Small, K., Penna, D.N., 2013. Technical Report Lac Des Iles Mine. Incorporating Prefeasibility Study Offset Zone Phase. Tetra Tech, Ontario, p. 594.

McDolald, I., Holwell, D.A., 2011. Geology of the northern Bushveld complex and the setting and genesis the Platreef Ni-Cu-PGE deposit. Economic Geology 17, 297–327.

Melcher, F., Grum, W., Simon, G., Thalhammer, T.V., Stumpfl, E.F., 1997. Petrogenesis of the ophiolitic giant chromite deposits of Kempirsai, Kazakhstan: a study of solid and fluid inclusions in chromite. Journal of Petrology 38 (10), 1419–1458.

Merwe, J.V.der, Cawthorn, R.G., 2005. Structures at the base of the upper group 2 chromitite layer, Bushveld complex, South Africa, on Karee Mine (Lonmin platinum). Lithos 83, 214–228. Elsevier at: www.sciencedirect.com.

Mohanty, D., Sen, A.K., 2008. PGE mineralisation in Kathpal Chromites, Sukinda Ultramafics complex, Orissa, India. The Indian Mineralogist 42 (1), 62–70.

Mohanty, J.K., Paul, A.K., Charchi, N., 2008. PGE carriers at Boula-Nausahi igneous complex, India. The Indian Mineralogist 42 (1), 53–61.

Mohanty, J.K., 2009. Field tour guide to Boula-Nausahi igneous complex. In: International Symposium on Magmatic Ore Deposits, pp. 17–21.

Mondal, S.K., Mathez, E.A., 2007. Origin of the UG2 chromitite layer, Bushveld complex. Journal of Petrology 48 (3), 495–510.

Mondal, S.K., Zhou, M.F., 2010. Enrichment of PGE through interaction of evolved boninitic magmas with early formed cumulates in a gabbro–breccia zone of the Mesoarchean Nuasahi massif (eastern India). Mineral Deposita 45, 69–91. http://dx.doi.org/10.1007/s00126-009-0264-0.

Mondal, S.K., Ripley, E.M., Li, C., Frei, R., 2006. The genesis of Archaean chromitites from the Nuasahi and Sukinda massifs in the Singhbhum craton, India. Precambrian Research 148, 45–66.

Mosier, D.L., Singer, D.A., Moring, B.C., Galloway, J.P., 2012. Podiform Chromite Deposits – Database and Grade and Tonnage Models. Scientific Investigations Report 2012-5157 US Geological Survey , p. 54.

Mossman, D.J., 2000. High-Mg Arc-Ankaramtic Dikes, Greenhills complex, South Land, New Zealand. The Canadian Mineralogist 38, 191–216.

Mudd, G.M., Jowitt, S.M., 2014. A detailed assessment of global nickel resource trends and endowments. Economic Geology 109, 1813–1841.

Mukwakwami, J., Lesher, C.M., Lafrance, B., 2014. Geochemistry of deformed and hydrothermally mobilized magmatic Ni-Cu-PGE ores at the Garson Mine, Sudbury. Economic Geology 109, 367–386.

Mungall, J.E., 2005a. Magmatic geochemistry of the platinum-group elements. Mineralogical Association of Canada Short Course Series 35, 1–34.

Mungall, J.E., 2005b. Exploration for platinum-group element deposits, mineralogical association of Canada. Short Courses Series 35, 512.

Mäkinen, J., Makkonen, H.V., 2004. Petrology and structure of the Palaeoproterozoic (1.9 Ga) Rytky nickel sulfide deposit, Central Finland: a comparison with the Kotalahti nickel deposit. Mineral Deposita 39, 405–421. http://dx.doi.org/10.1007/s00126-004-0413-4.

Naldrett, A.J., Li, C., 2007. The Voisey's Bay deposit, Labrador, Canada. In: Goodfellow, W.D. (Ed.), Mineral Deposits of Canada: A Synthesis of Major Deposit-Types, District Metallogeny, the Evolution of Geological Provinces, and Exploration Models: Geological Association of Canada. Mineral deposit Division, Special Publication No. 5, pp. 387–407.

Naldrett, A.J., Wilson, A., Kinnalrd, J., Chunnett, G., 2009. PGE Tenor and metal ratios within and below the Merensky Reef, Bushveld complex: implications for its genesis. Journal of Geology 50 (4), 625–659.

Naldrett, A.J., 2010. Secular variation of magmatic sulfide deposits and their source magmas. Economic Geology 105, 669–688.

Naldrett, A.J., Wilson, A., Kinnaird, J., Yudovskaya, A., Chunnett, G., 2011a. The origin of chromitites and related PGE mineralization in the Bushveld Complex: new mineralogical and petrological constraints. Mineral Deposita 1–24. http://dx.doi.org/10.1007/s00126-011-0366-3.

Naldrett, A., Kinnaird, J., Wilson, A., Yudovskaya, M., Chunett, G., 2011b. Genesis of the PGE-enriched Merensky reef and chromitite seams of the Bushveld complex. Economic Geology 17, 235–296.

Naldrett, A.J., 2004. Magmatic Sulfide Deposits – Geology, Geochemistry and Exploration. Springer Publication, p. 728.

Naldrett, A.J., 2011. Fundamentals of magmatic sulfide deposits. Economic Geology 17, 1–50.

OberthÜr, T., 2011. Platinum-group element mineralization of the main sulfide zone, Great Dyke, Zimbabwe. Reviews in Economic Geology 17, 329–349.

Pariser, G.D., 2013. Chromite: World Distribution, Uses, Supply and Demand, Future. PDAC Convention, Toronto, p. 23. www.boldventuresinc.com/…/2013-Mar3-PDAC.

Peng, R., Zhai, Y., Li, C., Ripley, E.M., 2013. The Erbutu Ni-Cu deposit in the central Asian Orogenic belt: a Permian magmatic sulfide deposit related to boninitic magmatism in an arc setting. Economic Geology 108, 1879–1888.

Perring, C.S., Barnes, S.J., Hill, R.E.T., 1995. The physical volcanology of Archaean komatiite sequences from Forrestania, Southern Cross Province, Western Australia. Lithos 34, 189–207.

Perring, C.S., 2015a. Volcanological and structural controls on mineralization at the mount Keith and Cliffs komatiite-associated nickel sulfide deposits, Agnew-Wiluna belt, western Australia—Implications for ore genesis and targeting. Economic Geology 110, 1669–1695.

Perring, C.S., 2015b. A 3-D geological and structural synthesis of the Leinster area of the Agnew-Wiluna belt, Yilgarn craton, western Australia, with special reference to the volcanological setting of komatiite-associated nickel sulfide deposits. Economic Geology 110, 469–503.

Peterson, J.A., 1984. Metallogenetic maps of the ophiolite belts of the western United States. Miscellaneous Investigation Series US Geological Survey 14.

Petrus, J.A., Ames, D.E., Kamber, B.S., 2014. On the track of the elusive Sudbury impact: geochemical evidence for a chondrite or comet bolide. Terra Nova 27 (1). http://dx.doi.org/10.1111/ter.12125 (Abstract).

Piña, R., Gervilla, F., Barnnes, S.J., Ortega, L., Lunar, R., 2013. Partition Coefficients of platinum group and chalcophile elements between Arsenide and sulfide phases as determined in the Beni Bousera Cr-Ni mineralization (North Morocco). Economic Geology 108, 935–951.

Platinum Australia Limited, 2014. Kalplat Projects. www.platinumaus.com.au/viewStory/Kalplats%20Projects.

Platinum Group Elements, 2011. www.australianminesatlas.gov.au/…/platinum_group_elements_10.jsp.

Prendergast, M.D., 2008. Archean komatiitic sill-hosted chromite deposits in the Zimbabwe craton. Economic Geology 103, 981–1004.

Prichard, H.M., Fisher, P.C., McDonald, I., Knight, R.D., Sharp, D.R., Williams, J.P., 2013. The distribution of PGE and the role of arsenic as a collector of PGE in the spotted Quoll nickel ore deposit in the Forrestania greenstone belt, western Australia. Economic Geology 108, 1903–1921.

Pripachkin, P.V., Rundkvist, T.V., Miroshnikova, Y.A., Chernyavsky, A.V., Borisenko, E.S., 2015. Geological structure and ore mineralization of the South Sopchinsky and Gabbro-10 massifs and the Moroshkovoe lake target, Monchegorsk area, Kola Peninsula, Russia. Mineral Deposita 20. http://dx.doi.org/10.1007/s00126-015-0605-0.

Proenza, J.A., Zaccarini, F., Escayola, M., Cábana, C., Schalamuk, A., Garuti, G., 2008. Composition and textures of chromite and platinum-group minerals in chromitites of the western ophiolitic belt from Pampean Ranges of Córdoba, Argentina. Ore Geology Review 33, 33–48. Elsevier www.sciencedirect.com.

Queffurus, M., Barnes, S.J., 2014. Selenium and sulphur concentrations in country rocks from the Duluth Complex, Minnesota, USA: implications for formation of the Cu-Ni-PGE Sulfides. Economic Geology 109, 785–794.

Radhakrishna, B.P., 1996. Mineral Resources of Karnataka. Geological Society of India, p. 471.

Reid, R., Keays, R.R., Lightfoot, P.C., 2015. Geochemical Stratigraphy of the Keweenawan Midcontinent rift volcanic rocks with regional implications for the genesis of associated Ni, Cu, Co, and platinum group element sulfide mineralization. Economic Geology 110, 1235–1267.

Ripley, E.M., Li, C.A., 2011a. A review of conduit-related Ni-Cu-(PGE) sulfide mineralization at the Voisey's Bay deposit, Labrador, and the Eagle deposit, northern Michigan. Economic Geology 17, 181–197.

Ripley, E.M., February 2011b. Magmatic Ni-Cu-PGE deposits – Sources and sulfur and metals. Advances in Ore Systems and Exploration Success 39 Perth 23.

Ripley, E.M., 2014. Ni-Cu-PGE mineralization in the Partridge river, South Kawishiwi, and eagle intrusions: a review of contrasting styles of sulfide-rich occurrences in the midcontinent rift system. Economic Geology 109, 309–324.

Robert, L., Thorne, R.L., Roberts, S., Richard Herrington, R., 2012. Climate change and the formation of nickel laterite deposits. Geological Society of America 40 (4), 331–334.

Rosengren, N.M., Grguric, B.A., Beresford, S.W., Fiorentini, M.L., Cas, R.A.F., 2007. Internal stratigraphic architecture of the komatiitic dunite-hosted MKD5 disseminated nickel sulfide deposit, Mount Keith Domain, Agnew-Wiluna Greenstone Belt, Western Australia. Mineral Deposita 1–25. http://dx.doi.org/10.1007/s00126-007-0139-1.

Rudashevsky, N.S., Mcdonald, A.M., Cabri, L.J., Nielsen, T.F.D., Stanley, C.J., Kretzer, Y.U.L., Rudashevsky, V.N., 2004. Skaergaardite, PdCu, a new platinum-group intermetallic mineral from the Skaergaard intrusion, Greenland. Mineralogical Magazine 68 (4), 615–632.

Saumur, B.M., Cruden, A.R., Evans-Lamswood, D., Lightfoot, P.C., 2015. Wall-Rock structural controls on the genesis of the Voisey's Bay intrusion and its Ni-Cu-Co magmatic sulfide mineralization (Labrador, Canada). Economic Geology 110.

Sciortino, M., Mungall, J.E., Muinonen, J., 2015. Generation of high-Ni sulfide and alloy phases during serpentinization of dunite in the Dumont Sill. Quebec. Economic Geology 110, 733–761.

Scoates, J.S., Jon Scoates, R.S., 2013. Age of the Bord river Sill, Southeastern Manitoba, Canada, with implication for the secular variation of layered intrusion-hosted stratiform chromite mineralization. Economic Geology 108, 895–907.

Seat, Z., Beresford, S.W., Grguric, B.A., Waugh, R.S., Hronsky, J.M.A., Mary Gee, M.A., Groves, D.I., Mathison, C.I., 2007. Architecture and emplacement of the Nebo–Babel gabbronorite-hosted magmatic Ni–Cu–PGE sulphide deposit, West Musgrave, western Australia. Mineral Deposita 1–31. http://dx.doi.org/10.1007/s00126-007-0123-9.

Seat, Z., Beresford, S.W., Grguric, B.A., Mary Gee, M.A., Grassineau, N.V., 2009. Reevaluation of the role of external sulfur addition in the genesis of Ni-Cu-PGE deposits: evidence from the Nebo-Babel Ni-Cu-PGE deposit, West Musgrave, western Australia. Economic Geology 104, 521–538.

Seat, Z., Mary Gee, M.A., Grguric, B.A., Beresford, S.W., Grassineau, N.V., 2011. The Nebo-Babel Ni-Cu-PGE sulfide deposit (West Musgrave, Australia): Pt. 1. U/Pb zircon Ages, whole-rock and mineral chemistry, and O-Sr-Nd isotope compositions of the intrusion, with constraints on petrogenesis. Economic Geology 106 (4), 527–556.

Silva, J.M., Filho, C.F.F., Bühn, B., Dantas, E.L., 2011. Geology, petrology and geochemistry of the "Americano do Brasil" layered intrusion, central Brazil, and its Ni–Cu sulfide deposits. Mineral Deposita 57–90 Springer.

Silva, J.M., Filho, C.F.F., Giustina, M.R.S.D., 2013. The Limoeiro deposit: Ni-Cu-PGE sulfide mineralization hosted within an ultramafic Tubular magma conduit in the Borborema province, Northeastern Brazil. Economic Geology 108, 1753–1771.

Singh, M.R., Manikyamba, C., Ray, J., Ganguly, S., Santosh, M., Saha, A., Rambabu, S., Sawant, S.S., 2015. Major, Trace and Platinum Group Element (PGE) Geochemistry of Archean Iron Ore Group and Proterozoic Malangtoli Metavolcanic Rocks of Singhbhum Craton, Eastern India: Inferences on Mantle Melting and Sulphur Saturation History. Elsevier, pp. 1–27. OREGEO-01509.

Smeeth, W.F., Iyengar, S.P., 1916. Mineral resources of Mysore. Bulletin Mysore Geological Department 7.

SNC-Lavalin, 2006. Ambatovy Project-Madagascar-Feasibility Study-Ni/Co Production, pp. 1–55.

Song, X.Y., Li, X.R., 2009. Geochemistry of the Kalatongke Ni–Cu–(PGE) sulfide deposit, NW China: implications for the formation of magmatic sulfide mineralization in a postcollisional environment. Mineral Deposita 303–327. http://dx.doi.org/10.1007/s00126-008-0219-x.

Song, X., Wang, Y., Chen, L., 2011. Magmatic Ni-Cu-(PGE) deposits in magma plumbing systems: features, formation and exploration. Geoscience Frontiers 2 (3), 375–384 Production and hosting by Elsevier.

Song, X.Y., Danyushevsky, L.V., Keays, R.R., Chen, L.M., Wang, Y.S., Tian, Y.L., Xiao, J.F., 2012. Structural, lithological, and geochemical constraints on the dynamic magma plumbing system of the Jinchuan Ni–Cu sulfide deposit, NW China. Mineral Deposita 277–297. http://dx.doi.org/10.1007/s00126-011-0370-7.

Song, X.Y., Yi, J.N., Chen, L.M., She, Y.W., Liu, C.Z., Dang, X.Y., Yang, Q.A., Wu, S.K., 2016. The giant Xiarihamu Ni-Co sulfide deposit in the East Kunlun Orogenic belt, northern Tibet plateau, China. Economic Geology 110, 29–55.

Stone, W.F., Archibald, N.J., 2004. Structural controls on nickel sulphide ore shoots in Archaean komatiite, Kambalda, WA: the volcanic trough controversy revisited. Elsevier Journal of the Structural Geology 26, 1173–1194.

Stowe, C.W., 1994. Compositions and Tectonic Settings of Chromite Deposits. Economic Geology 89, 528–546.

Suárez, S., Prichard, H.M., Velasco, F., Fisher, P.C., McDonald, L., 2010. Alteration of platinum-group minerals and dispersion of platinum-group elements during progressive weathering of the Aguablanca Ni–Cu deposit, SW Spain. Mineral Deposita 45, 331–350. http://dx.doi.org/10.1007/s00126-009-0275-x.

Sufriadin, I.A., Pramumijoyo, S., Warmada, I., Imai, A., 2011. Study on mineralogy and chemistry of the Saprolitic nickel ore from Soroako, Sulawesi, Indonesia: implication for the lateritic ore processing. Journal of Southeast Asian Applied Geology 3 (1), 23–33.

Sun, T., Qian, Z.Z., Deng, Y.F., Li, C., Song, X.Y., Tang, Q., 2013. PGE and isotope (Hf-Sr-Nd-Pb) constraints on the origin of the Huangshandong magmatic Ni-Cu sulfide deposit in the central Asian Orogenic belt, Northwestern China. Economic Geology 108, 1849–1864.

Tao, Y., Li, C., Hu, R., Ripley, E.M., Du, A., Zhong, H., 2007. Petrogenesis of the Pt–Pd mineralized Jinbaoshan ultramafic intrusion in the Permian Emeishan large igneous province, SW China. Mineral Deposita 321–337. http://dx.doi.org/10.1007/s00410-006-0149-5.

Thakurta, J., Ripley, E.M., Li, C., 2014. Platinum group element geochemistry of sulfide-rich Horizons in the Ural-Alaskan–type ultramafic complex of Duke Island, Southeastern Alaska. Economic Geology 109, 643–659.

Theart, H.F.J., De Nooy, C.D., 2001. The platinum group minerals in two parts of the Massive sulfide body of the Uitkomst complex, Mpumalanga, South Africa. South African Journal of Geology 104, 287–300.

Thorne, R., Roberts, S., Herrington, R., 2012. The formation and evolution of the Bitincke nickel laterite deposit, Albania. Mineralium Deposita 47, 933–947.

Törmänen, T., Konnunaho, J.P., Hanski, F., Moilanen, M., Heikura, P., 2015. The Paleoproterozoic komatiite-hosted PGE mineralization at Lomalampi, central Lapland greenstone belt, northern Finland. Mineral Deposita 1–20. http://dx.doi.org/10.1007/s00126-015-0615-y.

Tornos, T., Galindo, C., Casquet, C., Pevida, L.R., Martínez, C., Martínez, E., Velasco, F., Iriondo, A., 2006. The Aguablanca Ni–(Cu) sulfide deposit, SW Spain: geologic and geochemical controls and the relationship with a mid-crustal layered mafic complex. Mineral Deposita 737–769. http://dx.doi.org/10.1007/s00126-006-0090-6.

U.S. Geological Survey, 2015. Mineral Commodity Summaries 2015. U.S. Geological Survey. 196 p. http://dx.doi.org/10.3133/70140094.

Vaillant, M.L., Saleem, A., Barnes, S.J., Fiorentini, M.L., Miller, J., Beresford, S., Perring, C., 2015a. Hydrothermal remobilisation around a deformed and remobilised komatiite-hosted Ni-Cu-(PGE) deposit, Sarah's Find, Agnew Wiluna greenstone belt, Yilgarn Craton, Western Australia. Mineral Deposita 1–20. http://dx.doi.org/10.1007/s00126-015-0610-3.

Vaillant, M.L., Barnes, S.J., Fiorentini, M.L., Miller, J., McCuaig, T.C., Muccilli, P.A., 2015b. Hydrothermal Ni-As-PGE geochemical Halo around the Miitel komatiite-hosted nickel sulfide deposit, Yilgarn craton, western Australia. Economic Geology 110, 505–530.

Wang, C.Y., Zhou, M.F., 2006. Genesis of the Permian Baimazhai magmatic Ni–Cu–(PGE) sulfide deposit, Yunnan, SW China. Mineral Deposita 771–783. http://dx.doi.org/10.1007/s00126-006-0094-2.

Wilhelmij, H.R., Cabri, L.J., 2015. Platinum mineralization in the Kapalagulu intrusion, western Tanzania. Mineral Deposita 1. http://dx.doi.org/10.1007/s00126-015-0603-2.

Williams, D.A., Kerr, R.C., Lesher, C.M., 2011. Mathematical modeling of thermomechanical erosion beneath Proterozoic komatiitic basaltic sinuous rilles in the Cape Smith Belt, New Québec, Canada. Mineral Deposita 1–16. http://dx.doi.org/10.1007/s00126-011-0364-5.

Xia, M.Z., Jiang, C.Y., Li, C., Xia, Z.D., 2013. Characteristics of a newly discovered Ni-Cu sulfide deposit hosted in the Poyi ultramafic intrusion, Tarim craton, NW China. Economic Geology 108, 1865–1878.

Yang, K., Whitbourn, L., Mason, P., Huntington, J., 2013. Mapping the Chemical composition of nickel Laterites with Reflectance Spectroscopy at Koniambo, New Caledonia. Economic Geology 198, 1285–1299.

Yao, Y., Viljoen, M.J., Viljoen, R.P., Wilson, A.H., Zhng, H., Liu, B.G., Ying, H.L., Tu, G.Z., Luo, Y.N., 2001. Geological Characterization of PGE-bearing Layered Intrusions in Southern Sichuan Province, China. Information Circular No. 358. Economic Geology Research Institute. University of the Witwatersrand, Johannesburg, p. 21.

Yudovskaya, M.A., Kinnaird, J.A., Sobolev, A.V., Kuzmin, D.V., McDonald, J., Wilson, A.H., 2013. Petrogenesis of the lower zone olivine-rich cumulates beneath the Platreef and their correlation with recognized occurrences in the Bushveld complex. Economic Geology 108, 1923–1952.

Yumul Jr., G.P., 2001. The Acoje block Platiniferous dunite Horizon, Zambales ophiolite complex, Philippines: melt type and associated geochemical controls. Resource Geology 51 (2), 165–174.

Zhang, M., O'Reilly, S.Y., Wang, K.L., Hronsky, J., Griffin, W.L., 2008. Flood basalts and metallogeny: the lithospheric mantle connection. Earth Science Review 86, 145–174Elsevier, Science Direct.

Zhou, M.F., Robinson, P.T., Lesher, C.M., Keasys, R.R., Zhang, C.J., Malpas, J., 2005. Geochemistry, Petrogenesis and Metallogenesis of the Panzhihua Gabbroic layered intrusion and associated Fe–Ti–V Oxide deposits, Sichuan province, SW China. Journal of Petrology 46 (11), 2253–2280.

Zientek, M.L., Parks, H.L., 2014. A Geologic and Mineral Exploration Spatial Database for the Stillwater Complex, Montana Scientific Investigations Report 2014–5183. U.S. Department of the Interior. U.S. Geological Survey, p. 40.

Zientek, M.L., Corson, S.R., West, R.D., 2005. Discovery of the J-M Reef, Stillwater Complex, Montana: the role of soil and silt platinum and palladium geochrmical survey. In: Mungall, J.E. (Ed.), Exploration for Platinum-group Element Deposits, Mineralogical Association of Canada. Short Courses Series, vol. 35, pp. 391–407.

Zientek, M.L., 2010. Magmatic Ore Deposits in Layered Intrusions-descriptive Model for Reef-type PGE and Contact-type Cu-Ni-PGE Deposits USGS Open-File Report 2012-1010. 2012, p. 48.

Index

Printed in the United States
By Bookmasters